Lecture Notes in Physics

Edited by H. Araki, Kyoto, J. Ehlers, München, K. Hepp, Zürich
R. Kippenhahn, München, H.A. Weidenmüller, Heidelberg,
J. Wess, Karlsruhe and J. Zittartz, Köln

Managing Editor: W. Beiglböck

282

F. Ehlotzky (Ed.)

Fundamentals of Quantum Optics II

Proceedings of the Third Meeting on Laser Phenomena
Held at the Bundessportheim in Obergurgl, Austria
February 22–28, 1987

Springer-Verlag Berlin Heidelberg GmbH

Editor

Fritz Ehlotzky
Institute for Theoretical Physics, University of Innsbruck
Technikerstraße 25, A-6020 Innsbruck, Austria

ISBN 978-3-662-13657-7 ISBN 978-3-540-47716-7 (eBook)
DOI 10.1007/978-3-540-47716-7

Originally published by Springer-Verlag Berlin Heidelberg New York in 1987.
Softcover reprint of the hardcover 1st edition 1987

2153/3140-543210

FOREWORD

The Seminar on Fundamentals of Quantum Optics II was the third meeting on Laser Phenomena held at the Bundessportheim in Obergurgl. It was attended by 41 physicists from Austria, The Federal Republic of Germany, France, Great Britain, Hungary, Italy, Poland, Switzerland and The United States, who work actively in the rapidly developing field of quantum optics.

The first meeting in this series (Obergurgl, February 26 - March 3, 1984) also addressed the subject of quantum optics and was published as Fundamentals of Quantum Optics. Acta Physica Austriaca, Vol. 56, No. 1 - 2, 1984.

The present Seminar offered the opportunity to discuss at leisure problems of mutual interest to theoreticians and experimentalists who are working on various aspects of the field of quantum optics. The intention was to bring together people who are doing research on quantum chaos, squeezed states, quantum jumps, quantum electrodynamics in a cavity, cooling and trapping of particles, and on other fundamentals.

At the seminar 18 Invited Lectures were given by:

N.B. Abraham (Bryn Mawr)
Z. Bialynicka-Birula (Warsaw)
J. Dalibard (Paris)
W. Ertmer (Bonn)
E. Giacobino (Paris)
R. Graham (Essen)
F. Haake (Essen)
S. Haroche (Paris and New Haven)
J. Javanainen (Rochester)
H.J. Kimble (Austin)
P.L. Knight (London)
G. Leuchs (MPI Garching)
P. Meystre (Tucson)
J. Mlynek (Zürich)
J.M. Raimond (Paris)
H. Risken (Ulm)
A. Schenzle (Essen)
P.E. Toschek (Hamburg)

In addition, there were 10 contributed talks given at the meeting.

The following pages present the full text of the invited lectures and the abstracts of the contributed papers. The editor is grateful to the contributors for their collaboration in preparing their typescripts for rapid publication.

The active yet relaxed atmosphere of the Bundessportheim at Obergurgl, surrounded by the snow-capped peaks of the Ötztal Alps, provided a congenial setting for a very stimulating and rewarding meeting. It is a pleasure to thank all participants for their interest and enthusiasm. The most valuable secretarial assistance of Miss E. Merl is gratefully acknowledged.

Innsbruck, April 1987 F. Ehlotzky

ACKNOWLEDGEMENTS

The Seminar on Fundamentals of Quantum Optics II
has been supported by:

Bundesministerium für Wissenschaft und Forschung
Bundesministerium für Unterricht, Kunst und Sport
Amt der Tiroler Landesregierung
Magistrat der Stadt Innsbruck
Österreichische Akademie der Wissenschaften
Bundeskammer der Gewerblichen Wirtschaft
Österreichische Forschungsgemeinschaft

C O N T E N T S

I N V I T E D   L E C T U R E S

Part I: Chaos in Quantum Systems

Part II: Squeezed Quantum States

CONTRIBUTED PAPERS (ABSTRACTS)

I N V I T E D   L E C T U R E S

PART I: Chaos in Quantum Systems

QUANTUM CHAOS FOR KICKED SPINS

F. Haake, M. Kuś,* and R. Scharf
Fachbereich Physik, Universität-GHS Essen
Postfach 103 764, D-4300 Essen

The quasienergy statistics for kicked quantum systems displaying chaos in the classical limit fall into three universality classes. These correspond to the socalled orthogonal, unitary, and symplectic ensembles of random matrices. Realizations of all three kinds of dynamics with kicked spins will be presented. The universality of level repulsion will be demonstrated by dynamical and statistical arguments. The transition from regular (level clustering) to irregular behavior (level repulsion) will be discussed briefly.

Classically chaotic motion of spin systems has been observed by subjecting small crystalline magnets with an easy plane of magnetization to a magnetic field varying periodically in time [1]. The relevant dynamical variables are the components of a global spin vector J which moves with its length conserved,

$$J^2 = j(j+1) \ . \qquad (1)$$

The experiments in question involve quantum numbers j sufficiently large for quantum effects to be entirely negligeable.

Theoretical investigations have shown that for quantum numbers j of the order of several hundred very interesting quantum mechanical aspects of chaotic motion arise [2]. Present-day technology should allow the observation of these effects with small clusters of atoms.

In this talk the quasienergy statistics of kicked quantum spins obeying (1) will be

---

*Permanent address: Institute for Theoretical Physics, University of Warsaw, Hoza 69, 00-681 Warsaw, Poland

discussed. We think of a periodic sequence of delta-shaped kicks such that the unitary time evolution operator transporting the wave vector from kick to kick reads

$$U = e^{-i \frac{k}{2j} J_z^2} e^{-i p J_y} . \tag{2}$$

The factor in U obviously describes a precession of J around the y axis by an angle p; this precession can be realized by exposing the system to a magnetic field pointing in the y direction. The second factor in U might be due to a magnetic anisotropy and may be interpreted as a nonlinear torsion around the z axis by an angle proportional to kJ_z/j. We shall refer to the coupling constant k as to the kick strength. Powers of U, U^n with n = 1, 2, 3, ..., yield a stroboscopic description of the evolution of the quantum wave vector of the spin.

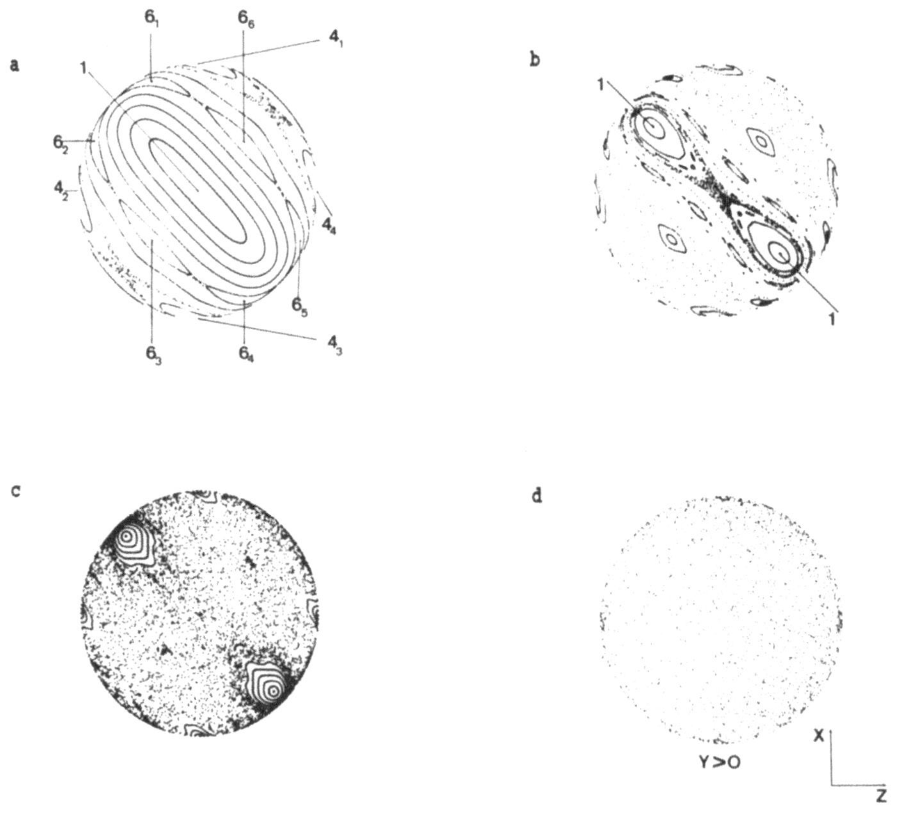

Fig. 1a-d. Classical motion on the unit sphere. Shown are some trajectories on the northern hemisphere (Y > 0). Fixed points are labeled by 1, points on n-cycles by n_i.
a: $p = \pi/2$, $k = 2$; **b:** $p = \pi/2$, $k = 2.5$;
c: $p = \pi/2$, $k = 3$; **d:** $p = \pi/2$, $k = 6$.

Classical behavior emerges in the limit $j \to \infty$; the classical vector $X = \lim_{j\to\infty} J/j$ moves on the unit sphere [2]. As a background to our discussion of finite j we present in Fig. 1 portraits of stroboscopic trajectories of the classical top for $p = \pi/2$ and various values of the kick strength k. The interesting message to be drawn is the predominance of regular motion for small k ($k \lesssim 2.5$) and of chaos for $k \gtrsim 3$.

The classical transition from mostly regular to mostly chaotic behavior is paralleled by a dramatic change in the quantum mechanical eigenvalue spectrum of the operator U.

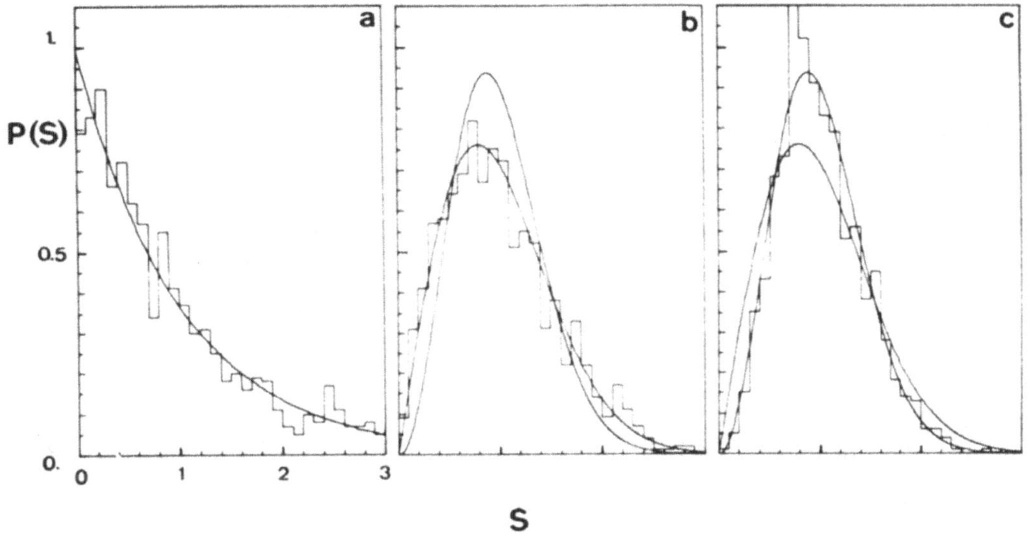

Fig. 2. Level spacing statistics, **a**: under conditions of classically regular motion . b - d: under conditions of classical chaos. **b**: j = 500, with generalized time reversal invariance T gives linear repulsion.
c: j = 500, no T, quadratic repulsion.
d: j = 499.5, with T, no parity, quartic repulsion (symplectic case) [4].

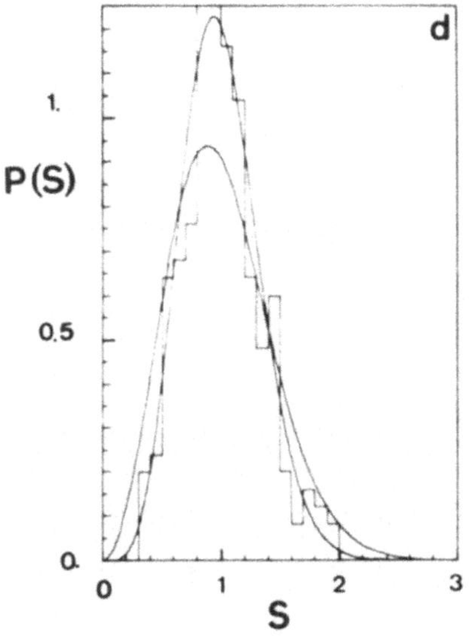

Since U is unitary the eigenvalues are unimodular and can be characterized by eigenphases φ_n,

$$U|n\rangle = e^{i\varphi_n}|n\rangle \,, \tag{3}$$

which all lie in the interval $0 \le \varphi_n < 2\pi$. Upon diagonalizing U for, say, j = 500 one finds that the φ_n tend to cluster for small k but display repulsion for large k. Numerically obtained level spacing histograms for the operator (2) are given in Figs. 2a (clustering limit) and 2b (repulsion limit). The smooth curve in Fig. 2b is an average level spacing distribution for Dyson's orthogonal ensemble of unitary (2j+1) by (2j+1) matrices generated from any one such matrix by arbitrary orthogonal transformations [3].
The level spacings of our U for large k are thus typical of any randomly chosen symmetric unitary matrix of like dimension. Symmetric unitary rather than general unitary matrices are of relevance here because of the following time reversal invariance of our U,

$$\begin{aligned} T\, U\, T^{-1} &= U^{-1} \\ T &= e^{ipJ_y}\, e^{i\pi J_z}\, K \end{aligned} \tag{4}$$

where K is the complex conjugation operation [2]. In fact, the invariance (4) places a restriction on the matrix elements of U so as to leave only one real parameter free in each off-diagonal element.

By breaking the time reversal (4) we obtain matrices U which have two real parameters free in each off-diagonal element and are thus genuinely unitary. Examples of such evolutions have been discussed in [2]. Fig. 2c shows a level spacing histogram pertaining to the level-repulsion limit of

$$U = e^{-i(k'/2j)J_x^2}\, e^{-i(k/2j)J_z^2}\, e^{-ipJ_y} \tag{5}$$

for which no time reversal holds. The smooth curve corresponds to Dyson's unitary ensemble of unitary (2j+1) by (2j+1) matrices [3]. The important difference between the "orthogonal" and the "unitary" case lies in the degree of level repulsion. In both cases the level spacing distribution P(S) rises like a power out of the origin,

$$P(S) \sim S^{\beta} . \tag{6}$$

The exponent β is unity for the evolution (2) and two for the evolution (5).

We have recently succeeded [4] in constructing an evolution operator for kicked spins the level spacing statistics of which pertains to the universality class of Dyson's symplectic ensemble [3]. Group theoretical arguments show that realization of that case requires half integer j, the presence of an antiunitary time reversal invariance and the absence of any other discrete rotation invariance [2c]. As shown in Fig. 2d the level repulsion in that case is quartic, $\beta = 4$. It is interesting to realize that this repulsion enhancement is related to Kramer's degeneracy. Each level φ_n is doubly degenerate, i.e. has two independent eigenvectors $|n\rangle$ and $|\tilde{n}\rangle$ associated with it. Each level pair φ_n, φ_m (with $\varphi_n \neq \varphi_m$) thus gives rise to eight offdiagonal matrix elements of U. The potentially sixteen real parameters in these elements are reduced in number to four by unitarity and time reversal.

The explanation of the success of random matrix theory with respect to the level spacing distribution is a long standing problem. A bit of progress was achieved when the mathematical equivalence of the quantum mechanical eigenvalue problem (3) to the classical Hamiltonian dynamics of a system of N interacting particles [5] in one spatial dimension was recognized [6]. Instead of formulating the equivalence in question for an evolution operator of the form (2), i.e.

$$U = e^{ikV} U_0 \tag{7}$$

with an exactly solvable part U_0 and a perturbation V (not commuting with U_0) we propose to study

$$U = e^{ikV/2} U_0 e^{ikV/2} . \tag{8}$$

Both operators U in (7) and (8) have identical eigenphases φ_n (as well as matrix elements $\langle n|V|m\rangle = V_{nm}$) and are thus equivalent for our purpose. The symmetrized version (8), however, is slightly more convenient to work with. Moreover, we first restrict ourselves to integer j and time reversal invariant dynamics. The eigenphases φ_n as well as the eigenvectors $|n\rangle$ of U depend on the weight k of the perturbation. That k dependence can be described by a set of differential equations obtained by differentiating the eigenvalue equation (3) with respect to k. These differential equations take the form of classical Hamiltonian equations if we associate

$$k \hat{=} \text{time}$$

$$\varphi_n \hat{=} \text{position on a unit circle}$$

$$\langle n|V|n\rangle \hat{=} \text{momentum } p_n$$

$$4 \sin\left[(\varphi_n-\varphi_m)/2\right] \langle n|V|m\rangle \hat{=} \text{angular momentum } \ell_{nm} ,$$

use angular momentum Poisson brackets for the ℓ_{ij} [7,8] and the N particle Hamiltonian

$$H = \frac{1}{2} \sum_{n=1}^{N} p_n^2 + \frac{1}{32} \sum_{n\neq m} \frac{\ell_{nm}^2}{\sin^2\left[(\varphi_n-\varphi_m)/2\right]} = \frac{1}{2} \operatorname{tr} V^2 . \tag{9}$$

Note that the interaction potential in (9) is repulsive.

For our purpose it is most convenient to write Hamilton's equations for the fictitious N particle system as differential equations for two N by N matrices L and F,

$$\frac{dL}{dk} = [M,L] , \quad \frac{dF}{dk} = [M,F] , \tag{10}$$

where [,] denote commutators. The matrix M occuring here plays the formal role of a generator of infinitesimal "time" translations. All three matrices L, F, and M are composed of the coordinates φ, the momenta p, and the angular momenta ℓ,

$$L_{ij} = p_i \delta_{ij} + \frac{1}{4} (1 - \delta_{ij}) \ell_{ij} \cot \left[(\varphi_i-\varphi_j)/2\right]$$

$$F_{ij} = \frac{1}{4} (1 - \delta_{ij}) \ell_{ij} \tag{11}$$

$$M_{ij} = \frac{1}{8} (1 - \delta_{ij}) \ell_{ij}/\sin^2 \left[(\varphi_i-\varphi_j)/2\right] .$$

Due to the commutator structure of the equations of motion (10) and the cyclic invariance of the trace of matrix products we immediately find the traces of arbitrary products of L and F,

$$C = \operatorname{tr} (L^\mu F^\nu L^\sigma F^\tau \ldots) , \tag{12}$$

to be constants of the motion [9]. Eqs. (10,11,12) remain valid even for halfinteger j and for dynamics without time reversal invariance.

We are now equipped for a statistical mechanical discussion. Let us first take our fictitious N particle system to be in equilibrium. Presumably, such a "time"-independent equilibrium behavior is reached for large k. Any normalizable function of the constants of the motion (12) is a potential candidate for an equilibrium phase space probability density of the φ, p, and ℓ. Were we concerned with real particles it would be natural to admit only the energy (9).Since H does not appear to have any clearcut physical meaning for the original quantum mechanical eigenvalue problem we see no reason to prefer it to other constants of the motion. We therefore propose to explore a generalized canonical ensemble in which all independent constants are admitted,

$$\rho\ (\varphi,p,\ell)\ =\ Z^{-1}\ \exp\left[-\sum_n \alpha_n C_n\right] \tag{13}$$

with Lagrange multipliers α_n determined by the ensemble means of the C_n.

The distributions (13) contain more information than we are interested in. Upon integrating out the momenta p and angular momenta ℓ we obtain probability densities for the coordinates, i.e. for the eigenphases of the evolution operator (7),

$$P(\varphi)\ =\ \int \left[\prod_{n=1}^{N} d\ p_n\right] \left[\prod_{n<m} d^{\beta}\ \ell_{nm}\right]\ \rho(\varphi,p,\ell)\ . \tag{14}$$

The index β in the angular momentum differentials indicates whether each particle pair (i.e. pair of eigenphases of U) is represented by one, two, or four independent real parameters in the angular momentum ℓ_{nm} (i.e. the offdiagonal elements V_{nm} of the perturbation V in U). The index β is determined by the symmetries of U as discussed above.

The positions φ_i enter the probability density $P(\varphi)$ only through the matrix L in (12). There would thus be no φ dependence of $P(\varphi)$ at all if only constants of the motion of the form tr F^{ν} were admitted in (13), a case we exclude from the following consideration. Incidentally, the energy $H = \frac{1}{2}$ tr (L^2-F^2), perhaps the most natural constant of the motion to be included in (13) provides $P(\varphi)$ with a nontrivial φ dependence. We should now appreciate that the constants C involving factors L and thus the canonical distribution (13) depend singularly on the spacings $\varphi_i-\varphi_j$. Obviously, this means level repulsion in the spectrum of the evolution operator (7) for large enough k. To bring about the level repulsion implied by the reduced distribution $P(\varphi)$ it is convenient to change integration variables in (14) as

$$\ell_{ij} = \tilde{\ell}_{ij} \sin\left[(\varphi_i - \varphi_j)/2\right] . \tag{15}$$

The corresponding change in the integration measure is

$$d^\beta \ell_{ij} = \left|\sin\left[(\varphi_i - \varphi_j)/2\right]\right|^\beta d^\beta \tilde{\ell}_{ij} \tag{16}$$

and we thus arrive at

$$P(\varphi) = \left[\prod_{i<j} \left|\sin\left[(\varphi_i - \varphi_j)/2\right]\right|^\beta\right] p(\varphi) \tag{17}$$

where

$$p(\varphi) = \int \left[\prod_i d\, p_i\right] \left[\prod_{i<j} d^\beta \tilde{\ell}_{ij}\right] \rho \tag{18}$$

no longer vanishes for $\Delta\varphi = 0$. The behavior of $P(\varphi)$ for $\Delta\varphi \to 0$, i.e. the degree of level repulsion in the quantum spectrum of U for large k is thus seen to be dominated by the first factor in (17) and to be independent of the choice of the constants C admitted in the ensemble (13). The remaining factor in (17), $p(\varphi)$, takes care of the behavior of $P(\varphi)$ at large spacings and does depend on the choice of the constants C. Further investigations are necessary to clarify the role of $p(\varphi)$ for the level spacing distribution and other spectral characteristics.

From the statistical mechanical point of view adopted here the transition from "regular" level clustering to "irregular" level repulsion with increasing weight k of the perturbation V in (7) appears as a relaxation into equilibrium. To shed some light on the equilibration process we consider two constants of the motion for the fictitions N particle system. Of the two constants of the motion we have in view one is

$$C_0 = \frac{1}{2}\sum_n p_n^2 + \frac{1}{32}\sum_{n\neq m} \frac{|\ell_{nm}|^2}{\sin^2\left[(\varphi_n - \varphi_m)/2\right]} = \frac{1}{2}\,\mathrm{tr}\,(L^2 - F^2) \tag{19}$$

(which for integer j and time reversal invariant dynamics is the Hamiltonian (9)) and the other

$$C_1 = \sum_{n\neq m} |\ell_{nm}|^2 = \mathrm{tr}\, F^2 . \tag{20}$$

Conservation of C_1 implies that the dynamics of the N particle system takes place on a hypersphere in the subspace spanned by the angular momenta ℓ. At $k = 0$, then, it is typical for the perturbation V to have nonvanishing matrix elements V_{nm} in the eigenrepresentation of $U(0) = U_0$ only close to the diagonal [10]. Consequently, the number of initially nonvanishing angular momenta ℓ_{nm} is not $N(N-1)$ but rather only proportional to N. As k increases, however, the matrix V tends to fill up with nonzero elements and more and more angular momenta ℓ_{nm} take on appreciable values. This is obvious from the equation of motion for ℓ_{nm} following from the equations (10),

$$\frac{d}{dk}\,\ell_{nm} = \frac{1}{8} \sum_{i(\neq n,m)} \ell_{ni}\,\ell_{im} \left\{\sin^{-2}\left[(\varphi_n-\varphi_i)/2\right] - \sin^{-2}\left[(\varphi_m-\varphi_i)/2\right]\right\} . \tag{21}$$

Therefore we expect $\sum_{n\neq m} |\ell_{nm}|$ to increase with k. From energy conservation for the fictitious particles we infer that close approach of two particles is possible for all pairs with vanishing ℓ_{nm}. Indeed, (19) implies a lower bound for the distance of any pair of particles,

$$|\varphi_n - \varphi_m| \geq \frac{1}{2}\,|\ell_{nm}|/\sqrt{C_0} . \tag{22}$$

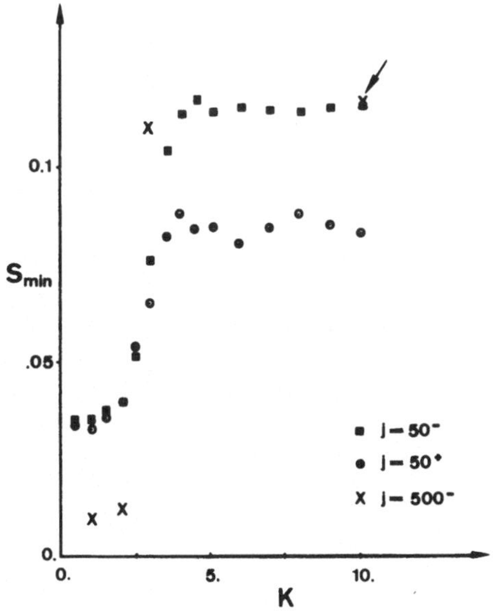

Fig. 3. Typical lower bound $S = \Delta\varphi/\overline{\Delta\varphi}$ for level spacings (defined in the text) versus kick strength, for $j = 50$ and $j = 500$.

At k = 0 only a tiny fraction (~ 1/N) of the particle pairs has close approach energetically forbidden. But as more and more angular momenta become finite in magnitude with increasing k the fraction of particle pairs with forbidden close encounters grows. Equilibrium will be reached when nonzero angular momenta ℓ_{nm} force all particles to stay apart from oneanother at distances of the order of the mean level spacing.

As a rough estimate for typical smallest level spacings we might take the average

$$\Delta\varphi = \frac{1}{N(N-1)} \sum_{n \neq m} \frac{1}{2} |\ell_{nm}|/\sqrt{\tau_0} \tag{23}$$

which has the mean level spacing

$$\overline{\Delta\varphi} = 2\pi/N \tag{24}$$

as upper bound. Our above arguments indicate that $\Delta\varphi$ should increase with k. Fig. 3 confirms that prediction for the kicked top described by the evolution operator (2). As one would expect in the vicinity of the integrable case (k = 0) the normalized quantity $S = \Delta\varphi/\overline{\Delta\varphi}$ goes to zero for increasing N = j. It is interesting to see S to display a rather pronounced growth in the range $2.5 \lesssim k \lesssim 3.5$ which is precisely the range in which the classical top makes the transition from dominantly regular to dominantly chaotic behavior (see Fig. 1). The saturation of S for $k \gtrsim 3.5$ corresponds to the practically complete coverage of the phase space of the classical top with chaotic trajectories. Although the saturation value seems to be independent of j, differences between the two parities are visible.

We gratefully acknowledge support by the Alexander-von-Humboldt-Stiftung and the Gesellschaft von Freunden und Förderern der Universität-Gesamthochschule Essen. Special thanks for discussion and help are due to G. Eilenberger.

REFERENCES

1) F. Waldner, D. R. Barberis, and H. Yamazaki, Phys. Rev. **A31**, 420 (1985).

2)a) F. Haake, M. Kuś, J. Mostowski, and R. Scharf in "Coherence, Cooperation, and Fluctuations", F. Haake, L. Narducci, and D. Wall (eds). p. 220, Cambridge University Press 1986,

b) F. Haake, M. Kuś, and R. Scharf, Z. Physik **B65**, 381 (1987),

c) M. Kuś, R. Scharf, and F. Haake, Z. Physik **B66**, 129 (1987).

3) F. J. Dyson, J. Math. Phys. **B3**, 1199 (1962).

4) F. Haake, M. Kuś, and R. Scharf to appear in "Quantum Chaos", E. R. Pike (ed.), (Plenum, New York).

5) $N = 2j+1$ if there is no conserved k independent parity. Otherwise N is the dimension of an irreducible representation of U_0 and V.

6)a) P. Pechukas, Phys. Rev. Lett. **51**, 943 (1983),

b) T. Yukawa, Phys. Rev. Lett. **54**, 1883 (1985),

c) T. Yukawa, Phys. Lett. **116**, 227 (1986).

7) $\{\ell_{nm}, \ell_{ik}\} = \ell_{nk}\,\delta_{mi} + \ell_{in}\,\delta_{mk} + \ell_{km}\,\delta_{ni} + \ell_{mi}\,\delta_{nk}$

8) S. Wojciechowski, Phys. Lett. **111A**, 101 (1985).

9) Clearly, not all of these constants are independent; we do not know whether all constants of the motion can be written in this form.

10) In the case (2) $V_{nm}(k = 0)$ is tridiagonal because there exists a conserved k independent parity.

LOCALIZATION AND DELOCALIZATION IN A DISSIPATIVE QUANTUM MAP

Th. Dittrich and R. Graham
Fachbereich Physik, Universität Essen GHS
D-4300 Essen, F.R. Germany

Abstract

The time evolution of the density matrix generated by a quantum version of the 'standard map' (kicked rotator) with dissipation is studied. The quantum ensemble corresponding to the classical invariant measure on the strange attractor is obtained and discussed. The influence of weak dissipation on localization phenomena in the quantized standard map is assessed qualitatively and studied quantitatively in numerical experiments. Complete delocalization is found for sufficiently strong dissipation. For extremely weak dissipation a remnant of quantum localization is found to survive even in the steady state.

1. Introduction

In recent years there has been a strong interest in the behavior of quantum systems under the influence of intense externally applied periodic fields (cf. reviews given in [1], [2],[3]). Examples which have been investigated both theoretically and experimentally include molecules vibrationally excited by strong infrared laser fields and Rydberg atoms in strong microwave fields. From these studies it has become clear (i) that the classical dynamics of these systems is chaotic, and (ii) that this property alone already accounts for the main experimental results, such as the 'fluence dependence' of the average number of absorbed quanta in infrared excitation of molecules [3], and the observed strong multi-photon ionization of Rydberg atoms in non-resonant microwave fields below the ionization threshold in corresponding static fields [4].

However, given the fact that one is really dealing with quantum systems in these experiments the question has naturally been asked what the quantum corrections to the chaotic classical behavior should be. This question has been dealt with in a vast literature, an important part of which has concerned itself with two main subjects: (i) the investigation of a simple model system, the 'periodically kicked planar rotator' which is classically equivalent to Chirikov's 'standard map' [5] and

(ii) the analysis of a physically more realistic model of a 1-dimensional hydrogen atom in an external microwave field [6]. From this work one has learned that important quantum effects in periodically driven classically chaotic systems can, in fact, exist, arising from subtle coherence effects in the quantum wave function describing a system which is classically chaotic.

So far, these effects are best understood in the example of the kicked rotator to which we shall confine our discussion in the following. A rather close physical realization of this model by the microwave excitation of rotational bands in a diatomic molecule has recently been proposed [7]. Let us briefly recall some of the important results for the kicked rotator. Its Hamiltonian takes the form

$$H = \frac{p^2}{2} - \frac{K}{(2\pi)^2} \cos 2\pi q \sum_{n=-\infty}^{\infty} \delta(t-n) , \tag{1.1}$$

where we have chosen units in which the moment of inertia of the rotator and the period of the kicks is unitary. The parameter K is a measure of the strength of the kicks. Integrating the canonical equations of motion following from (1.1) between two times n , $n+1$ immediately preceding two subsequent kicks we obtain the 'standard map'

$$\begin{aligned} p_{n+1} &= p_n - \frac{K}{2\pi} \sin 2\pi q_n , \\ q_{n+1} &= (p_{n+1} + q_n) \pmod 1 , \end{aligned} \tag{1.2}$$

where $q_n := q(n)$, $p_n := p(n-\varepsilon)$, and $\varepsilon \to 0^+$. Eqs. (1.2) may be read as classical or as quantum mechanical Heisenberg equations of motion with the Poisson bracket or commutator

$$\{p,q\} \mathrel{\hat{=}} \frac{i}{\hbar}[p,q] = 1 . \tag{1.3}$$

The quantum map in the Schrödinger picture takes the form

$$|\psi_{n+1}\rangle = U|\psi_n\rangle = e^{-i\frac{p^2}{2\hbar}} e^{\frac{i}{\hbar}\frac{K}{4\pi^2}\cos 2\pi q} |\psi_n\rangle . \tag{1.4}$$

The classical map for $K>0$ behaves like a typical non-integrable Hamiltonian system which is neither integrable nor fully chaotic [8]. For $0<K<4$ the classical map has stable 1-cycles (0, m) where m are the integers. For $0<K<2$ there are also stable 2-cycles (0, m + 1/2), (1/2, m + 1/2) whose angular momentum is in between that of the 1-cycles. For $K< K_c \simeq 0.9716$ the 1-cycles and the 2-cycles are separated by a continuous KAM curve $p=p(q)$ $(0\leq q\leq 1)$. In this domain chaos exists only locally in phase space. For $K>K_c$ the last KAM torus has broken down and there is chaos on a large scale in phase space. Typically, the mean square angular momentum grows diffusively, i.e. linear in n, in this regime $\langle p^2\rangle = D(K)n$. Nevertheless for arbitrarily large values of K there also exist windows of stability for cycles of period 2 or higher and for 'accelerator modes' ($\frac{1}{2\pi}\mathrm{Arctan}\frac{l}{K}$, $m+ln$) for integer l whose angular momentum increases linearly with the number of kicks.

Quantum mechanically, there is strong numerical and theoretical evidence [5] that the diffusive motion of $\langle p^2\rangle$ is destroyed by coherence effects of the wavefunction. The Floquet states of (1.4) satisfying

$$e^{-i\frac{p^2}{2\hbar}}\, e^{\frac{i}{\hbar}\frac{K}{4\pi^2}\cos 2\pi q}\,|U_\lambda\rangle = e^{-i\omega_\lambda}\,|U_\lambda\rangle \qquad (1.5)$$

for typical values of $K>K_c$, seem to be exponentially localized in the angular momentum representation due to a mechanism similar to Anderson localization in disordered systems in real space [9]. As a result any initial wavefunction $|\psi_0\rangle$ localized in p has appreciable overlap only with a finite number of Floquet states spanning a finite dimensional subspace and remains in that subspace under the action of (1.4). $|\psi_n\rangle$ therefore remains localized in the angular momentum representation in the course of time precluding diffusion of angular momentum. There are exceptions to this typical case, however. In the case of 'quantum resonances' where in the chosen units

$$2\pi\hbar \in \mathbb{Q}\,, \qquad (1.6)$$

the spectrum of Floquet exponents ω_λ in (1.5) is continuous, the Floquet states (1.5) are extended, and the mean square angular momentum for large times increases quadratically in n [10]. Again, this is a quantum mechanical coherence phenomenon, which supersedes the classical diffusive behavior. Furthermore, it has been shown that there is an infinite set of measure zero of irrational values of $2\pi\hbar$ where (1.5) has a possibly singular continuous spectrum [11]. The physical consequences of this mathematically interesting result seem not quite clear at present.

The quantum effects which have been mentioned occur as a consequence of coherence and are therefore very fragile against any perturbations reducing coherence in a wavepacket [12]. Damping mechanisms are very important perturbations of this kind. It is the purpose of the present contribution to review our recent work concerning the influence of quantum effects in the dissipative standard map [13 - 16]. A physical system approximately realizing this case is a diatomic molecule in a surrounding medium inducing frictional effects excited rotationally by a microwave field. The classical map, in this case, takes the form

$$p_{n+1} = \lambda p_n - \frac{K}{2\pi}\sin 2\pi q_n ,$$
$$q_{n+1} = (q_n + p_{n+1}) \,(\mathrm{mod}\, 1), \tag{1.7}$$

with $0 \leq \lambda < 1$. Under the map (1.7) phase space volume is reduced by a factor λ in each step. Therefore the ω-limit sets of eq. (1.7) (i.e. the invariant manifolds approached for $n \to \infty$) have zero volume in phase space and are either fixed points, limit cycles, or strange attractors with fractal dimension $d < 2$. As is easily seen from the first of eqs. (1.7), the ω-limit sets must all lie in the compact region

$$|p| \leq \frac{K}{2\pi(1-\lambda)} . \tag{1.8}$$

For $\nu := (1-\lambda) \to 0$ the ω-limit sets must approach invariant manifolds of the conservative systems and the bound (1.8) is removed to infinity. A discussion of the periodic orbits of the classical map has been given in [17]. In the following sections we first describe a quantum version of the map (1.7) in the form of a master equation (section 2). Then we consider the stationary state of the quantum map (section 3), i.e. quantum effects on the invariant probability distribution around a strange attractor. Finally, in section 4 we turn to dynamical results for the mean square angular momentum of the rotator and consider the fate of localization and quantum coherence as a function of the dissipation parameter $\nu = (1-\lambda)$.

2. The Quantum Map

A quantum map defines the transformation of a state at the discrete time n to a state at the subsequent time n+1. For a conservative quantum system a state is described by the wavefunction $|\psi_n\rangle$ and the quantum map is given by a unitary operator U as in eq. (1.4). The state of a dissipative quantum system, on the other hand, is defined by the density matrix ρ_n. A quantum map $\rho_n \to \rho_{n+1}$ can still be defined in this case, provided the knowledge of ρ_n is sufficient, in principle, to predict the state ρ_{n+1}. This is only possible, if memory effects are negligible, i.e. if the time-evolution of the density matrix is Markovian, which we assume in all that follows. Then an operator (more precisely a superoperator) $\underline{G}$ in the linear space of density matrices exists which defines the map

$$\rho_{n+1} = \underline{G}\,\rho_n . \tag{2.2}$$

The operator $\underline{G}$ must preserve the normalization, positivity and hermiticity of the density matrix ρ and must be consistent with the uncertainty principle, which has e.g. the consequence

$$\mathrm{Tr}\,\rho_n^2 \leq 1, \tag{2.3}$$

where equality holds for a pure state.

An operator $\underset{\sim}{G}$ with these and the additional property that the classical limit of the quantum map reproduces the classical map (1.7) has been given in [13], but, as stated there, even within these restrictions, it is not unique. For simplicity the operator $\underset{\sim}{G}$ was chosen as a product $\underset{\sim}{G} = \underset{\sim}{U} \cdot \underset{\sim}{D}$ where $\underset{\sim}{U}$ is a unitary operator in the space of density matrices equivalent to (2.1) if applied to a pure state, and $\underset{\sim}{D}$ is a real operator describing a purely dissipative process.

This factorization of the quantum map corresponds to a factorization of the classical map into a purely dissipative step

$$\begin{aligned} p_{n+\frac{1}{2}} &= \lambda p_n, \\ q_{n+\frac{1}{2}} &= q_n, \end{aligned} \tag{2.4}$$

and the purely conservative map

$$\begin{aligned} p_{n+1} &= p_{n+\frac{1}{2}} - \frac{K}{2\pi} \sin 2\pi q_{n+\frac{1}{2}}, \\ q_{n+1} &= \left(q_{n+\frac{1}{2}} + p_{n+1}\right) (\mathrm{mod}\, 1), \end{aligned} \tag{2.5}$$

which, together, give back the dissipative map (1.7). Quantum mechanically, the dissipative step (2.4) is described by the solution of the initial value problem of the Markovian master equation

$$\frac{\partial \rho}{\partial t} = \underset{\sim}{L}\rho = \frac{1}{2} |\ln \lambda| \left([\Gamma, \rho\Gamma^+] + [\Gamma\rho, \Gamma^+] \right), \tag{2.6}$$

with $\Gamma^+ := (\Gamma)^+$ and Γ defined by

$$\Gamma := \sum_{l \geq 0} \sqrt{|l|}\, |l-1\rangle\langle l| + \sum_{l \leq 0} \sqrt{|l|}\, |l+1\rangle\langle l|. \tag{2.7}$$

In (2.7) we use the angular momentum representation defined by

$$p|l\rangle = 2\pi\hbar l |l\rangle. \tag{2.8}$$

Eq. (2.6) describes the absorption of quanta of the absolute value of $|p|$ by a reservoir. We use the solution of eq. (2.6) to define $\underset{\sim}{D}$ by

$$\rho_{n+\frac{1}{2}} = \underline{D}\,\rho_n = \exp(\underline{L})\,\rho_n\,, \qquad (2.9)$$

and $\underline{U}$ by

$$\rho_{n+1} = \underline{U}\,\rho_{n+\frac{1}{2}} = U\rho_n U^+\,, \qquad (2.10)$$

where the unitary operator U in the space of state vectors is given by eq. (1.4). Explicitly, the quantum map in angular momentum representation then takes the form

$$\langle l'|\rho_{n+1}|m'\rangle = \sum_{l,m} G(l',m'|l,m)\langle l|\rho_n|m\rangle\,, \qquad (2.11)$$

with

$$G(l',m'|l,m) = \lambda^{\frac{1}{2}(|l|+|m|)}\Big[U(l',m'|l,m) + \\ +\Theta_{lm,0}\sum_{j=1}^{\min(|l|,|m|)}\left(\frac{1-\lambda}{\lambda}\right)^j\sqrt{\binom{|l|}{j}\binom{|m|}{j}}\;U\!\left(l',m'|l-\frac{l}{|l|}j,\,m-\frac{m}{|m|}j\right)\Big]\,, \qquad (2.12)$$

$$U(l',m'|l,m) := \langle l'|U|l\rangle\langle m|U^+|m'\rangle\,. \qquad (2.13)$$

Here $\Theta_{p,0}$ is unity for $p\geq 0$ and vanishes otherwise. The operator $\underline{G}$, according to eq. (2.12) is given by a sum

$$\underline{G} = \sum_j \underline{G}_j\,,$$

where each of the operators $\underline{G}_j$ describes the propagation from time n to n+1 with the absorption of j quanta of $|p|$ by the reservoir. The superposition of the $\underline{G}_j$ in eq. (2.12) shows that processes with different values of j are mutually incoherent, while coherence is preserved for all amplitudes from time n to n+1 with the same value of j. This competition between coherent and incoherent processes is the main feature of $\underline{G}$. Details of course depend on the assumptions we made on the form of the master equation. However, the general fact that some amplitudes add coherently while others add incoherently is independent of these assumptions, and we can therefore hope that the map (2.12) can give us insight in the qualitative consequences of this competition. In the following it is useful to introduce the representation of the density matrix by the Wigner quasi-probability density in the phase space $0\leq q<1, -\infty<p<\infty$. It is defined by

$$W_n(p,q) := \sum_{l=-\infty}^{\infty} W_n^{(l)}(q)\,\delta(p-\pi\hbar l), \tag{2.14}$$

$$W_n^{(l)}(q) = \sum_{l'=-\infty}^{\infty} \frac{1+(-1)^{l+l'}}{2}\, e^{2\pi i l' q} \langle \tfrac{l+l'}{2} | \varrho | \tfrac{l-l'}{2} \rangle . \tag{2.15}$$

Two somewhat unusual features of this representation are worth pointing out. One is the fact that $W(p,q)$ has support in phase space not only on the quantized values of angular momentum $p_m = 2\pi\hbar m$ with integer m but also at momenta p_m where m is half-integer, as is shown by eq. (2.14). A second related peculiarity follows from eq. (2.15) which shows that $W_n^{(l)}(q)$ for $l=2m$ is periodic in q with period 1/2 rather than 1, while $W_n^{(l)}(q)$ with $l=2m+1$ is antiperiodic in q (i.e. changes sign) with period 1/2. It follows that the unphysical values of angular momentum drop out if $W_n(p,q)$ is integrated over the unit interval of q to yield the probability distribution of angular momentum. The Wigner representation is particularly useful for a discussion of the classical limit of the quantum map and the leading quantum corrections. Detailed derivations are given in [13] and we merely present the result here. Instead of presenting the map for the Wigner quasi-probability distribution [13] it is rather more transparent to write down an equivalent 'quasi-stochastic' map [18]. It takes the form

$$\begin{aligned} p_{n+1} &= \lambda p_n - \frac{K}{2\pi}\sin\left(2\pi(q_n+\xi_{n+1})\right)+\eta_{n+1}, \\ q_{n+1} &= \left(q_n + p_{n+1} + \xi_{n+1}\right) (\mathrm{mod}\ 1), \end{aligned} \tag{2.16}$$

where the 'quasi-noise' terms η_{n+1} and ξ_{n+1} are uncorrelated for different values of n (expressing the basic Markov assumption we made) and have non-vanishing second and third order cumulants

$$\begin{aligned} \langle \xi_n^2 \rangle &= \frac{(1-\lambda)\hbar}{32\pi^3\lambda|p_{n-1}|}, \\ \langle \eta_n^2 \rangle &= \frac{\hbar}{2\pi}\lambda(1-\lambda)|p_{n-1}|, \\ \langle \xi_n \eta_n \rangle &= 0, \end{aligned} \tag{2.17}$$

$$\langle \eta_n^3 \rangle = -\frac{\hbar^2}{2^{11}\pi^3} K \sin 2\pi q_{n-1} \,. \tag{2.18}$$

Higher order cumulants occur also, but are proportional to higher powers of $\hbar$ and hence negligible for $\hbar$ sufficiently small. The fact that third order cumulants of η_n appear has the consequence that η_n is distributed with a quasi-probability density which is not everywhere non-negative. Hence, η_n cannot be simulated by classical noise, except in an approximation where the third order cumulant is neglected. In the latter approximation the quantum map is indistinguishable from a classical map with noise.

The map (2.16) is stochastically equivalent to a map for a quasi-probability density $\tilde{W}_n(p,q)$ from which the Wigner quasi-probability density (2.15) can be generated by

$$W_n^{(l)}(q) = \frac{1}{2}\left[\tilde{W}_n(\pi\hbar l, q) + (-1)^l\, \tilde{W}_n(\pi\hbar l, q+\tfrac{1}{2})\right]. \tag{2.19}$$

Hence, neglecting third and higher order cumulants, eq. (2.16) can be used to generate a semi-classical approximation for the Wigner function by stochastic simulation.

Semi-classical approximations of this kind have been used in earlier work on quantum effects in chaotic optical systems like the complex Lorenz model (single-mode laser) [19] and second harmonic generation [20]. In these more realistic systems it has so far not been possible to improve on this approximation, while this <u>is</u> possible in our present quantum map. Therefore, it is interesting to compare exact and semi-classical results in some cases in the following.

3. Steady State

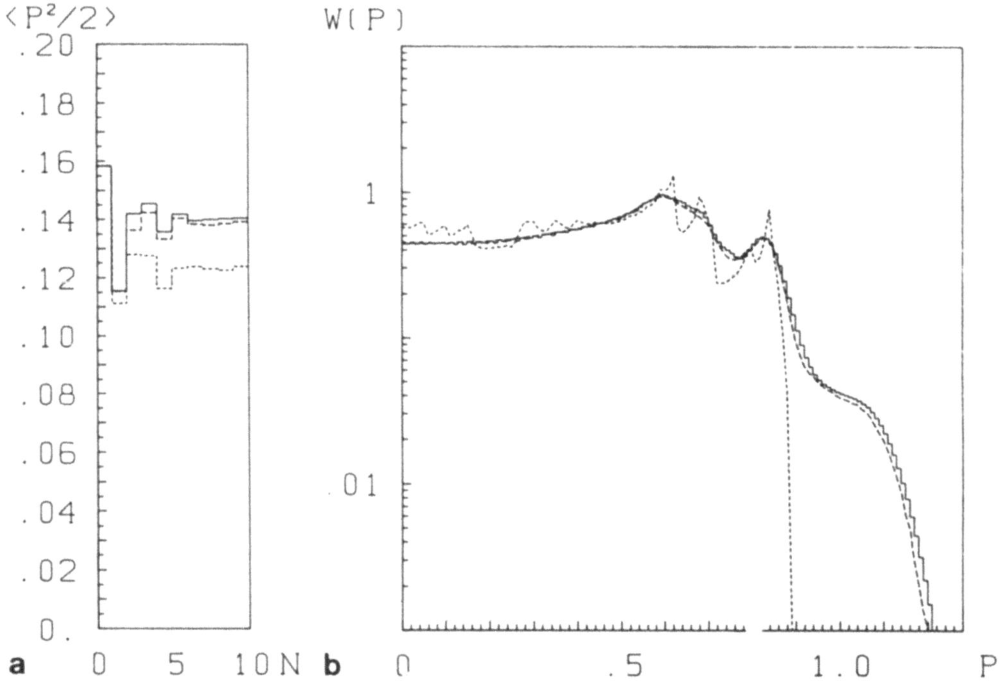

Fig. 1

Time evolution of the total energy over the first ten iterations (a) and reduced angular momentum distribution in the stationary state (b), shown for the quantum mechanical (solid line), semiclassical (coarse-dashed line), and classical (fine-dashed line) cases. The parameter values are K = 5.0, λ = 0.3, and for the quantum mechanical and the semiclassical cases $\hbar = 0.01/2\pi$.

In this section we present some numerical results [14] for the quantum map (2.11) for the parameter values K = 5.0, λ = 0.3, $\hbar = 0.01/2\pi$. (We recall that $\hbar$ is given in units in which the moment of inertia of the rotator and the kicking period are unity). The initial state chosen in the numerical work is always the zero angular momentum eigenstate $\rho_0 = |0\rangle\langle 0|$. For the value of $\hbar$ chosen one expects to reach a steady state after a relaxation time $\Delta n \simeq (1-\lambda)^{-1}$, i.e. after only a few time steps. In fig. 1a we plot the averaged kinetic energy of the rotator as a function of the discrete time n for the classical map (fine-dashed line), the semi-classical map neglecting third order cumulants (coarse-dashed line) and the full quantum map (full line). A steady state is apparently reached after a relaxation time which is not influenced by quantum effects. There are quantum effects visible in the size of the averaged kinetic energy, but it can also be seen that these effects are very well approximated by the semi-classical noisy map. In fig. 1b we also show the full probability distribution of angular momentum in the steady state in the three cases. Only the region $p>0$ is shown because of symmetry. Again quite drastic quantum effects are visible, but they are well described by the noisy map. We now return to results for the full phase-space distribution in the steady state. In figs. 2a-d various phase-space functions are plotted against q for fixed values of p over a base-line whose vertical position gives the respective value of p. This method was found to produce quite clear pictures of the phase-space functions. In fig. 2c the exact Wigner distribution $W^{(l)}(q)$ is plotted in this way, fig. 2b gives the corresponding semi-classical result $\tilde{W}(p,q)$, fig. 2a shows the classical phase-space distribution. The (in principle infinitely) nested structure of the classical strange attractor is seen to be smoothened by the quantum noise already in the semi-classical distribution. The full Wigner function $W(p,q)$ is seen to differ from $\tilde{W}(p,q)$ by the obvious 'kinematical' effects of quantization (only $p=\pi\hbar l$ with integer l occurs in the Wigner function) and doubled periodicity described by eq. (2.19). In addition, however, wavy patterns are seen to occur in $W(p,q)$ (making $W(p,q)$ non positive-definite) which are not at all present in the semi-classical result.These patterns are therefore produced by higher order cumulants in the 'quasi-stochastic' map. Finally, in fig. 2d we show the corresponding result for the positive Q-function [21]

$$Q(p,q) := \langle \alpha | \varrho | \alpha \rangle, \tag{3.1}$$

where $|\alpha\rangle$ is a coherent state with amplitude $\alpha = \sqrt{2\hbar}\,(\,q(\mathrm{mod}\ 1) + ip)$ is related to $W(p,q)$ by

$$Q(p,q) = \frac{4\hbar}{\pi} \int d^2\alpha' \, e^{-2|\alpha-\alpha'|^2} W(p',q') \tag{3.2}$$

and appears correspondingly broadened. In contrast to $W(p,q)$, the function $Q(p,q)$ is, in principle, observable. It is the best locally resolved measurable phase-space distribution in quantum theory and therefore the best one can do, in principle, to observe local features of the Wigner distribution. We notice that all wavy features of the Wigner distribution have disappeared in Q . In fact the substitution of $\tilde{W}(p,q)$ for $W(p,q)$ produces a practically indistinguishable result. We see therefore that observable results in the present case are well reproduced by the semi-classical approximation. We can expect this to be true whenever the condition

$$\sqrt[3]{|\langle \eta^3 \rangle|} \Big/ \sqrt{\langle \eta^2 \rangle} \ll 1 \tag{3.3}$$

is satisfied for the cumulants in eqs. (2.17), (2.18). This condition is well satisfied for the parameter values chosen here.

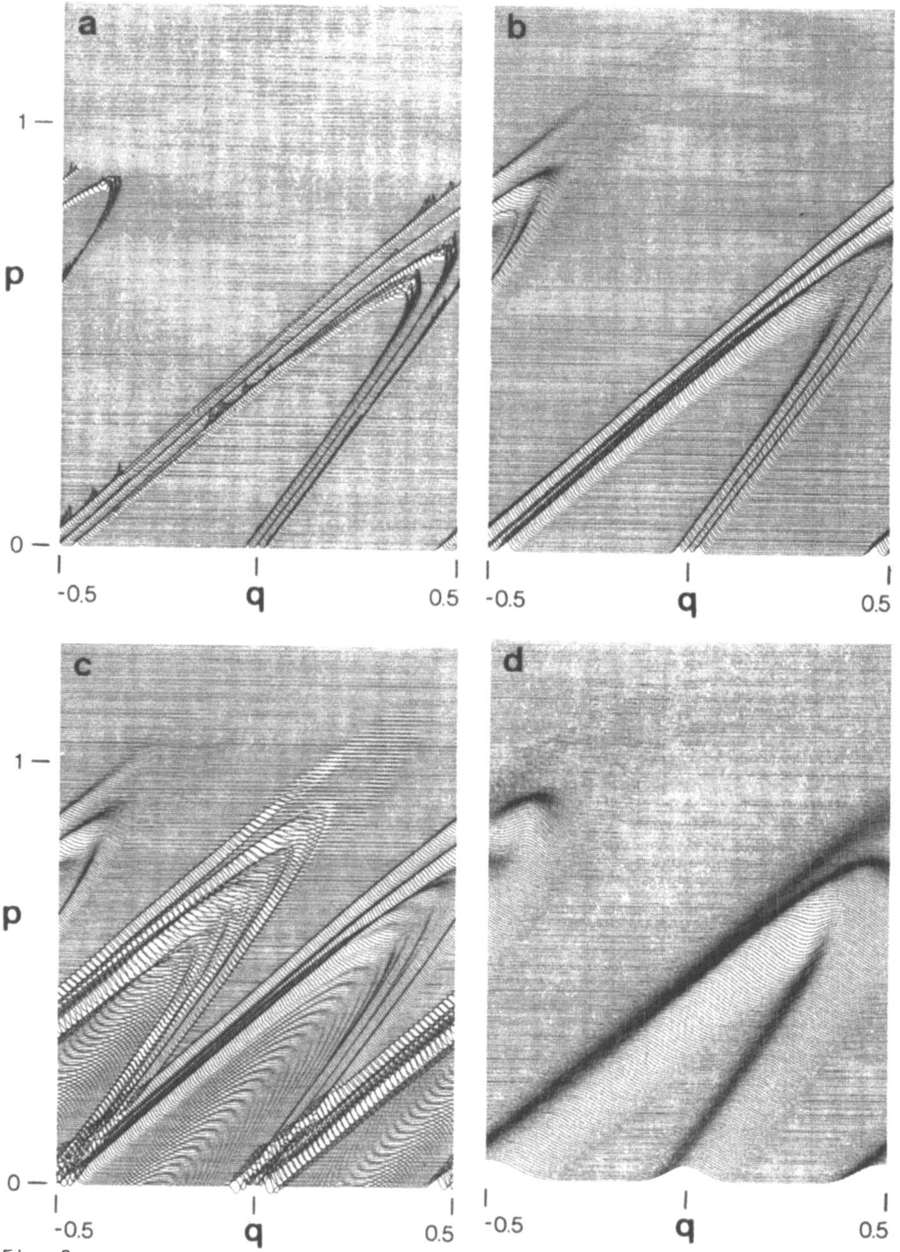

Fig. 2

Phase space distribution functions in the stationary state. Parts (c,d) results of quantum mechanical calculations, parts (a,b) results of iterations of the classical map, in (b) with quantum noise simulated in the semiclassical approximation. Part (d) is a coarse-grained version of part (c).
The parameter values are $K = 5.0$, $\lambda = 0.3$, $\hbar = 0.01/2\pi$ (b-d), $\hbar = 0$(a).
The distribution functions are point symmetric with respect to the origin and periodic in phase; only the upper half of one period is shown.

4. Localization and Delocalization

As has been mentioned in the introduction, the Floquet (quasi-energy) states of the conservative quantum map determined from eq. (2.5) localized in the action variable for typical values of $\hbar$. Floquet states $|U_\lambda\rangle$ which are neighboured in the action variable within twice the localization length L have finite overlap and their eigenphases repel each other. Hence their mean spacing of eigenvalues can be estimated as $\delta\omega_\lambda \simeq 2\pi/2L$. For times $n<n^*=2\pi/\delta\omega_\lambda$ the uncertainty principle tells us that the quantization of Floquet states cannot yet be resolved in the time evolution of an initially localized wave packet. Hence, there is classical chaotic diffusion of the action variable for times $n<n^*$ and one estimates from eq. (1.2) (if $p_0=0$)

$$\langle p_n^2\rangle \simeq \frac{1}{2}\left(\frac{K}{2\pi}\right)^2 n. \tag{4.1}$$

One can use this to estimate the localization length L of the conservative map in a self-consistent fashion as the momentum scale $p^*=2\pi\hbar L$ where the classical diffusion has to break down [22]. Hence

$$(2\pi\hbar L)^2 \simeq \frac{1}{2}\left(\frac{K}{2\pi}\right)^2 n^* = \frac{1}{2}\left(\frac{K}{2\pi}\right)^2 \frac{2\pi}{\delta\omega_\lambda} \propto L\left(\frac{K}{2\pi}\right)^2, \tag{4.2}$$

and one concludes that

$$L \simeq n^*/2 \simeq \left(\frac{K}{4\pi^2\hbar}\right)^2, \tag{4.3}$$

as an estimate of order of magnitudes. How is this simple picture changed by dissipation [13], [15]? New important time scales are introduced by dissipation, namely the classical relaxation time to the steady state $n_d=(1-\lambda)^{-1}$ and the mean life-time of a quasi-energy state $|U_\lambda\rangle$ due to dissipation $n_\lambda=[(1-\lambda)\langle U_\lambda||p||U_\lambda\rangle/2\pi\hbar]^{-1}$. The latter life-time can be computed from the master equation. If we insert in the latter the momentum scale $2\pi\hbar L$ we obtain the time scale

$$n_c = \frac{1}{(1-\lambda)L} \simeq \frac{(4\pi^2\hbar)^2}{(1-\lambda)K^2} \tag{4.4}$$

on which incoherent transitions due to dissipation occur in a wavepacket initially started at $\rho_0 = |0\rangle\langle 0|$ and propagating under the dissipative map. If $n_c<n^*$, i.e.

$$1-\lambda > \frac{1}{2}\left(\frac{4\pi^2\hbar}{K}\right)^4, \tag{4.5}$$

incoherent transitions dominate over coherent propagation and we cannot expect to see even transient signatures of localization. Another way of stating the condition (4.5) is

$$\frac{2\pi}{\delta\omega_0} > n_0 , \qquad (4.6)$$

where $\delta\omega_0$ and $2\pi/n_0$ are, respectively, the average level spacing and level widths of Floquet states near $p=0$ within a distance L . Condition (4.5) was satisfied for the parameter values chosen in the preceding section. We now define the parameter domain of 'weak dissipation' by the condition

$$\frac{2\pi}{\omega_0} < n_0 \qquad \text{(weak dissipation)} . \qquad (4.7)$$

The disruption of coherence by incoherent processes then does not occur sufficiently frequently to inhibit localization altogether, but incoherent transitions now move the wavepacket diffusively from one localized state of size L to the next. After such incoherent transitions the action variable therefore spreads diffusively according to

$$\frac{\langle \Delta p^2 \rangle}{(2\pi\hbar)^2} \simeq L^2 N . \qquad (4.8)$$

The time-scale $\partial N/\partial n$ for incoherent processes also depends linearly on the momentum scale (cf. the expression for n_λ given above) and we can therefore estimate self-consistently

$$\frac{\partial N}{\partial n} \simeq \frac{1-\lambda}{2\pi\hbar}\sqrt{\langle \Delta p^2 \rangle} \simeq (1-\lambda) L\sqrt{N} , \qquad (4.9)$$

which gives

$$N \simeq \frac{n^2}{4n_c^2} , \qquad (4.10)$$

and with eq. (4.8)

$$\frac{\langle \Delta p^2 \rangle}{(2\pi\hbar)^2} \simeq \left(\frac{K}{4\pi^2\hbar}\right)^8 \frac{(1-\lambda)^2}{4} n^2 . \qquad (4.11)$$

Eq. (4.11) remains valid only as long as the widths of the Floquet states involved (which grow proportionally with the scale of angular momentum) are small compared to their average level spacing $\delta\omega_\lambda \simeq 2\pi/2L$. The angular momentum scale in the steady state can then be estimated from eq. (4.11) inserting $n_d = (1-\lambda)^{-1}$ for n . Then we find

$$\langle \Delta p^2 \rangle \simeq \frac{1}{4}\left(\frac{K}{4\pi^2\hbar}\right)^8 (2\pi\hbar)^2 , \qquad (4.12)$$

and, self-consistently, the condition

$$(1-\lambda) < \left(\frac{4\pi^2\hbar}{K}\right)^6 \qquad (4.13)$$

for the validity of eq. (4.11) until the steady state is reached. We use eq. (4.13) as a definition of the parameter regime of very weak dissipation. The condition (4.13) just guarantees that the angular momentum scale (4.12) in the quantum steady state is smaller than the corresponding scale

$$\langle \Delta p^2 \rangle \simeq \frac{K^2}{16\pi^2} \frac{1}{1-\lambda} . \tag{4.14}$$

in the classical steady state. In other words eq. (4.13) ensures that effects of localization are visible even in the steady state. If eq. (4.7) but not eq. (4.13) is satisfied, eq. (4.11) must break down before the steady state is reached. The breakdown occurs once the angular momentum scale $|p|$ is reached where the widths and average spacings of quasi-energy levels are comparable, which yields

$$\begin{aligned} |p| &\approx \left[2(1-\lambda)(K/4\pi^2\hbar)^2\right]^{-1}(2\pi\hbar) \\ n_b &\simeq \left[(1-\lambda)^2 (K/4\pi^2\hbar)^6\right]^{-1} \end{aligned} \tag{4.15}$$

for the angular momentum scale and the time scale, respectively, where the transition occurs. For longer times and larger angular momenta the system returns to classical diffusion. Therefore, in the region of weak dissipation (but not very weak dissipation) localization is a transient phenomenon only, whose signature is the n-dependence of eq. (4.11).

Let us finally consider the preliminary numerical evidence for the various regimes of dissipation which we have obtained so far.

In figs. 3,4 we show results for the long-time development of the mean rotational energy $\langle p^2/2 \rangle$ of the rotator for a kick strength K = 10.0 far above $K_c \simeq 1$, with $\hbar = 0.15(\sqrt{5}+1)/4\pi$ (the golden mean factor serving to avoid the influence of quantum resonances). For these parameter values, the conservative case (dashed line in figs. 3a,4a) is characterized by an initial diffusive increase of the energy with an estimated rate $\langle p_n^2/2 \rangle \simeq 0.6n$ and an onset of localization after $n^* \simeq 86$ time steps, leading to a quasi-periodic behavior with the energy fluctuating around $\langle p^2/2 \rangle \simeq 13.6$. There is reasonable agreement of order of magnitudes with the respective numerical values $n^* \simeq 40$, $\langle p_n^2/2 \rangle \simeq 0.4n$, $\langle p^2/2 \rangle \simeq 20$.

The full line in fig. 3a corresponds to a dissipation rate $1-\lambda = 5\cdot 10^{-6}$ in the regime of extremely weak dissipation $0 < 1-\lambda \lesssim 1.3\cdot 10^{-5}$. We observe a slow increase of the mean energy compared to the conservative case, the fluctuations still following closely those in the conservative system. Moreover, we indeed find a nearly quadratic time dependence for the difference of the two energies. In the log-log plot fig. 3b the actual exponent is compared with the theoretically estimated value 2

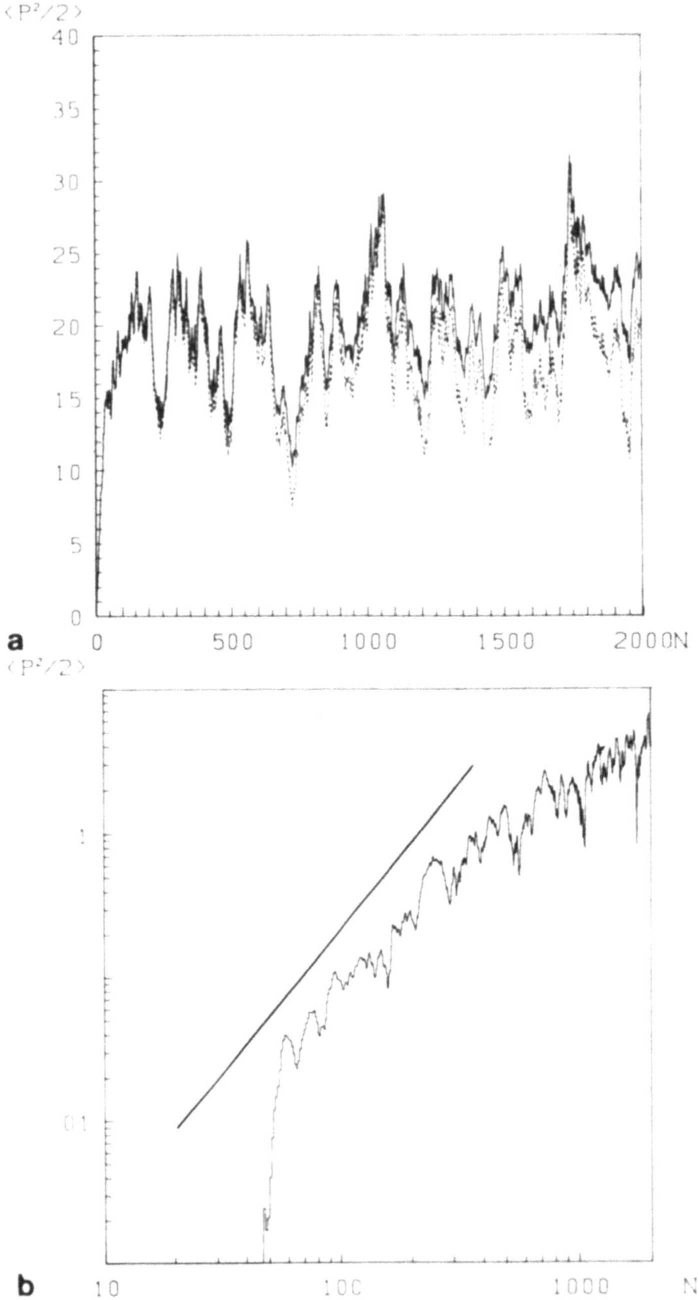

Fig. 3

Time evolution of the total energy over the first 2000 iterations (a), shown for extremely weak dissipation (solid line) and for the conservative case (dashed line), and difference of the two functions in log-log representation (b). The straight line in part (b) has the slope 2, corresponding to a quadratic time dependence.

The parameter values are $K = 10.0$, $\lambda = 1.0\text{-}5 \cdot 10^{-6}$ (damped case), $\lambda = 1.0$ (conservative case), $\hbar = 0.03863$.

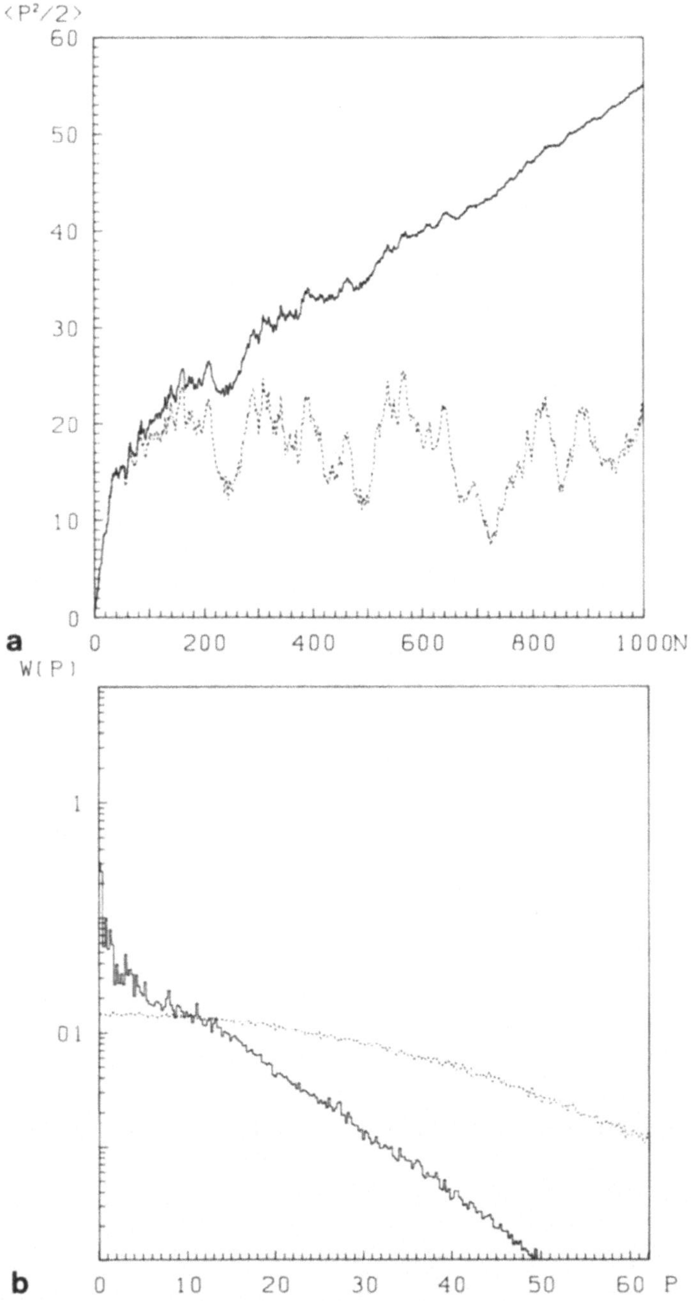

Fig. 4

Time evolution of the total energy over the first 1000 iterations (a), shown for weak dissipation (solid line) and for the conservative case (dashed line), and reduced angular momentum distribution (b), shown for the quantum mechanical (solid line) and classical (dashed line) cases with weak dissipation as in part (a). The parameter values are $K = 10.0$, $\lambda = 1-1\cdot 10^{-4}$ (a, damped case, and b), $\lambda = 1.0$ (a, conservative case), $\hbar = 0.03863$ (a and b, quantum mechanical case), $\hbar = 0$ (b, classical case).

(straight line). The absolute value of the increase rate $\langle p_n^2/2 \rangle \simeq 1 \cdot 10^{-6} n^2$ is in reasonable agreement with the estimated value $\langle p_n^2/2 \rangle \simeq 0.6 \cdot 10^{-6} n^2$. With $1-\lambda = 1.0 \cdot 10^{-4}$, the result shown in fig. 4a (full line) corresponds to the regime of weak dissipation ($1.3 \cdot 10^{-5} \lesssim 1-\lambda \lesssim 2.7 \cdot 10^{-4}$). In this case, the onset of localization still remains visible. Disruptions of coherence (whose time scale is $n_c \simeq 230$ here), however, dominate the behavior of the system soon thereafter. The increase of energy remains far below the classically expected rate, which we attribute to the fact that the average $\langle p^2/2 \rangle$ receives contributions from angular momentum scales both in the localized region at small $|p|$ and the delocalized region at large $|p|$; in any case there is no more evidence for a quadratic time dependence. In fig. 4a the fluctuations cease to be correlated with those of the conservative quantum system after $\simeq$ 1000 iterations, which is consistent with the estimated value of $n_b \simeq 1300$.

In fig. 4b we present the distribution over angular momentum obtained after 1000 time-steps for the same parameter values as in fig. 4a. For small angular momenta, where localization of quasi-energy states is efficient and diffusion is prohibited the distribution is different in form and less smooth than for large angular momenta where localization of quasi-energy states is destroyed and diffusion occurs. The momentum scale estimated from (4.15) is $|p| \simeq 28$ and somewhat larger than but of the same order of magnitude as the momentum scale $|p| \simeq 10$ where the distribution changes its form.

In the experiments of figs. 3,4 the increasing angular momentum scale prevented us from following the system into the steady state, which should be reached after $n_d \simeq 2 \cdot 10^5$ and after $n_d \simeq 1 \cdot 10^4$ iterations for the cases of extremely weak and weak dissipation, respectively.

In summary, the numerical results demonstrate the existence of the various regimes of dissipation strength $\nu = (1-\lambda)$ and localization or delocalization which have emerged from our qualitative analytical estimates. Furthermore, these estimates are found to give useful orders of magnitude which help to interpret the numerical results.

References

1. 'Chaotic Behavior in Quantum Systems', ed. G. Casati, Plenum (New York 1985)
2. 'Proceedings of the Workshop on Quantum Chaos Trieste 1986' Physica Scripta, to be published
3. J.R. Ackerhalt, P.W. Milonni, M.L. Shih, Phys. Rep. 128,205 (1985)
4a. J.E. Bayfield, P.M. Koch, Phys. Rev. Lett. 33,258 (1974); J.E. Bayfield, L.D. Gardner, P.M. Koch, Phys. Rev. Lett. 39,76 (1977); P.M. Koch, J. Physique 43,C2 (1982); K.A.H. von Leeuven, G.v. Oppen, S. Renwick, J.B. Bowlin, P.M. Koch, R.V. Jensen, O. Rath, D. Richards, J.G. Leopold, Phys. Rev. Lett. 55,2231 (1985); J.N. Bardsley, B. Sundaram, L.A. Pinnaduwage, J.E. Bayfield, Phys. Rev. Lett. 56, 1007 (1986)

4b. N.B. Delone, B.P. Krainov, D.L. Shepelyansky, Usp. Fiz. Nauk. 140,355 (1983) (Sov. Phys. Usp. 26,551 (1983)); R.V. Jensen, Phys. Rev. Lett. 49,1365 (1982); Phys. Rev. A30,386 (1984); J.G. Leopold, D. Richards, J. Phys. B18,3369 (1985); J. Phys. B19,1125 (1985); D.L. Shepelyansky, in ref. 1 ; G. Casati, B.V. Chirikov, I. Guarneri, D.L. Shepelyansky, Phys. Rev. Lett. 56,2437 (1986); G. Casati, B.V. Chirikov, D.L. Shepelyansky, I. Guarneri, Phys. Rev. Lett. 57,823 (1986); R. Blümel, U. Smilansky, Phys. Rev. Lett. 52,137 (1984); Phys. Rev. A30,1040 (1984); preprint WIS 87/1/ph.
5. G. Casati, B.V. Chirikov, F.M. Izrailev, J. Ford, in Lecture Notes in Physics 93, Springer (Berlin, 1979) p. 334; F.M. Izrailev, D.L. Sheoelyansky, Theor. Math. Phys. 43,553 (1980); Sov. Phys. Dokl. 24,996 (1979); T. Hogg, B.A. Huberman, Phys. Rev. Lett. 48,711 (1982); Phys. Rev. A28,22 (1983); S. Fishman, D.R. Grempel, R.E. Prange, Phys. Rev. Lett. 49,509 (1982); D.R. Grempel, R.E. Prange, S. Fishman, Phys. Rev. Lett. A29,1639 (1984); S.J. Chang, K.J. Shi, Phys. Rev. Lett. 55,269 (1985); J.D. Hanson, E. Ott, M. Antonsen, Phys. Rev. A29,819 (1984); D.L. Shepelyansky, Physica D8,208 (1984)
6. references given in 4b
7. R. Blümel, S. Fishman, U. Smilansky, J. Chem. Phys. (1986) in press
8. A.J. Lichtenberg, M.A. Lieberman, Regular and Stochastic Motion, Springer (Berlin 1983)
9. See the references by S. Fishman, D.R. Grempel, R.E. Prange in 5.
10. See F.M. Izrailev, D.L. Shepelyansky in 5.
11. G. Casati, I. Guarneri, Comm. Math. Phys. 95,121 (1984)
12. E. Ott, T.M. Antonsen, J.D. Hanson, Phys. Rev. Lett. 53,2187 (1984)
13. T. Dittrich, R. Graham, Z. Physik B62,515 (1986)
14. T. Dittrich, R. Graham, preprint
15. R. Graham, in 'Quantum Chaos', ed. F. Sarkar, E.R. Pike, Plenum (New York 1987)
16. T. Dittrich, R. Graham, to be published
17. G. Schmidt, B.W. Wang, Phys. Rev. A32,2994 (1985)
18. R. Graham, Europhys. Lett., to appear
19. R. Graham, Phys. Rev. Lett. 53,2020 (1984)
20. M. Dörfle, R. Graham, in 'Optical Instabilities', ed. R.W. Boyd, M.G. Raymer, L.M. Narducci, Cambridge University Press (Cambridge 1986), p. 352)
21. C.W. Gardiner, 'Handbook of Stochastic Methods' Springer (Berlin 1983)
22. B.V. Chirikov, F.M. Izrailev, D.L. Shepelyansky, Sov. Sci. Rev. C2,209 (1981)

COHERENCES AND CORRELATIONS IN CHAOTIC OPTICAL SIGNALS

N.B. Abraham, A.M. Albano, B. Das and M.F.H. Tarroja
Department of Physics, Bryn Mawr College
Bryn Mawr, PA 19010 USA

Although it is common to analyze fluctuating optical signals on a statistical basis and to use correlation functions and spectral analysis to find bandwidths, pulse shapes and short term memory features, it is now possible to perform certain specific numerical studies which can uncover characteristics of the causes of the fluctuations. That is, by analysis of the time series of the signals, we can distinguish differences between broadband fluctuations caused by deterministic effects and those caused by stochastic processes. We are very familiar with the notion that many random processes cause Gaussian amplitude fluctuations and the corresponding negative exponential intensity probability distribution function. Among such processes is spontaneous emission from a large and dilute collection of incoherently excited atoms. In contrast, semiclassical laser light can be described by a delta function distribution of intensity and only a slow phase drift for the complex electric field amplitude.

Recently simple algorithms have been offered which aid in determining whether the broadband fluctuations of a signal have their origin in stochastic or deterministic processes [1]. Deterministic processes can be easily identified if they generate constant or periodic signals. However, when the deterministic process generates seemingly random fluctuations, then the situation may well be that which is now called "deterministic chaos", a particular form of evolution in the phase space of the system. Chaotic behavior often generates broadband spectra that are essentially indistinguishable from those caused by stochastic noise. In particular for lasers, chaotic dynamics can cause the laser to have an optical power spectrum which closely resembles the spectrum of spontaneous emission.

Work in the last five years has identified methods of analysis of seemingly random signals which can distinguish some types of chaotic behavior from stochastic behavior. The basic method is to take a digitized record of one fluctuating variable, use it to reconstruct a topological equivalent to the evolution of the system in its variable space, and then to measure characteristics of the set of points forming the solution. For low dimensional chaotic systems, the solution set forms a low dimensional fractal of dimension larger than two. By careful analysis of the dimension of the reconstructed attracting set, one can find fractional values which are clearly different from, and smaller than, the integer values that would be characteristic of stochastic noise filling of a variable space of the same number of dimensions.

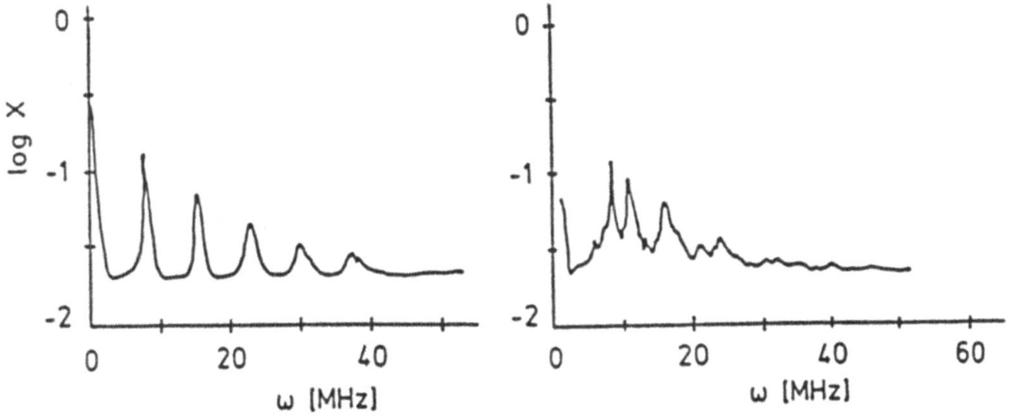

Figure 1: Two examples of experimentally measured intensity power spectra for a spontaneously pulsing, single mode, xenon-helium (inhomogeneously broadened) laser [1].

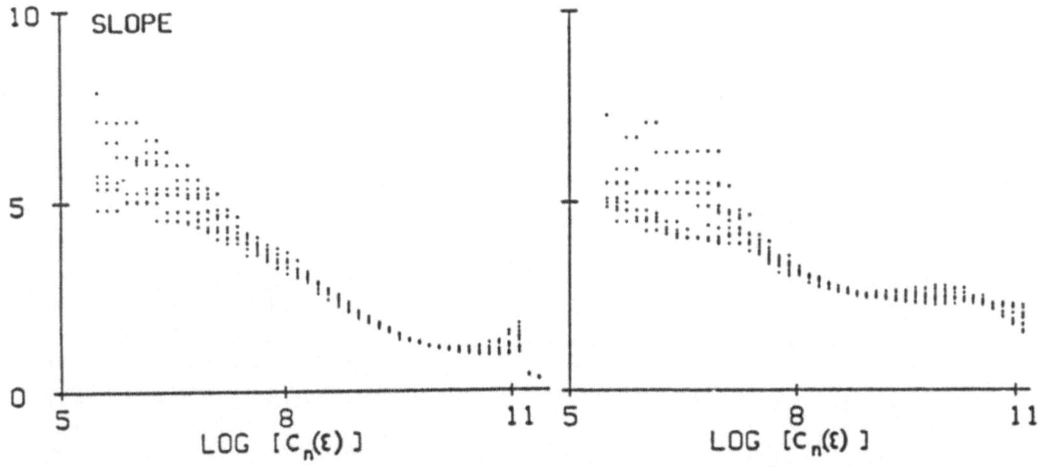

Figure 2: Slopes of the Correlation integral for different embedding dimensions for digitized intensity time series corresponding to the power spectra shown in Fig. 1: dimensions of order one for the periodic signal and of order 2.2 for the quasiperiodic signal with broadband features [1].

These techniques have been applied to the analysis of laser signals and there is evidence that broadband features of the optical power spectrum are in certain cases representative of chaotic rather than stochastic behavior. An example is shown in Figures 1 and 2, where we show the intensity power spectra from experimental measurements and the slopes of correlation sums for the Grassberger-Procaccia method of calculating the correlation dimension [see Refs. 1 and 2]. The plateau regions indicate that a fractal structure exists over a certain length scale in the attracting set and can be taken as strong evidence for chaos. Nearly the same values for the dimension were found when we analyzed the numerically generated time-series

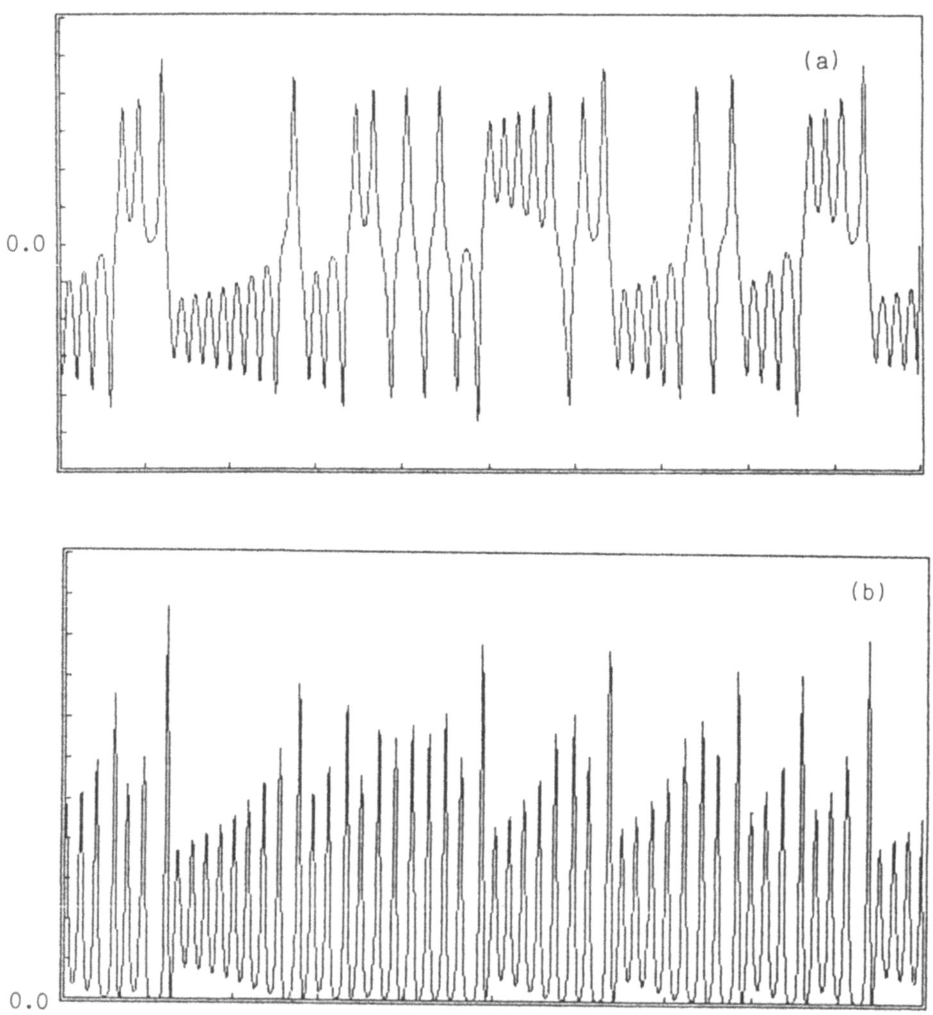

Figure 3: Temporal fluctuations of the amplitude (a) and the intensity (b) for the Lorenz-Haken model in the chaotic regime.

that represent solutions for the intensity output of an inhomogeneously broadened laser [1].

We have more recently inquired about other properties of a chaotic signal in order to probe the possibility that the measurement of coherence or correlation functions might reveal chaotic dynamics either more clearly or with greater computational ease.

As examples of work of this nature, we show first the results for the Lorenz-Haken model for a single mode, homogeneously broadened laser in resonance [3]. In Figure 3, we show examples of the characteristic time evolution of the field amplitude and the intensity in the chaotic region.

In Figure 4, we show the corresponding power spectra. Note that the amplitude spectrum is smoothly broadband while the intensity spectrum shows

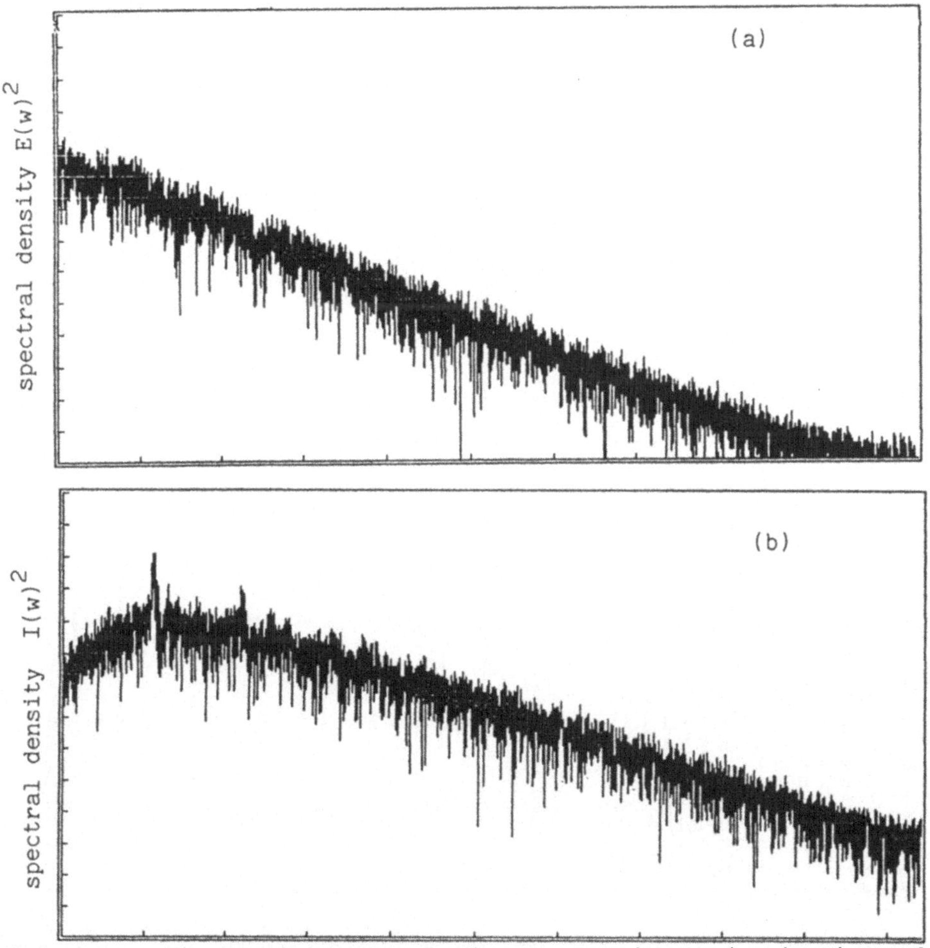

Figure 4: Power spectra corresponding to the time series in Figure 3. Vertical scale is 10 dB per division.

distinct peaks, over 30 dB above the broadband noise as a result of the elimination of the phase noise part of the amplitude fluctuations. The low frequency portion of the power spectrum of the intensity is also significantly reduced. As both signals are derived from the same chaotic dynamics, we see that the degree of chaos is not equally revealed by different variables or by nonlinear functions of those variables.

Similar results have been found for the model of an inhomogeneously broadened laser which we have used to successfully model experimental measurements [4]. The major distinction is that in this case, chaos is

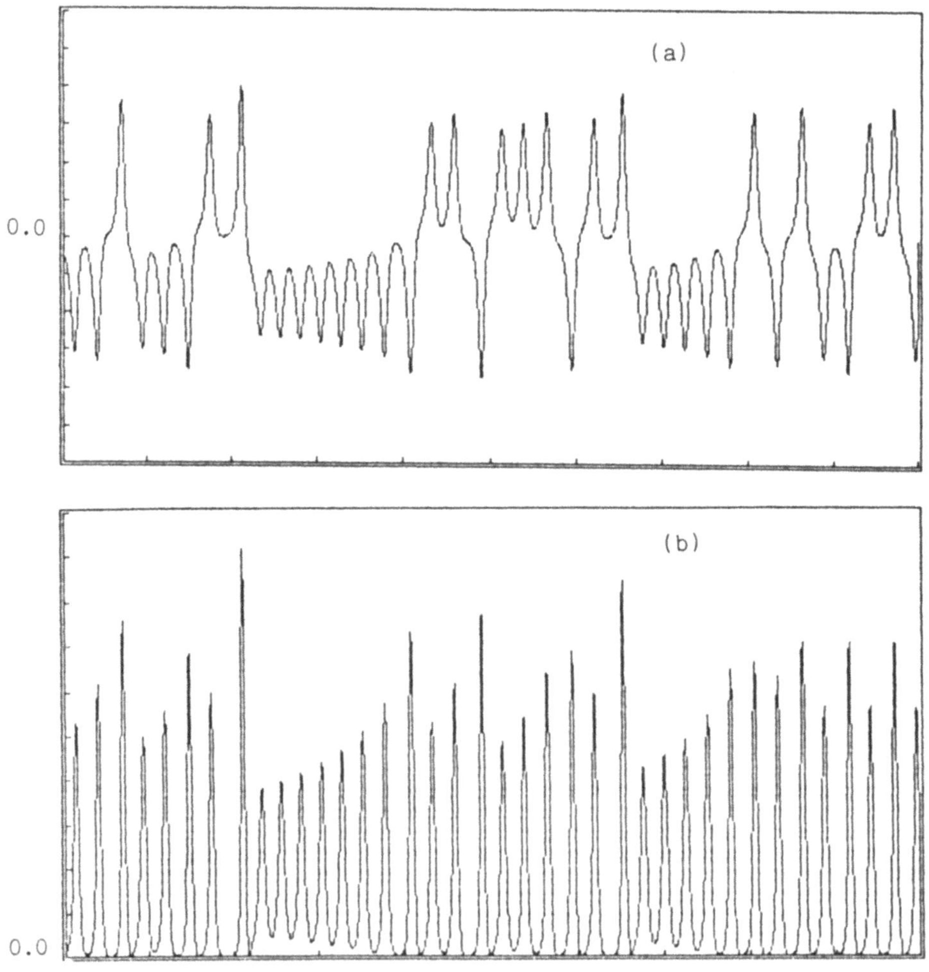

Figure 5: Electric field amplitude (a) and intensity (b) pulsations from numerical integration of a model for an inhomogeneously broadened ring laser.

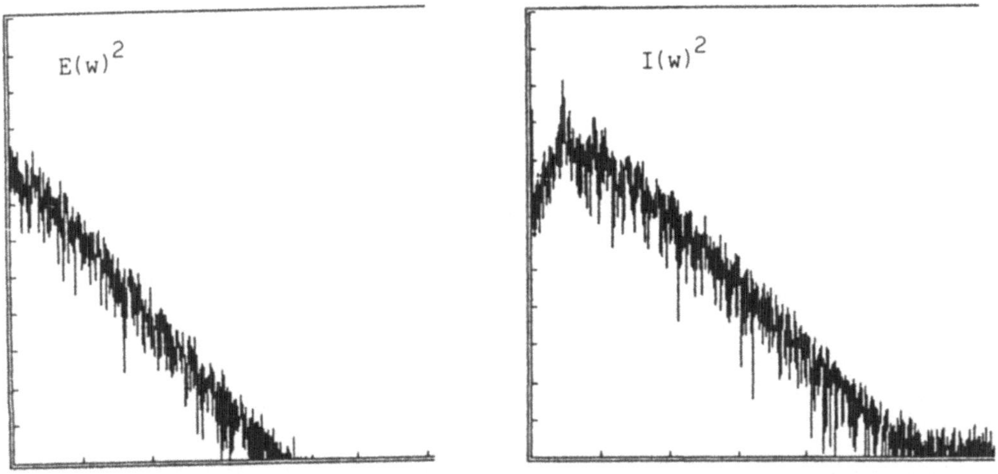

Figure 6: Amplitude (a) and Power Spectra (b) for the signals shown in Figure 5. Vertical scale is 10 dB per division.

found very close to the threshold for laser action.

Some of the information contained in the contrast between the intensity and amplitude spectra can be useful in interpreting experimental data. We recognize that chaotic evolution may cause a large amount of broadband phase noise while the effect on the intensity is predominantly periodic modulation with a lesser amount of chaotic fluctuations. If we return to some experimental data of several years ago for a Fabry-Perot He-Ne laser (similarly inhomogeneously broadened on the 3.39 micron line and unstable), we can compare this information with the high-resolution spectra that could be obtained because of the intrinsic quiescence of the He-Ne gas discharge (in contrast to the noisiness of the He-Xe discharge). In Figure 7 we display several samples of our He-Ne data [5] which show the kind of intensity power spectra we find for the model calculations for fully-developed chaos. Here also, the periodic modulation is more than 20dB above a broadband portion of the spectrum. The broadband spectrum is peaked at the frequency of the periodic modulation and there is less broadband noise near zero frequency. This suggests to us that we should repeat the He-Ne experiments to obtain clear heterodyne spectra which should reflect the non-periodic and broadband nature of the amplitude spectrum obtained from the models.

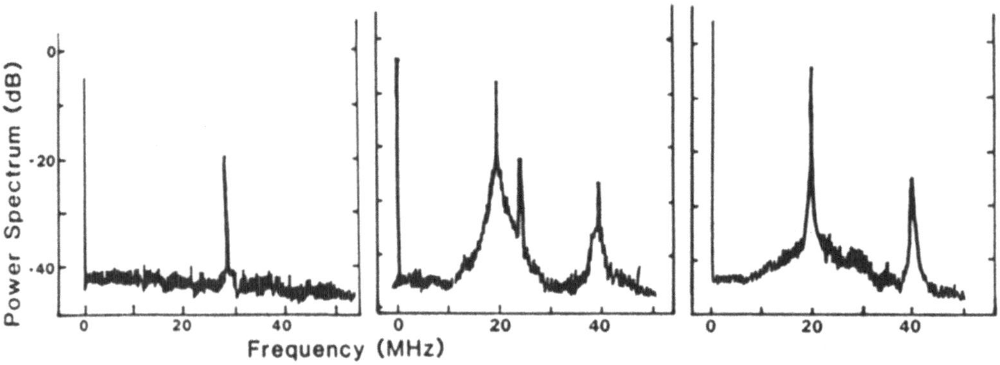

Figure 7: Intensity power spectra for a chaotically pulsing He-Ne laser after [5].

Our alternative method of analysis is to calculate correlation functions. The simplest correlation function, the autocorrelation function, is calculated from the data (from which the mean has been subtracted) by the relation:

$$C_{11}(\tau) = \langle A(t)A(t+\tau)\rangle / \langle A(t)^2\rangle \ , \qquad (1)$$

which is related by the Wiener-Khintchine Theorem to the power spectrum of A(t) by a simple Fourier Transform.

Higher order correlation functions can be calculated. However, they are then functions of multiple temporal coordinates as different time delays can be used in the different factors. We have found that additional and helpful information already appears in the simplest form of the third-order correlation function

$$C_{21}(\tau) = \langle A(t)A(t)A(t+\tau)\rangle / (\langle A(t)^2\rangle)^{3/2} \ . \qquad (2)$$

Results for the Lorenz-Haken model are shown in Figs. 8 and 9. Figure 8 gives the autocorrelation for the signals shown in Figs. 3 and 4, and Fig. 9 gives the simple third-order correlation function for the same signals.

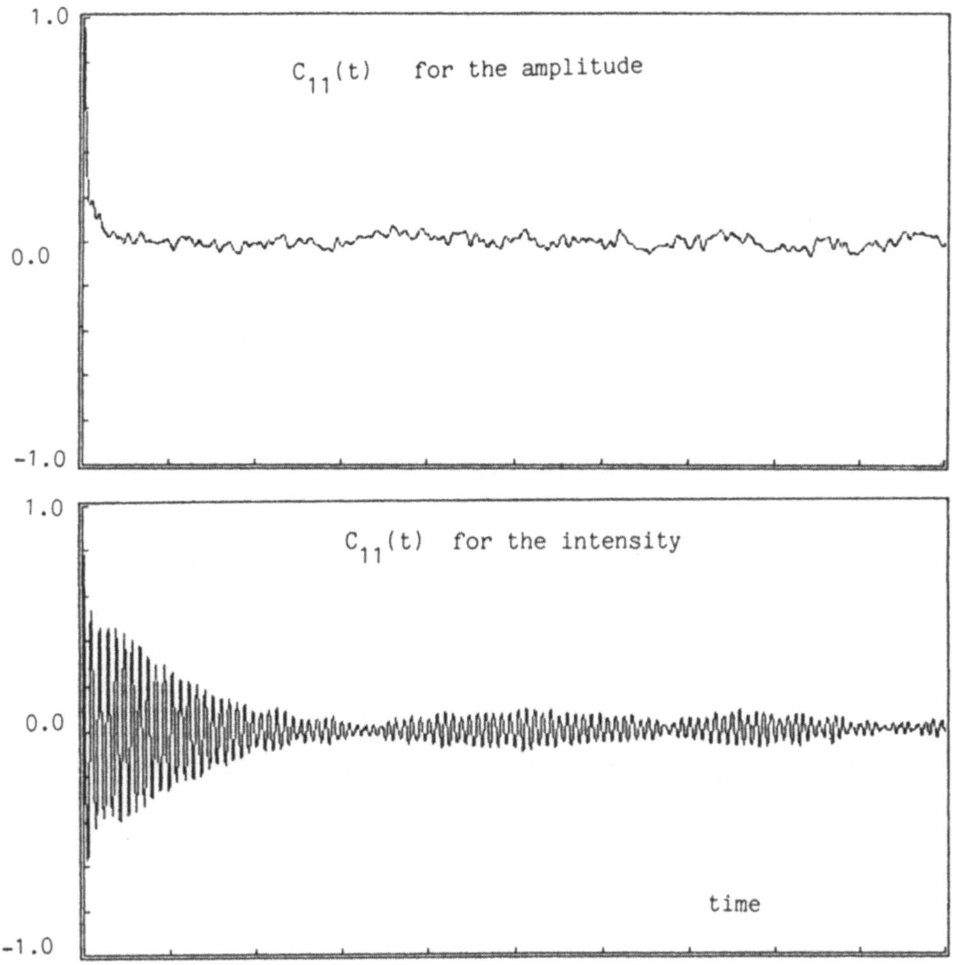

Figure 8: Autocorrelation function of the chaotic signals from the Lorenz-Haken model shown in Figs 3 and 4.

The third order correlation function for the amplitude shows little information because the symmetry of the signal about zero amplitude gives a null result for all delays. The third-order function for the intensity shows considerably more information with two characteristic decay times for the largest part of the signal. One is of the order of the period of the dominant intensity pulsation frequency (indicated by the rapid decay from $C_{11}(0) = 1.0$ to about $C_{11}(\tau) = 0.5$), while the other is of order of ten periods of the pulsation frequency. Other longer term correlations are also visible after more than 100 periods of the fundamental intensity

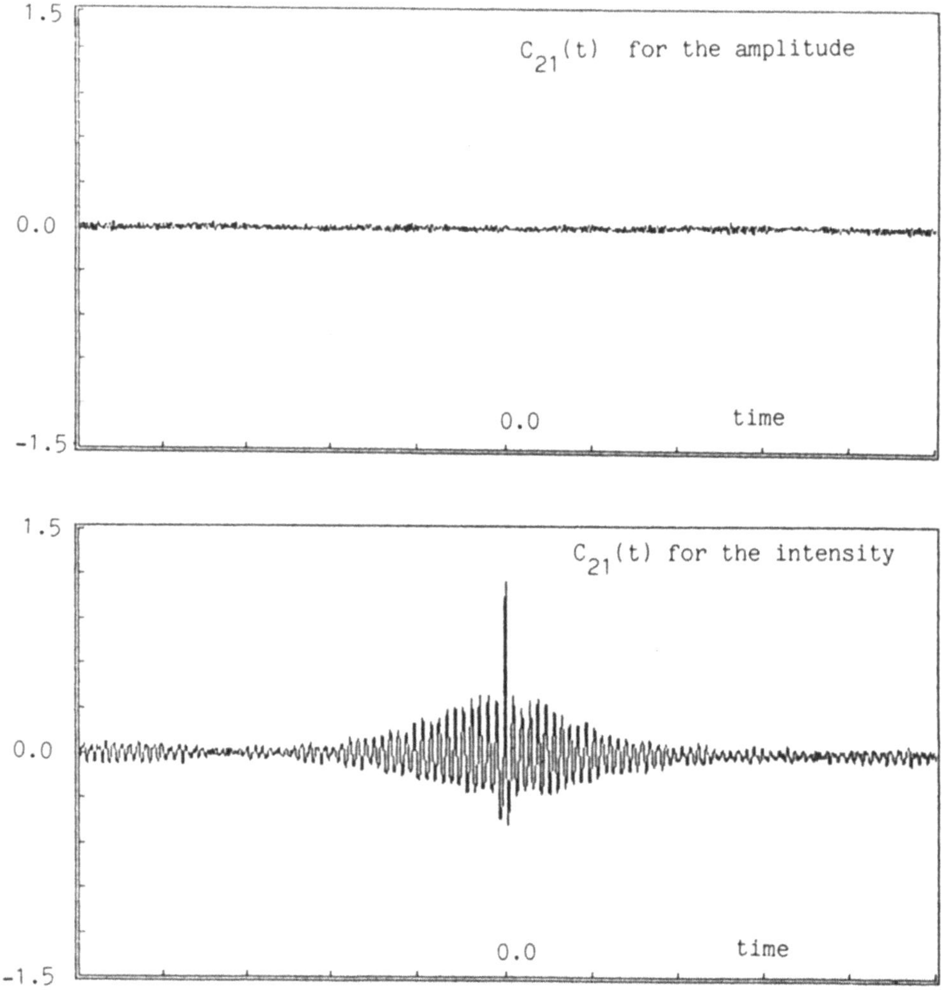

Figure 9: Third-order correlation function for the chaotic signals from the Lorenz-Haken model.

pulsation frequency.

Similarly distinctive features have been found for experimental data from the unstable He-Xe laser and they differ from characteristics we find in the fluctuations of amplified spontaneous emission from a heavily saturated source.

A major difference found here is that the ASE shows a single peak at zero delay time for the third order correlation function, which indicates that there is no significant memory or characteristic evolution after each pulse. Observing the intensity pulsations in real time we notice that it

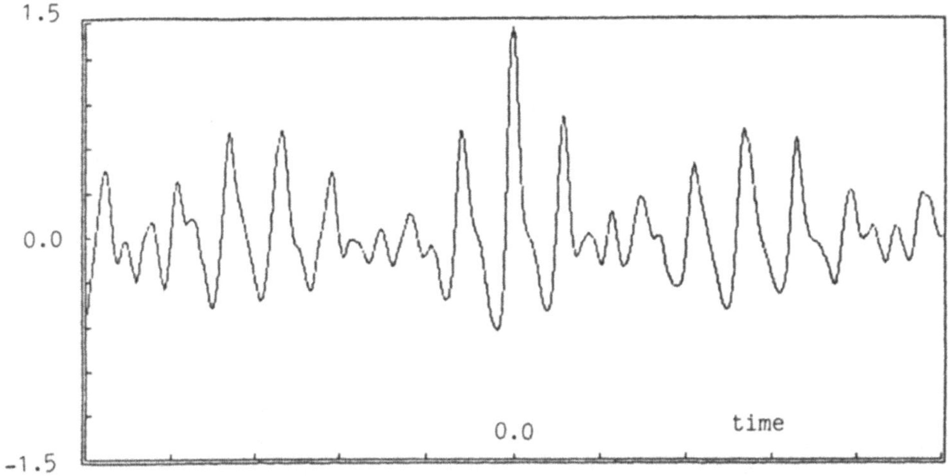

Figure 10: Third-order intensity correlation function for experimental data from a He-Xe ring laser as described in [4]. (Data record of 500 pts.)

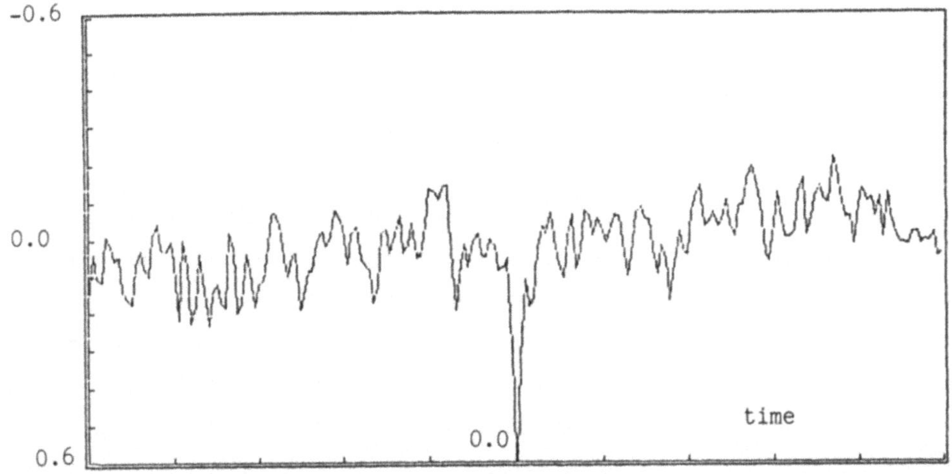

Figure 11: Third-order intensity correlation function for experimental data from a heavily saturated source of amplified spontaneous emission using the same He-Xe 3.51 micron transition used in the laser studies (discharge length: 2.0 m; Pressure: 193 mTorr Xe, 4.0 Torr He). Data stream of 500 points. The function is inverted because of an inverting amplifier in the experimental setup. The signal analyzed here is raw data including amplifier noise of about 5%.

is a common occurence that the intensity remains near zero after a large pulse. We have also found that the fluctuations in copropagating or counterpropagating, orthogonally polarized, beams in the ASE source are anticorrelated with the strong pulsations in this beam [6]. For this preliminary data we believe that there is no significance to the function shown in Fig. 11 other than the large peak near zero delay time. The fluctuating baseline for longer times is presumed to arise from the very short data record (see Figs. 13 for results for longer data records).

While it is common to presume that amplified spontaneous emission is stochastic in origin and output, we have checked this for our heavily saturated source and confirmed that dimensionality tests are unable to discern a dimension (and thus unable to discern the existence) for an attractor [1]. An improved algorithm for projecting the embedded (reconstructed) attractor onto its principal axes, using the singular value decomposition techniques espoused by Broomhead and King, was also applied to the data and again a dimension or characteristic structure for an attractor failed to emerge [7]. We might have expected some deterministic features caused by the evolution of intense pulses in a superflourescent manner.

Thus the ASE represents an optical signal which has a broadband spectrum equivalent to that of a chaotic laser, yet the origin of the fluctuations in the two cases is quite different. In fact, this is immediately evident in the intensity power spectra. Unlike the chaotic laser which has a broadband amplitude spectrum associated with a predominantly periodic intensity pulsation, the ASE reveals no underlying dynamical structure as its intensity power spectrum is as smoothly broadband as the amplitude spectrum. The intensity power spectrum in this case is the natural homodyne convolution of amplitude spectrum with itself as would be expected for a stochastic source. We recall that the periodic features of the intensity of the chaotic laser signal, which could not be predicted by convolution of the amplitude power spectrum, indicate that this is not similarly stochastic.

Other stochastic features of the ASE signal are revealed by measuring the intensity probability distribution function which is shown in Figure 12, and the higher order correlation functions defined by

$$g_{nm}(\tau) = \langle (A(t))^n (A(t+\tau))^m \rangle / (\langle A(t)^2 \rangle)^{(n+m)/2} \tag{3a}$$

and

$$C_{nm}(\tau) = g_{nm}(\tau) - g_{nm}(\infty) \tag{3b}$$

which are shown in Figure 13. Assuming that the amplifier and detector noise is independent of the signal and is added to the photocurrent signal, we have subtracted away the corresponding cumulants of the noise determined

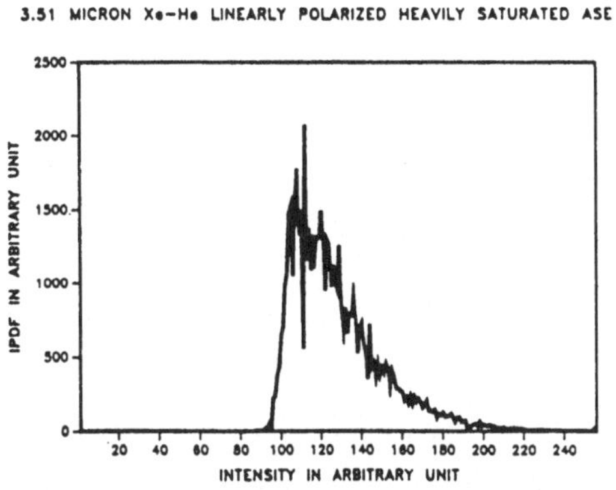

Operating Conditions:

Discharge length: 3.0 m

Pressure: 192 mTorr Xe

2.0 Torr He

Current 5.5 mA

Figure 12: Intensity probability distribution function for heavily saturated amplified spontaneous emission.

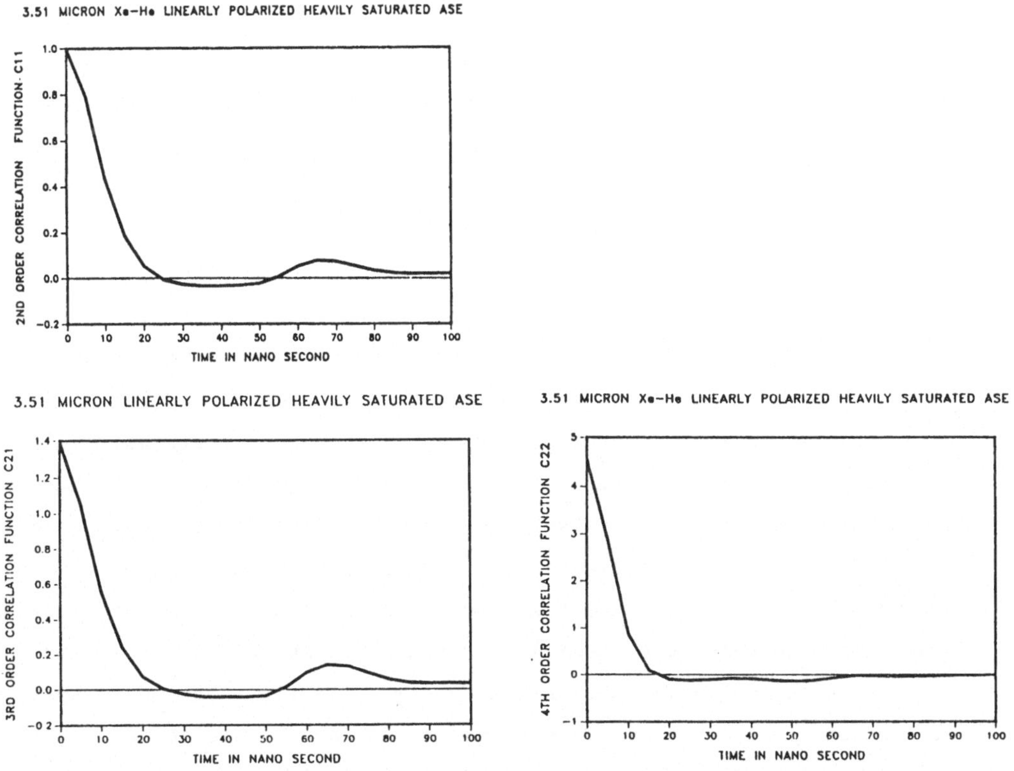

Figure 13: Correlation functions of the digitized ASE intensity using data records of 60,000 points. The gain is somewhat more heavily saturated.

by measuring the data record with the optical signal blocked.

Here we see another indication of the lack of deterministic correlations in that the correlation functions die out quickly without long term ringing. The second and third order correlation functions do show a secondary peak at about 70 ns, coincidentally close to the characteristic duration of the anticorrelation effects seen earlier [6]. However, as this data is preliminary, we are not yet certain of the reliability of this feature and we will be investigating it further. The higher order correlation functions are narrower in time, indicating that the pulses of greater height are also narrower as we could infer from direct observation of the intensity time series.

Though not shown here, the intensity and amplitude spectra are both broadband and without peaks within the resolution limits of our rf spectrum analyzers.

CONCLUSIONS

It appears that intensity power spectra, intensity correlation functions, and heterodyne spectra (if available) can be used together to build a strong case for stochastic or deterministically chaotic origins for fluctuating optical signals. When the intensity power spectrum contains peaks that are not predicted by the differences between peaks in the amplitude spectrum, it is likely that there is deterministic evolution causing the signal. When, instead, the intensity power spectrum is simply the homodyne convolution of the amplitude spectrum and when the spectra and higher order correlation functions show no peaks other than the ones near zero, then it is likely that the signal has stochastic origins.

There is clear evidence in this dynamical single-mode laser chaos that the phase of the field is a "key player" in the chaotic motion. However, all of the variables are dynamically coupled and at least three must be actively involved to produce the chaos. In the case of the Lorenz-Haken model, we have taken the laser in resonance, in which case the field is one of only three real variables in the system. Clearly a broadband spectrum of frequencies is involved in the amplitude chaos by a combination of switching sign and pulsing magnitude. It is rather surprising that the modulation of the intensity is so relatively periodic. The projection of the chaotic attractor onto an intensity phase space shows relatively weak blurring due to the chaos. One can partially understand the chaos as predominantly frequency modulation while the intensity modulation remains predominantly periodic. Nevertheless it is, of necessity, a matter of degree. The chaotic modulation must be present in all three of the variables.

The distinction between phase and intensity is all the more puzzling because the field variable in this case is real. It would be easier to understand if the field amplitude were complex, as it could then be described by two real variables (a normative amplitude and a phase). The phase noise we see corresponds to the hopping between the two basins of attraction in the regions of positive and negative field amplitude. "Phase" in this case means the sign of the electric field and it switches telegraphically in a chaotic manner.

We are not certain how the emphasis on chaotic switching relates to the different kind of behavior when the laser is detuned (requiring five equations in a generalized Lorenz-Haken model). Both experimentalists and theoreticians have pointed out the the detuned laser is much more likely to exhibit periodic rather than chaotic pulsations.

The importance of phase chaos is not universal in optical systems as in many models and experiments, such as for the modulated laser discussed in this volume by Tredicce, only the intensity is involved (and not the field amplitude) as one of the dynamical variables.

Continuing studies are directed at improving the distinctions that can be made and testing other correlation and coherence measures for their relative aid in discerning features that are characteristic of stochastic or deterministic causes of fluctuations.

ACKNOWLEDGEMENTS

This work was supported in part by a Research Contract DAAG29-85-K-0256 from the U.S. Army Research Office. The work of M.F.H.T. was supported by a Newport Research Award administered by the Optical Society of America for the Newport Research Corporation.

REFERENCES

1. N.B. Abraham, A.M. Albano, B. Das, T. Mello, M.F.H. Tarroja, N. Tufillaro and R.S. Gioggia, in Optical Chaos, ed. J. Chrostowski and N. B. Abraham, SPIE Proceedings Volume 667 (SPIE, Bellingham, 1986) p. 2.

2. N.B. Abraham, A.M. Albano, T.H. Chyba, L.M. Hoffer, M.F.H. Tarroja, S.P. Adams and R.S. Gioggia, in Instabilities and Chaos in Quantum Optics, eds. F.T. Arecchi and R.G. Harrison, Springer Series in Synergetics, Volume 34, (Springer, Berlin-Heidelberg, 1987), to be published.

3. H. Haken, Phys. Lett. 53A, 77 (1975).

4. M.F.H. Tarroja, N.B. Abraham, D.K. Bandy, L.M. Narducci, Physical Review A 34, 3148-3158 (1986).

5. R.S. Gioggia and N.B. Abraham, Optics Communications, 47, 278-282 (1983).

6. S.P. Adams and N.B. Abraham, in Coherence and Quantum Optics V, eds. L. Mandel and E. Wolf, (Plenum, NY, 1984) pp. 233-242.

7. A.M. Albano, G.C. DeGuzman, A.E. Mees, and P.E. Rapp, "Data Requirements for Reliable Estimation of Correlation Dimension", Proceedings of the 1986 Wales Conference on Chaos in Biology (to be published); N.B. Abraham, A.M. Albano and M.F.H. Tarroja, "Dynamics and Chaos in Optical Systems", in Chaos and Related Nonlinear Phenomena, ed. I. Procaccia, (Plenum, NY, 1987) to be published.

PART II: Squeezed Quantum States

SQUEEZED STATE GENERATION BY THE NORMAL MODE SPLITTING OF TWO-LEVEL ATOMS IN AN OPTICAL CAVITY

H.J. Kimble, M.G. Raizen, L.A. Orozco, Min Xiao and T.L Boyd
Department of Physics
University of Texas at Austin
Austin, Texas 78712

In recent years there has developed a rapidly growing interest in the manifestly quantum or nonclassical features of the electromagnetic field. From the perspective of quantum optics, such nonclassical features are of great intrinsic interest; however, a widening audience is now being attracted to problems in this area because of potential applications to measurement science, to optical communication, and to atomic spectroscopy. Squeezed states of the electromagnetic field are one example of quantum states that are the subject of intense activity(1), with several observations of squeezing now having been reported(2-7). In at least one case, a sufficient degree of squeezing to warrant serious attention to possible application has been demonstrated(4). Squeezing refers to the phase dependent redistribution of the quantum fluctuations of the field such that the variance in one of two orthogonal quadrature operators drops below the level of fluctuations set by the vacuum state of the field (the zero-point level).

While a diverse set of processes has been identified for squeezed state generation(8), it has proven to be somewhat difficult for a variety of scientific and technological reasons to find experimental systems that actually fulfill the potential indicated by many model calculations. In this paper we wish to describe a new regime for squeezed state generation involving a collection of two-level atoms coupled to a single mode of a high finesse resonator. We demonstrate both theoretically and experimentally that significant degrees of squeezing can be achieved by employing the normal-mode structure of this coupled system in a domain in which the oscillatory exchange of excitation between the cavity mode and the atoms in the cavity provides the requisite phase sensitive amplification and deamplification of quantum fluctuations. With few exceptions (9, 11), previous investigations of this system have focused on the *"good-cavity"* limit in which the atomic variables have been adiabatically eliminated and with them the very structure that we have identified(10, 12-15).

To gain qualitative insight into the nature of the processes responsible for squeezed state generation in this new regime, we begin with an analysis of the interaction of a collection of N two-level atoms with a single mode of a high finesse interferometer. The model Hamiltonian H for this system has been extensively studied in quantum optics and is taken to be of the form(16,17)

$$H = H_0 + H_a + H_c ,$$
$$H_0 = (\hbar\omega_a/2)\, J_z + \hbar\omega_c \quad a^+ a + \hbar g[iJ\_a^+ + H.c.] \tag{1}$$

The coherent coupling of the atomic polarization to the cavity field is described by H_0. $\{J_z, J_\pm\}$ are collective atomic operators for the N atoms of transition frequency ω_a, and $\{a, a^+\}$ are the annihilation and creation operators for the single cavity mode of resonant frequency ω_c. The atoms and field mode are coupled through an assumed dipole interaction with coupling coefficient

$$g = \left(\omega_c \mu^2 / 2\hbar\varepsilon_0 V \right)^{1/2}.$$

Decay of the atomic inversion is assumed to be purely radiative at rate γ, while the atomic polarization decay rate is designated by $\gamma_\perp$. Both decay processes are described by H_a. The field amplitude decays at a rate κ via coupling at the cavity mirrors to a set of continuum input-output modes as described by H_c ,which also includes the possibility of excitation by an external field of frequency ω_L .

While an incredibly diverse set of phenomena is described by Eq. (1), we wish to focus on a feature of this Hamiltonian that is well known within the context of cavity QED with Rydberg atoms(18-21) but which is often overlooked in optical physics (22). For $\omega_a = \omega_c$ and for a weak intracavity field $x<<1$, with $x=(<a^+ a>/n_0)^{1/2}$, $n_0 = \gamma_\perp \gamma / 4g^2$, the atomic system can be replaced by an atomic oscillator obeying a boson algebra $[b,b^+]=1$(11, 21). In this case H_0 can be rewritten as a sum of two uncoupled harmonic oscillations,

$$H_0 = \hbar(\omega + g\sqrt{N})C^\dagger_+ C_+ + \hbar(\omega - g\sqrt{N})C^\dagger_- C_- , \tag{2}$$

where the normal mode operators $C_\pm$ have been introduced, $C_\pm = 1/\sqrt{2}(a \pm ib)$, with eigenvalues $\lambda_\pm = i(\omega \pm g\sqrt{N})$. This lifting of the degeneracy of the first excited state of the atom-field system has been termed a *"vacuum-field Rabi splitting"*(19) and is a result of the mutual coupling of the two systems of N atoms and cavity mode. If decay is included, Carmichael has shown more generally that $\lambda_\pm = 1/2(\gamma/2+\kappa) \pm [1/4\, (\gamma/2 - \kappa)^2 - g^2 N]^{1/2} \equiv \beta \pm i\upsilon$, where we have transformed to a rotating frame of frequency ω and have assumed radiative damping, $2\gamma_\perp = \gamma$(11). We see that $\lambda_\pm$ contains an imaginary part only for $1/2|\gamma/2-\kappa|<g\sqrt{N}$. That is, since a and b are independently coupled to separate reservoirs, a periodic exchange of excitation occurs only if the decay rates of the two oscillators are not too dissimiliar. It is this exchange that is crucial to our analysis. (In experiments reported elsewhere (23), we have observed this oscillatory exchange in transient decay.)

For describing squeezed state generation, one must extend the above discussion to include nonzero detunings and larger intracavity fields. We have carried out such an analysis by linearization about the steady state of the Maxwell-Bloch equations that result from Eq (1). For the sake of brevity, in the remainder of our theoretical discussion we restrict attention to the case $2\gamma_{\perp} = \gamma$. The eigenvalue structure is found to consist of five eigenvalues, which in general comprise a set of one real value and two pairs of complex conjugates corresponding to the eigenfunctions formed from the atomic inversion, atomic polarization, and cavity field. Figure 1 shows the dependence of the imaginary parts (υ_1,υ_2) of these eigenvalues on intracavity field x for a fixed value of the atomic cooperativity parameter $C=Ng^2/\kappa\gamma$. In Fig.1a the cavity detuning $\theta = (\omega_c - \omega_L)/\kappa$ and atomic detuning $\Delta = (\omega_a - \omega_L)/\gamma$ are both set to zero, while in Fig.1b $\theta = -0.65$ and $\Delta = 10.7$. The limiting value of (υ_1, υ_2) as $x\to0$ in Fig. 1a is precisely the vacuum-field Rabi splitting described above, generalized to include nonzero detunings in Fig. lb. While for $x\to0$ the normal mode operators corresponding to these eigenvalues are roughly equal admixtures of cavity and atomic operators, for x>>1 we see two rather

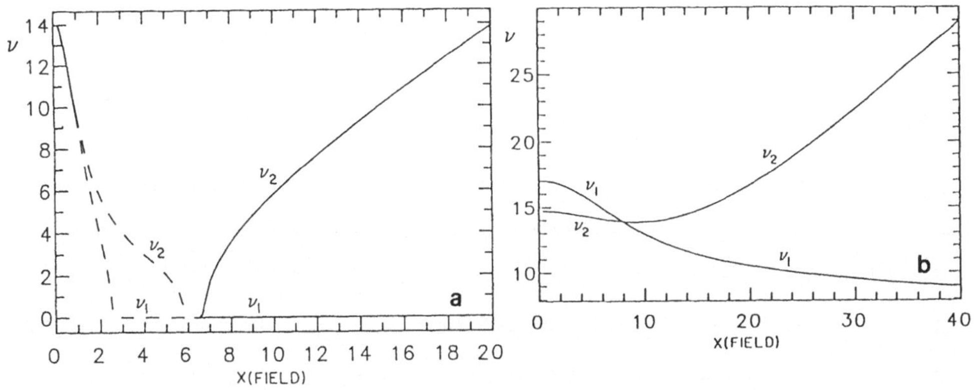

Figure 1--Imaginary parts (υ_1, υ_2) of the eigenvalues resulting from a linearization of the Maxwell-Bloch equations versus intracavity field x. The frequency υ is in units of γ (a) atomic detuning $\Delta = 0 =$ cavity detuning θ (b) atomic detuning $\Delta = 10.7$, cavity detuning $\theta = -0.65$. In both plots the ratio μ of cavity loss to atomic decay rate is 11.8 and cooperativity parameter C = 20. In (a) the region x = 1.05, 6.07 is bistable while in (b) no bistability is present.

disparate frequencies in Fig. 1, which are just the usual free-space Rabi frequency $\delta^2 = x^2 + 4\Delta^2$(with damping given by γ) and the cavity ringing frequency θ (with damping given by κ). Thus in the limit $x>>(1,\Delta)$, the eigenvalue problem decouples with two sets of eigenfunctions of character determined predominantly by atomic or cavity properties. In our work we concentrate on the region around the crossing shown near $x=8$ in Fig.1b. At this point the eigenvalue spectrum results in a phase sensitivity to fluctuations through an eigenmode structure that conspires cooperatively to produce squeezing. We note that this approach is quite different from the usual perspective in which the atomic variables are adiabatically eliminated by taking $\mu = \kappa/\gamma \rightarrow 0$(12,14,15) and for which the weak field splitting shown in Fig.1 is lost.

Of course to translate this qualitative discussion into quantitative predictions for squeezed state generation, a detailed analysis beginning with the Hamiltonian Eq. (1) must be carried out. Given the existing literature in which a generalized Fokker-Planck equation in the positive-P representation has been derived for this problem(24), it is straightforward to arrive at an equation for the spectral density $A(\phi,\Omega)$ describing the fluctuations of the quadrature amplitude $y(t,\phi)=a(t)e^{-i\phi}+a^+(t)e^{i\phi}$ of the intracavity field. Complexity in the current problem arises from keeping the full set of five dynamical variables without adiabatic elimination. Squeezing of the field emitted through the output mirror of the cavity is described by the spectrum of squeezing $S(\phi,\Omega)=2\kappa A(\phi,\Omega)$ (25).

Our results for the spectrum of squeezing S are displayed in Fig. 2. $S(\phi,\Omega)$ is defined such that $S=0$ corresponds to the vacuum state while $S=-1$ corresponds to perfect squeezing of one quadrature amplitude. Note that the frequency Ω at which optimum squeezing occurs in curve (i) is approximately given by the frequency of the crossing point in Fig.1b; that is, optimum squeezing occurs near the frequency associated with the vacuum-field Rabi splitting. This behavior is not peculiar to Fig.1b and 2(i), but it is rather a common feature found in our numerical studies over quite large regions of the parameter space. Another general feature that emerges from our analysis is that the value of the intracavity field x at which optimum squeezing occurs is such that $x \sim \Delta$; that is, the field needs to be increased to the point of the onset of saturation. In Fig. 2 the phase ϕ of the output quadrature amplitude examined has been optimized at each value of Ω to maximize the squeezing. We denote the resultant spectrum as $S_-(\Omega)$; the spectrum of squeezing of the orthogonal quadrature amplitude with ϕ increased by 90° is designated $S_+(\Omega)$ and is not displayed.

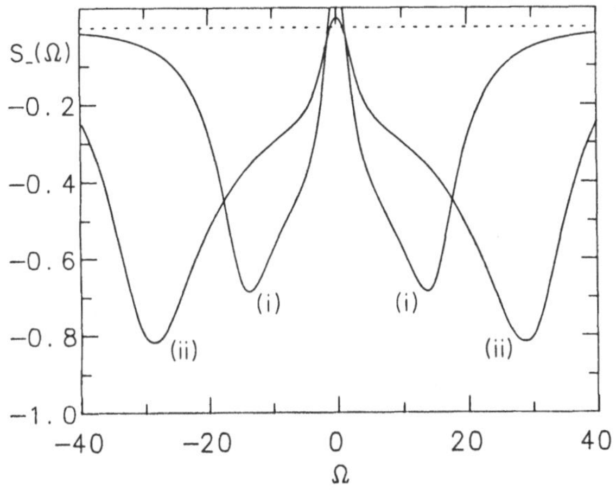

Figure 2--Spectrum of squeezing S_- versus offset frequency Ω in units of γ, with $S_- = 0$ corresponding to the vacuum state. The ratio μ of cavity loss to atomic decay rate is 11.8 (i) Cooperativity parameter $C = 20$, $x = 12.1$, $\Delta = 10.7$, $\theta = -0.65$ (ii) $C = 100$, $x = 17.7$, $\Delta = 18.2$, $\theta = -1.2$. Note that at each Ω the phase ϕ is varied to minimize S_-.

A more global picture of the nature of squeezing in this system is obtained from Fig. 3, which shows the dependence of S_- on atomic cooperativity parameter C. In Fig. 3 each point results from a search for optimum squeezing over all (Δ,θ, Ω,x) for given values of (C,μ). This search is carried out in a four-dimensional space with the value of S_- at any given point in the space requiring the inversion of five-dimensional matrices. Note that substantial degrees of squeezing are predicted for relatively modest values of (C, x), relative to earlier treatments of squeezed state generation with two-level atoms. In addition to the results presented in Fig.1-3, we have also constructed *"isosqueezing"* contours that indicate that squeezing in this new regime persists over relatively broad regions of the parameter space so that in practice the predicted effects should be reasonably robust with respect to deviations between actual experiment and the model calculation.

These results from our numerical evaluation of the spectrum of squeezing together with an analysis of the eigenvalue spectrum leads to the following simple picture for squeezed state generation via the normal mode splitting of the coupled atom-field system. (1)The splitting in the eigenvalue structure (υ_1 , υ_2) must be large as compared to the associated width of the spectral components, which for modest values of μ and Δ is a condition that $g\sqrt{N} \gg (\kappa,\gamma)$. (2) The cavity damping rate κ must be much larger than the atomic decay rate γ so that the dominant decay route of the spectral

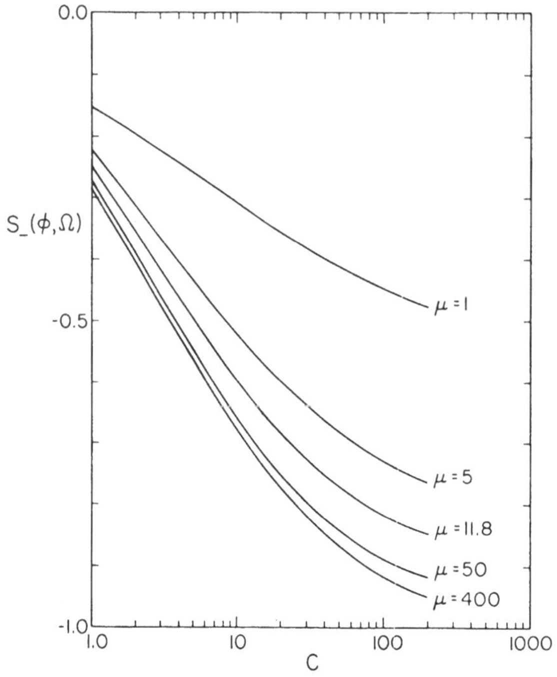

Figure 3--Dependence of the minimum value of the spectrum of squeezing S_ on atomic cooperativity C for fixed μ. Note that at any point on a curve the values of (Δ, θ, Ω, x) have been chosen for optimum squeezing.

features of A(Ω) is through the cavity output coupler, $\mu >> 1$(Note that conditions (1) and (2) imply $C >> \mu >> 1$). (3) The intracavity field x should be increased to the point $x \sim \Delta$. The first two conditions are necessary for forming a coupling induced structure in the system's normal mode spectrum appropriate for multiwave mixing. The third condition then ensures sufficient excitation for nonlinear processes to occur among the low lying (N-atom)+(cavity mode) states. A clear description of *"four-wave mixing"* in this system has been given by Varada et al.(26).

From the perspective of experimental design, the two critical considerations for the observation of squeezed states via the mechanism that we have identified are the requirements for large splittings $\upsilon = g\sqrt{N}$ and large cavity damping κ as compared to the atomic rate γ. Both conditions drive one to small cavity volumes-the lower limit on cavity waist being set by transit broadening and the lower limit on cavity length being the requirement of large optical density within the cavity. The cavity finesse is determined by a tradeoff between high finesse F for reaching large values of atomic cooperativity C for a given maximum attainable intracavity optical density ($C = \alpha l F/2\pi$, with αl=small signal absorption of the intracavity medium) and lower finesse for $\mu >> 1$ and for achieving a good approximation to a single-sided cavity for given absorption and scatter losses in the mirror coatings.

A diagram of the experimental arrangement that we have employed to achieve these ends is shown in Fig. 4. The cavity is formed by a pair of mirrors of radius of curvature 1m separated by 0.83mm. The transmission coefficients of the two mirrors are $T_1 = 0.0075$ and $T_2 = 0.0002$. The measured cavity finesse is F=680, while that inferred from the value of T_1 is $F_1 = 840$. Hence the ratio of output loss through m1 to loss by all other avenues is given by $\sigma = F/F_1 = 0.81$, which implies a 19% reduction in squeezing as compared to an ideal single-ended cavity. The intracavity medium is composed of optically prepumped beams of atomic sodium prepared in the ($3\,^2S_{1/2}, F=2, m_F=2$) state and excited with circularly polarized light to the ($3\,^2P_{3/2}, F=3, m_F=3$) state of the D_2 line. Collimation is provided by a 0.5mm aperture in the source oven and by a 0.3mm aperture located 250mm downstream from the oven and 15 mm upstream from the cavity waist. The maximum absorption αl for this configuration is $\alpha l = 0.2$. The fluorescence from the optical pumping beam together with the recorded hysteresis cycle in absorptive optical bistability provide a measure of the intracavity density and hence of C during the experiments. In separate measurements we have confirmed that to a good approximation an arrangement such as the one outlined above can be viewed as a single-mode cavity

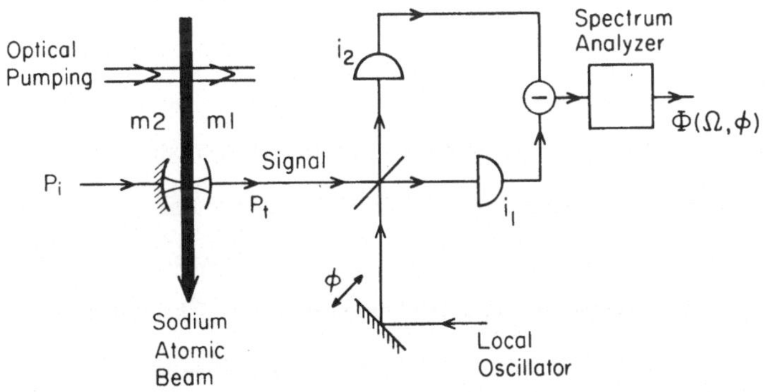

Figure 4--Diagram of the essential elements of the experiment for generating and detecting squeezed states.

containing a collection of two-level atoms within the mean-field theory of optical bistability(27). The excitation source in the experiments is a commercial frequency-stabilized cw dye laser pumped by an argon-ion laser operating at 515nm.

Detection of the fluctuations in the quadrature amplitude of the signal beam emitted through the mirror m1 is accomplished with the balanced homodyne detector indicated in Fig.4(28). The photodiodes are EGG-FFD-060 with the glass windows removed and with the reflection from the diode surface collected and refocussed onto the photodiode resulting in a quantum efficiency $\eta = 0.85 \pm 0.04$. The homodyne efficiency is measured to be approximately $\varepsilon = 0.93 \pm 0.07$ for each channel. Over the range 100-200MHz the *"shot noise"* associated with the 5mA dc photocurrent produced by the local oscillator exceeds the amplifier noise level by greater than 7dB. That the local oscillator is indeed at the vacuum level and does not carry appreciable excess amplitude noise is confirmed by a comparison of the noise levels observed when the two photocurrents i_1 and i_2 are combined first with 0° and then with 180° phase shift. With the exception of coherent lines at multiples of the 85MHz longitudinal mode spacing of the ion laser, we conclude that the local oscillator fluctuations are within ±1% of the vacuum level over the spectral range of interest in the current experiment. Furthermore, with the 180° phase shift actually employed in the squeezing measurements, any excess local oscillator noise is reduced by greater than 15dB.

An example of our observation of noise reductions below the vacuum level is given in Fig. 5. The figure displays the spectral density of photocurrent fluctuations, $\Phi(\phi,\Omega)$ on a logarithmic scale at fixed frequency $\Omega/2\pi$ =200MHz versus local phase ϕ. The trace marked (i) is the vacuum level obtained by blocking the signal beam. The trace labelled (ii) is with the signal beam present and clearly exhibits noise reductions R_- below the vacuum level. Note that the periodicity of these reductions with local oscillator phase ϕ is π rather than 2π , which is the periodicity of intensity fringes in the system. Also note that noise reduction is achieved with incident laser power of only a few hundred microwatts. After correction is made for the nonzero noise level of the amplifier, the observed reduction of -0.8dB below the shot noise level becomes -1.0dB. This figure represents a 20% noise reduction below the level set by the vacuum state of the field at the signal port of the balanced homodyne detector. We have also explored the dependence of the phase sensitive noise on offset frequency Ω. In qualitative terms the observed noise reductions extend over the same broad regions of frequency predicted by theory. For low frequency ($\Omega < 50$MHz), very large noise enhancements R_+ above the vacuum level are observed, again in qualitative accord with theory.

By including the propagation loss (1-T), detection quantum efficiency η, heterodyne efficiency ε, and escape efficiency σ from the cavity, we can relate the spectrum of squeezing S_- to the observed noise reduction R_- (4,29,30)

$$R_-(\Omega) = 1 + \sigma T \eta \varepsilon S_-(\Omega). \tag{3}$$

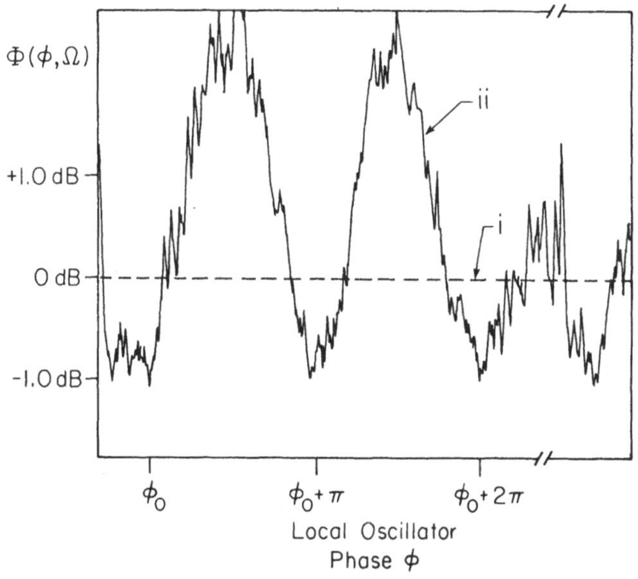

Figure 5--Spectral density $\Phi(\phi,\Omega)$ of fluctuations of the difference photocurrent $(i_1\text{-}i_2)$ versus local oscillator phase ϕ at fixed analysis frequency $\Omega/2\pi = 200$ MHz. (i) Signal input blocked to define vacuum level (ii) Phase sensitive fluctuations with signal beam present drop below the vacuum level. Operating conditions--$C = 17 \pm 2$, $\mu = 13$, $\Delta = 3.5 \pm 0.5$.

By separately measuring the quantities (σ, T, η, ε), we thus infer S_- from measurement of R_-. In the current arrangement, R_-=0.80 corresponds to S_-=-0.33, or to a 33% decrease in fluctuations relative to the vacuum level before degradation by the various loss mechanisms associated with escape and detection. Unfortunately a precise comparison with our theory is hampered at present by a lack of quantitative knowledge of (x,θ). However with the recorded values of (C=17±2, μ = 13) Fig.3 leads to an optimum prediction S_-=-0.66, which is a degree of squeezing considerably larger than that actually inferred from the data. Possible causes of this discrepancy include stray absorption due to background sodium vapor in the vacuum chamber (2), a reduction in optical pumping efficiency at large atomic detunings (a single laser was employed for pumping of the F=2 $\leftrightarrow$ F=3 transition), and saturation of the detection electronics by the noise power contained in the broad detection bandwidth. Another difficulty is that at present we do not have a well defined search procedure for optimization of the squeezing on the five-dimensional space of experimental parameters. We are currently working to eliminate each of these difficulties in future experiments. Of a more fundamental nature is the fact that our theoretical analysis is carried out for a traveling-wave interferometer with plane waves while the experiments are conducted in a standing-wave cavity with a Gaussian-transverse profile. While we are reasonably confident that the deterministic physics is adequately described by a single-transverse-mode theory (27) the precise role of a nonuniform cavity mode in altering the quantum fluctuations is yet to be understood, although we have carried out a full quantum treatment in the good cavity limit (31).

In summary, we have identified both theoretically and experimentally a new regime for squeezed state generation associated with the coupling of a collection of two-level atoms to an optical cavity. The physical process responsible for the squeezing is a coupling-induced mode splitting in the eigenvalue spectrum of the system, which for weak fields and zero detunings is just the vacuum-field Rabi splitting. We have presented a theoretical analysis of the squeezing in this system based upon the formalism developed in Ref.(24,25), and have predicted that large degrees of squeezing should be attainable with rather modest values of atomic density and intracavity field. An experiment to confirm these ideas has been carried out and noise reductions comparable to the best yet achieved in atomic vapors (2) have been recorded. Improvements in our apparatus should lead to observed noise reductions of greater than 50%. Apart from squeezed state studies, the investigation of a number of problems in cavity QED with the optical system we have constructed should be of some interest.

This work was supported by the Venture Research Unit of BP and by the Office of Naval Research. L.A. Orozco is supported by an IBM Graduate Fellowship. We gratefully acknowledge interactions with H.J. Carmichael and D.F. Walls which stimulated the current research program.

References

1) *"Squeezed States of the Electromagnetic Field,"* Feature Issue of JOSA B, eds. H.J. Kimble and D.F. Walls, August (1987).

2) R.E. Slusher, L.W. Hollberg, B. Yurke, J.C. Mertz and J.F. Valley, Phys. Rev. Lett. 55, 2409 (1985).

3) R.M. Shelby, M.D. Levenson, S.H. Perlmutter, R.G. DeVoe, and D.F. Walls, Phys. Rev. Lett. 57, 691 (1986).

4) Ling-An Wu, H.J. Kimble, J.L. Hall, and Huifa Wu, Phys. Rev. Lett. 57, 2520 (1986).

5) H.J. Kimble, J.L. Hall, JosaA3, 37, (1986); M.G. Raizen, L.A. Orozco, H.J. Kimble, JOSA A3, 46 (1986).

6. M.W. Maeda, P. Kumar, and J.H. Shapiro, Opt. Lett. 12 (1987), to be published.

7. Y. Yamamato, S. Machida, N. Imoto, M. Kitagawa, and G. Bjork, JOSA B, to be published (1987).

8) D.F. Walls, Nature (London) 306, 141 (1983).

9) M.D. Reid, A. Lane, D.F. Walls, in Quantum Optics IV, eds, J.D. Harvey and D.F. Walls (Springer-Verlag, 1986), p. 31.

10) L.A. Lugiato and G. Strini, Optics Commun. 41, 67 (1982).

11) H.J. Carmichael, Phys. Rev. A33, 3262 (1985).

12) M.D. Reid and D.F. Walls, Phys. Rev. A32, 396 (1985).

13) M.D. Reid and D.F. Walls, Phys. Rev. A34, 4929 (1986).

14) D.A. Holm, M. Sargent III, and B.A. Capron, in Optical Bistability III, eds. H.M. Gibbs, P. Mandel, N. Peyghambarian and S.D. Smith (Springer-Verlag, Berlin) 1986.

15) J.R. Klauder, S.L. McCall and B. Yurke, Phys. Rev. A33, 3204 (1986).

16) L.A. Lugiato, in Progress in Optics Vol. XXI, E. Wolf, ed., North-Holland, Amsterdam (19 84).

17) H.J. Carmichael, Quantum-Statistical Methods in Quantum Optics (Springer, Berlin) to be published.

18) Y. Kaluzny, P. Goy, M. Gross, J.M. Raimond, and S. Haroche, Phys. Rev. Lett. 51, 1175 (1983).

19) H.I. Yoo and J.H. Eberly, Phys. Reports 118, no. 5 (1985).

20) D. Meschede, H. Walther and G. Müller, Phys. Rev. Lett. 54, 551 (1985).

21) G.S. Agarwal, J. Opt. Soc. Am. B2, 480 (1985).

22) H.J. Kimble, in Frontiers in Quantum Optics, eds, E.R. Pike and Sarben Sarkar (Adam Hilger, Bristol and Boston) 1986.

23) L.A. Orozco, R.J. Brecha, M.G. Raizen, Min Xiao and H.J. Kimble, talk presented at the meeting of the Division of Atomic, Molecular and Optical Physics of the APS, Eugene, Oregon (1986).

24) P.D. Drummond and D.F. Walls, Phys. Rev. A 23, 2563 (1981).

25) C.W. Gardiner and M.J. Collett, Phys. Rev. A31, 3761 (1985).

26) G.V. Varada, M. Sanjay Kumar and G.S. Agarwal, to be published.

27) L.A. Orozco, H.J. Kimble and A.T. Rosenberger, Opt. Commun. (to be published) 1987.

28) Yuen, H.P., and Chan, V.W.S., Opt. Lett. 8, 177 and errata Opt. Lett 8, 345 (1983).

29) L. Mandel in Optics in Four Dimension-1980, edited by M.A. Machado and L.M. Narducci, AIP Conference Proceedings No. 65 (American Institute of Physics, New York, 1981).

30) J.H. Shapiro, IEEE J. Quantum Electron. 21, 237 (1985).

31) Min Xiao, H.J. Kimble and H.J. Carmichael, Phys. Rev. A (to be published) (1987).

QUANTUM NOISE REDUCTION VIA TWIN PHOTON BEAM GENERATION

E. Giacobino, C. Fabre, A. Heidmann, R. Horowicz and S. Reynaud
Laboratoire de Spectroscopie Hertzienne de l'ENS (Laboratoire associé au CNRS), Université Pierre et Marie Curie, F- 75252 PARIS CEDEX 05.

I. INTRODUCTION :

Considerable interest has been devoted lately to squeezed and non classical states of the electromagnetic field. As opposed to the coherent states introduced by Glauber (1) they are characterized by different quantum fluctuations on conjugate quantities (2) like either the two quadrature components or the phase and amplitude of the electric field. In principle the fluctuations on one of these quantities can be made arbitrarily small. Such states have been recently produced experimentally (3). Numerous ideas have been proposed to generate them in various experimental schemes.

Most of these ideas rely on the use of a non linear Hamiltonian acting on the vacuum field, to transform it into a non classical state. The Hamiltonian usually includes terms corresponding to the creation of pairs of correlated photons indicating a close connection between pair production and squeezing.

In this paper we shall focus our attention on parametric down conversion, in which a non linear crystal pumped at frequency ω_0 emits two signal fields at frequencies ω_1 and ω_2 such that $\omega_0 = \omega_1 + \omega_2$. The correlation between the "twin" photons produced by parametric process has been experimentally demonstrated in the seventies (4) and recently confirmed by more precise measurements (5), in agreement with quantum mechanical calculations (6). Recently it was used in the degenerate case to create squeezed states of light (7) and in the non-degenerate case to investigate the production of sub-Poissonian statistics via optoelectronic feedback (8). Only the non-degenerate case, in which the signal fields can be distinguished by either their polarizations or their frequencies will be considered here.

In contrast with previous experiments, we shall mainly concentrate on the study and production of "macroscopic" twin beams, namely intense laser-like beams with strong intensity correlation. Indeed, with the available C.W. lasers and non linear crystals the parametric process does not generate intense beams because the pump power is spread out into an infinity of twin modes. In order to favor only a few pairs of modes, the crystal can be inserted in an optical cavity having mirrors with a high reflectivity for the signal frequencies. Above some pump threshold, the system oscillates (9) and yields two intense light beams. As the photons are created by pairs, it can be expected that the noise on the difference $I_1 - I_2$ between the signal beam intensities is reduced. In section II, we theoretically investigate the noise characteristics of an ideal two-mode optical parametric oscillator (TMOPO) above threshold. In section III, we describe our experimental set-up and we discuss the predicted phenomena including the effect of losses in the cavity and the quantum efficiency of the detectors.

II. THEORY

In this section, we present a simple theoretical model for computing the characteristics of the TMOPO emission. We consider that the signal fields are confined in an optical ring cavity (Fig. 1) having the same damping rate for the two signal modes. We assume perfect phase matching and we neglect any loss mechanism other than the transmission of the coupling mirror. In addition we suppose that the one pass gain and losses are small. The mirrors are perfectly transmissive for the pump beam.

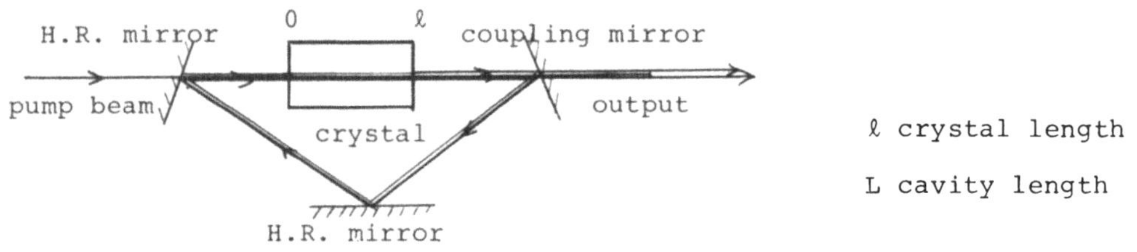

Figure 1: OPO cavity

Assuming a linear depletion of the pump field inside the crystal, the classical equations for the propagation of the fields in a parametric medium (9) lead to :

$$\alpha_i(\ell) = e^{ik_i\ell}(\alpha_i(0) + \xi\,\tilde{\alpha}_0\,\alpha_j^*(0)) \quad (i,j : 1,2) \quad i \neq j \qquad (1)$$

$\alpha_i(0)$ and $\alpha_i(\ell)$ $(i = (1,2))$ are the classical amplitudes of the signal field i (with wave vector k_i) respectively at the input and output of the crystal of length. $\tilde{\alpha}_0$ is the mean pump field amplitude :

$$\tilde{\alpha}_0 = \alpha_0(0) - \frac{\xi}{2}\,\alpha_1(0)\,\alpha_2(0) \tag{2}$$

and $\xi = 2c\chi\ell$

where χ is the second order non linear coefficient of the crystal. The fields α_i arriving on the coupling mirror are related to the fields α_i' leaving the coupling mirror (cf. Fig. 2) through the propagation in the cavity and in the crystal :

$$\alpha_i\, e^{-ik_iL} = \alpha_i' + \xi\alpha_0\,\alpha_j'^{*} \quad (i,j = 1,2;\; i \neq j) \tag{3}$$

where $\alpha_0 = |\tilde{\alpha}_0|$ (a proper choice of the phase of the pump field has been made).

On the coupling mirror of the cavity the signal fields inside the cavity, α_i and α_i' and outside the cavity, α_i^{in} and α_i^{out} (Fig. 2) are connected by

$$\alpha_i' = r\alpha_i + t\alpha_i^{in} \tag{4a}$$

$$\alpha_i^{out} = -r\alpha_i^{in} + t\alpha_i \tag{4b}$$

where r and t are the reflexion and transmission coefficients of the coupling mirror.

Figure 2:

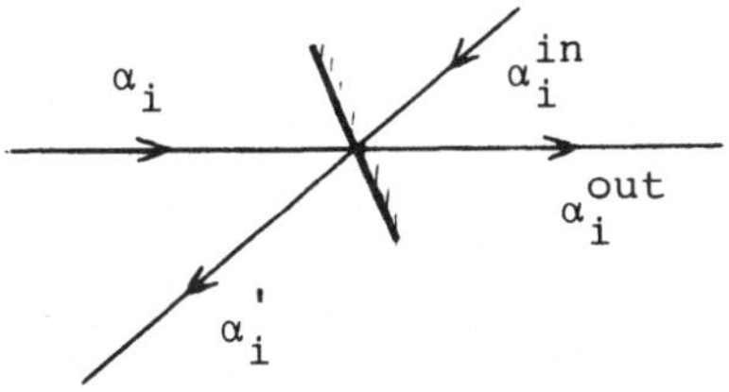

At resonance for the signal fields in the cavity ($\exp(-ik_iL) = 1$ for i = 1,2), and with no input signal fields, one gets

$$\gamma\alpha_1 = \xi\,\alpha_0\,\alpha_2^{*} \tag{5a}$$

$$\gamma\alpha_2 = \xi\,\alpha_0\,\alpha_1^{*} \tag{5b}$$

where $\gamma = 1-r$

Equations (6) have a solution $\alpha_1 = \alpha_2 = 0$ corresponding to no oscillation. A non zero solution can be found :

$$\bar{\alpha}_1^{\,2} = \bar{\alpha}_2^{\,2} = \left(|\alpha_{in}| - \frac{\gamma}{\xi}\right)\frac{2}{\xi} \tag{6}$$

if the input field $|\alpha_{in}|$ satisfies the threshold condition $|\alpha_{in}| > \gamma/\xi$ the solution for the mean pump field is :

$$\bar{\alpha}_0 = \frac{\gamma}{\xi} \tag{7}$$

Now, our purpose is to calculate the fluctuation spectrum of the intensity difference I between the two signal fields. In order to determine the field fluctuations we will use a semi-classical approach. We linearize the classical equations in the vicinity of the above solution for modes $\alpha_i(\omega)$ detuned by ω from the oscillating modes. Furthermore we consider that these modes are driven by the vacuum fluctuations entering the cavity through the coupling mirror.

Above threshold the intensity fluctuations are proportional to the amplitude fluctuations. We thus have to compute the noise spectrum of the variable

$$\alpha(\omega) = \alpha_1(\omega) - \alpha_2(\omega) \tag{8}$$

From Eqs (3) and (4) we obtain the equations giving $\alpha(\omega)$

$$\alpha(\omega)(e^{-i\omega\tau} - 1 + \gamma) = -\xi\, \bar{\alpha}_0\, \alpha^*(-\omega) + t\alpha^{in}(\omega) \tag{9a}$$

$$\alpha(\omega)(e^{-i\omega\tau} - 1 - \gamma) = -\xi\, \bar{\alpha}_0\, \alpha^*(-\omega) - t\alpha^{out}(\omega) \tag{9b}$$

where $\tau = L/c$ is the cavity round trip time and α^{in} and α^{out} are the input and output fluctuations corresponding to the quantity α. Let us note that the fluctuations of α are not coupled to the pump fluctuations due to the fact that $\bar{\alpha}_1 = \bar{\alpha}_2$. Solving the set of the linear equations (9) and their complex conjugates, leads to the following expression for the output fluctuations

$$\alpha^{out}(\omega) + \alpha^{out*}(-\omega) = -\frac{e^{-i\omega\tau} - 1}{e^{-i\omega\tau} - 1 + 2\gamma}\,(\alpha^{in}(\omega) + \alpha^{in*}(-\omega)) \tag{10}$$

the spectrum of the intensity difference is proportional to :

$$S_I(\omega) \propto \bar{\alpha}_1^{\,2}\, |\alpha^{out}(\omega) + \alpha^{out*}(-\omega)|^2 \tag{11}$$

Using Eq(10), we finally obtain :

$$S_I(\omega) = S_0\left(1 - \frac{T^2}{1+R^2 - 2R\cos\omega\tau}\right) \tag{12}$$

where $R = r^2 = 1-2\gamma$ and $T = t^2$ are the intensity reflection and transmission coefficients of the coupling mirrors and S_0 is the usual shot noise.

Eq. (12) clearly shows that photon noise is completely suppressed for $\omega = 0$. Suppression remains effective for frequencies ω inside the Airy peaks, that is for frequencies inside the cavity bandwidth $(\omega_c \sim (1-R)/\tau)$ This can be well understood by recalling that the photons are emitted by correlated pairs, but that the pair correlation is degraded by the cavity : the intensities I_1 and I_2 on the two detectors are expected to be nearly equal only when measured during a time longer than the cavity storage time $^1/\omega_c$.

This physical interpretation can be justified by a simple model where the field is described in terms of photons only. The key assumption of such a model is that the two photons of a pair are emitted simultaneously, but at random times. The mean rate of pair emission is denoted E. When a photon hits the cavity coupling mirror, it has the probability R for being reflected, the probability T for being transmitted $(R+T = 1)$. For a given pair of photons, the difference I of the intensities I_1 and I_2 detected by two photodetectors at time t is given by

$$I(t) = \delta(t - (t_0+m_1\tau)) - \delta(t - (t_0+m_2\tau)) \tag{13}$$

where t_0 is the first possible detection time and m_1 and m_2 are the numbers of round trips of photon 1 and 2 before they exit the cavity. The contribution of this pair emission to the Fourier transform $|I(\omega)|^2$ is

$$\begin{aligned} |I(\omega)|^2 &= |\exp(-i\omega(t_0+m_1\tau) - \exp(-i\omega(t_0+m_2\tau))|^2 \\ &= 2(1-\cos((m_1-m_2)\omega\tau)) \end{aligned} \tag{13}$$

The emission spectrum $S_I(\omega)$ is obtained by averaging (13) over the contributions of the various pair emissions :

$$S_I(\omega) = \frac{1}{\Delta t} < |I(\omega)|^2 > \tag{14}$$

where the integration time Δt is much longer than any other characteristic time. Noting that each pair contribution has to be weighted by the probability $TR^{m_1}\,TR^{m_2}$ that photons 1 and 2 undergo respectively m_1 and m_2 reflections we get :

$$S_I(\omega) = \frac{1}{\Delta t} E \, \Delta t \sum_{m_1,m_2} TR^{m_1} \, TR^{m_2} \, (1 - \cos(m_1-m_2)\omega\tau)$$

$$= 2E(1 - \frac{T^2}{1 + R^2 - 2R \cos\omega\tau}) \qquad (15)$$

We see that by a mere corpuscular approach we recover the preceding result that photon noise is reduced inside the Airy peaks of the cavity. The latter calculation is also valid for a low finesse cavity, whereas a high reflection coefficient had been assumed in the preceding calculation.

On the other hand, let us stress that the semi-classical calculation allows one to derive the fluctuations of other characteristics of the fields (10) and in particular to show that the sum of the phases of the twin beams is squeezed. But dealing with quantum fluctuations by a semi-classical method could seem questionable. However, we can first remark that it gives the right results below threshold : we have checked that we obtain the same results by using the standard quantum methods (11) for our two-mode-OPO assuming the case of "ideal noise" (12). On the other hand, well above threshold, it is well known that the mean quantum fields are adequately given by the classical equations, and that the quantum fluctuations have very small relative amplitudes. The linearized classical equations are also very frequently used to describe the dynamics of the fluctuations (cf. stability analysis). When the vacuum fluctuations are the only source of noise, our method seems to be a quite natural one. In addition the semi-classical calculation and the photon pair model are in remarkable agreement for the fluctuations of I_1-I_2 with fully quantum calculations above threshold (13)(14).

III. EXPERIMENT

Experimental set-up

The experimental set-up is sketched in Fig. 3. The optical parametric oscillator is pumped by the 528 nm line of a single mode Ar^+ laser, stabilized on an external Fabry-Perot cavity. The non-linear medium is a 7 mm KTP crystal, which is used in a type II phase matching configuration : the green pump beam propagates along an extraordinary ray and is phase matched with an ordinary and an extraordinary beam at the infrared frequencies, with incident and output angles close to zero. The KTP crystal is inserted in a 3 cm long cavity, closed by mirrors having a 2 cm radius of curvature. The input mirror has a high transmission for the green light and a maximum reflectivity for the infrared. The output mirror has a 1 % transmission for the infrared light and also a high transmission for the green light. An acousto-optic modulator serving as an optical isolator prevents the light back-reflected by the cavity from interacting with the Ar^+ laser.

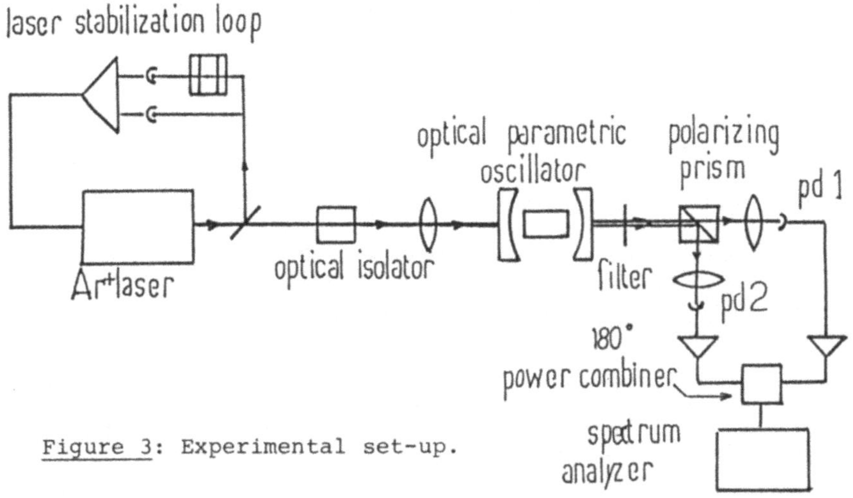

Figure 3: Experimental set-up.

Above a pump threshold power of about 100 mW, the system oscillates and two co-propagating cross-polarized beams with wavelengths close to 1.06 µ are emitted. The crystal is cut for exact phase matching at 0° incidence between two waves at 1.064 µm and one wave at 0.532 µm (for the purpose of YAG lasers doubling) ; the phase matching conditions for 0° incidence for pumping at 528 mm give emission of twin beams at 1.067 µm and 1.048 µm respectively. The stray infrared beam emitted in the backward direction through the high reflectivity mirror is used to stabilize the cavity length for maximum parametric emission. In the forward direction, a filter blocks the pump beam and the two infrared beams are separated with a polarizing beam-splitter and focused on two InGaAs photodiodes having a quantum efficiency of 0.9 at 1.06 µ. All the optical elements located after the output of the cavity are antireflection coated for the infrared light. The two photocurrents are amplified by low noise 40 dB amplifiers and subtracted by a 180° power combiner. The difference signal is monitored with a spectrum analyser.

The interest of type II phase matching is that the twin photons are distinguished by their polarizations even if their freauencies are equal or very close to each other ; this enables us to easily monitor the difference intensity of the twin beams, and its noise spectrum which is expected to be below the quantum noise.

Expected signal

As exposed above, the various theoretical models predict a reduction of noise below shot noise in a frequency band which is the frequency pass-band of the cavity. The noise should be completely suppressed near the zero frequency. However, this result has been obtained by assuming that the only loss mechanism comes from the coupling mirror. Actually other losses have to be taken into account. Absorption and stray reflections on the two faces of the crystal will be represented (Fig. 4) by losses coefficients R_1 and R_2 ($T_1 = 1-R_1$, $T_2 = 1-R_2$). The quantum efficiency of the detection will be accounted for by a transmission coefficient T_3.

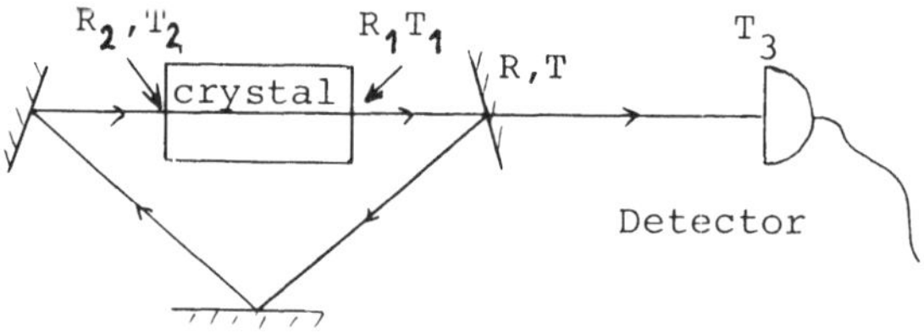

Figure 4

With these assumptions, the probability for one photon to undergo 0 reflection and be detected is $T_1\ T\ T_3$. The probability for one photon to undergo m reflections and be detected is $T_1\ T\ T_3\ (RT_2T_1)^m$. Inspection of the various possible events at the detectors leads to the following result:

$$S_I(\omega) = 2E\left\{\sum_m T_1T_3T\ (RT_2T_1)^m - \sum_{m_1,m_2} T_1T_3T\ (RT_2T_1)^{m_1}\ T_1T_3T\ (RT_2T_1)^{m_2} \cos(m_1-m_2)\omega\tau\right\}$$

$$= 2E\eta\left[1 - \frac{\eta\ (1-R')^2}{1+R'^2-2R'\cos\omega\tau}\right]$$

where $R' = RT_1T_2$ is the effective reflection coefficient of the cavity

and $\eta = \dfrac{TT_1T_3}{1-RT_1T_2}$ is the effective detection efficiency

The shot noise suppression is thus degraded by a factor η.
With $T_3 = 0.90$, $R_1=R_2 = 0.005$, $1-T=R = 0.99$
one gets

$\eta = 0.45$

Even with losses in the cavity which amount to about half the transmission of the coupling mirror, a sizeable reduction of the quantum noise should be observed.

IV. CONCLUSION

Such highly correlated intense beams could have applications in very various demains : first, it may enhance the sensitivity of absorption measurements (15) : if one inserts an absorption cell on one arm and scans the frequency around an absorption frequency, the signal-to-noise ratio of the absorption dip recorded on the signal $I_1 - I_2$ is no longer shot noise limited. Second, in a way analogous to ref. (16), one can monitor the I_1 intensity only and use this signal to react on I_2 (or on the pump intensity). This would provide an intensity squeezed laser like beam, i.e. an approximation of a Fock state $|N\rangle$, which has so far never been obtained in the laboratory.

ACKNOWLEDGEMENTS

High efficiency InGaAs photodiodes have been generously provided by Mr. de Cremoux, from THOMSON-CSF Company.

REFERENCES

(1) R.J. Glauber, Phys. Rev. 130, 2529 (1963) : 131, 2766 (1963).

(2) D.F. Walls, Nature 306, 141 (1983).

(3) R. Slusher, L. Hollberg, B. Yurke, J. Mertz, J. Valley, Phys. Rev. Lett. 55, 2409 (1985)
R. Shelby, M. Levenson, S. Perlmutter, R. De Voe, D. Walls, Phys. Rev. Lett. 57, 691 (1986).
Ling-An Wu, H.J. Kimble, J.L. Hall, Huifa Wu, Phys. Rev. Lett. 57 2520 (1986).

(4) D. Burnham, D. Weinberg, Phys. Rev. Lett. 25, 84 (1970).

(5) S. Friberg, C. Hong, L. Mandel, Phys. Rev. Lett. 54, 2011 (1985).

(6) B. Mollow, R. Glauber, Phys. Rev. 160, 1097 (1967).
B. Mollow, Phys. Rev. A 8, 2684 (1973).
C. Hong, L. Mandel, Phys. Rev. A 31, 2409 (1985).

(7) Ling-An Wu, H.J. Kimble, J.L. Hall, Huifa Wu, Phys. Rev. Lett. 57, 2520 (1986).

(8) E. Jakeman, J. Walker, Opt. Com. 55, 219 (1985).
R. Brown, E. Jakeman, E. Pike, J. Rarity, P. Tapster, Europhys. Lett. 2, 279 (1986).

(9) J. Giordmaine, R. Miller, Phys. Rev. Lett. 14, 973 (1965).
S. Harris, Proc. IEEE 57, 2096 (1969).
R. Smith : Optical Parametric Oscillators in "Laser Handbook I", T. Arecchi, E. Schultz-Dubois editors, North Holland (1973).

(10) S. Reynaud, C. Fabre, E. Giacobino, J.O.S.A. To be published.

(11) B. Yurke, Phys. Rev. A 29, 408 (1984) ; Phys. Rev. A 32, 300 (1985).
M. Collett, D. Walls, Phys. Rev. A 32, 2887 (1985).

(12) M. Reid, D. Walls, Phys. Rev. A 31, 1622 (1985).

(13) S. Reynaud, preprint, 1987

(14) C.M. Savage and D.F. Walls, J.O.S.A. To be published.

(15) N.C. Wong, J. Hall, J. Opt. Soc. Am. B 2, 1527 (1985).
M. Gehrtz, G. Bjorklund, E. Whittaker, J. Opt. Soc. Am. B 2, 1510 (1985).

(16) Y. Yamamoto, N. Imoto, S. Machida, Phys. Rev. A 33, 3243 (1986).

DYNAMICS OF THREE-LEVEL ATOMS: JUMPING AND SQUEEZING

P.L.Knight, S.M.Barnett, B.J.Dalton[1], M.S.Kim, F.A.M.de Oliveira[2] and K.Wodkiewicz[3]

Optics Section, Blackett Laboratory, Imperial College, London SW7 2BZ, U.K.

The competition between transitions in a driven three-level atom generates interesting dynamical correlations in the radiated electromagnetic field. We first review the problem of a three-level atom in a high-Q cavity excited by a quantized radiation field. We examine the phase sensitivity and squeezing properties of this two-photon system. We then discuss the quantum fluctuations in the fields radiated by a three-level atom in free-space and show how the three-level dynamics modify the phase-sensitive squeezing in the fluorescence. Intensity correlations between the emitted photons are studied and evidence for quantum jumps in the three-level dynamics examined. The description of sequential photoemission correlations using probability amplitudes is compared with Bloch equation treatments based on the quantum regression theorem.

---

permanent address: 1: Physics Department, University of Queensland, St.Lucia, Queensland 4067, Australia.

2: Physics Department, Universidade Federal do Rio Grande do Norte, 59000, Natal, RN Brazil.

3: Institute for Theoretical Physics, University of Warsaw, Hoza 69, Warsaw 00 681, Poland)

1. INTRODUCTION

In this paper we examine some of the fundamental processes by which three-level atoms can radiate light with manifestly quantum properties. These atoms have three energy levels, one of which is coupled to both of the others by electric dipole transitions. Three level-configurations are possible. The "ladder"-configuration has the ground state coupled to the most excited state via an intermediate third level. The "lambda"-configuration has two lower energy states, both of which are coupled to a common excited state. Finally, the "vee"-configuration has two excited states both of which are coupled to a common ground state.

In section two we consider the production of squeezed superposition states by the resonant interaction of a three-level atom in a ladder-configuration with one or two cavity modes [1]. In section three we discuss the possibility of squeezing by resonance fluorescence from a driven three-level atom in a lambda-configuration [2]. Finally, in section four we consider the quantum jumps and intensity correlations in fluorescence from a driven three-level atom in a vee-configuration [3].

2. CAVITY FIELD SQUEEZING: THREE-LEVEL JAYNES-CUMMINGS MODEL

The intra-cavity spontaneous emission from a specially prepared two or three-level atom can generate field states that are a superposition of the vacuum and the one- or two- photon states [1,4]. These states can exhibit squeezing, that is fluctuations in one quadrature of the electric field that are reduced below the level associated with the vacuum [5]. These superposition states are fundamentally different from the usual squeezed or two-photon coherent states in that they are not minimum uncertainty states of the uncertainty relation

$$\Delta a_1 \Delta a_2 \geq 1/4$$

where a_1 and a_2 are the hermitian quadratures of the annihilation operator ($a=a_1+ia_2$).

We consider a three-level atom with states $|1\rangle$, $|2\rangle$ and $|3\rangle$ in a ladder configuration resonantly excited by a quantized field consisting of one or two cavity modes. We imagine the cavity Q-factor to be sufficiently high that dissipative interactions may be neglected, and ignore temperature-dependent thermal effects. Such a system is experimentally approachable in the Rydberg atom maser [6,7]. We will demonstrate that a suitably-prepared atom interacting with a single cavity mode generates a maximally-squeezed two-photon superposition state, whereas if a pair of field modes is involved, multimode squeezing in superpositions of the modes is generated by the (reversible) spontaneous emission [1].

For a single-mode interaction, we take as our Hamiltonian the rotating-wave (RWA), interaction picture form

$$H=\hbar\{a^{+}(g_1\,|1\rangle\langle 2| + g_2|2\rangle\langle 3|) \\ +(g_1\,|2\rangle\langle 1| + g_2|3\rangle\langle 2|)a\} \tag{2.1}$$

where the coupling constants g_1, g_2 are taken as real. The three-level atom is prepared in a linear superposition of the ground, and uppermost states

$$|A(t=0)\rangle=\cos(\theta/2)\,|3\rangle+\exp(i\phi)\sin(\theta/2)\,|1\rangle \tag{2.2}$$

by coherent excitation with suitably chosen fields prior to t=0. The initial atom-field state is a product of the atomic superposition state and the vacuum field

$$|\psi(t=0)\rangle = |A(t=0)\rangle|0_f\rangle. \tag{2.3}$$

The Hamiltonian (2.1) couples the atom-field state $|3\rangle|0_f\rangle$ to the states $|2\rangle|1_f\rangle$ and $|1\rangle|2_f\rangle$ but in RWA the state $|1\rangle|0_f\rangle$ is uncoupled. At time t, the interaction picture wavefunction is then

$$|\psi(t)\rangle=\cos(\theta/2)\{C_3(t)|3\rangle|0_f\rangle + C_2(t)|2\rangle|1_f\rangle+C_1(t)|1\rangle|2_f\rangle\}$$
$$+\exp(i\phi)\sin(\theta/2)|1\rangle|0_f\rangle. \tag{2.4}$$

The time-dependent Schrödinger equation gives for the state amplitudes

$$C_3(t)=[2g_1^2+g_2^2\cos\Omega t]/[2g_1^2+g_2^2] \tag{2.5}$$

$$C_2(t)=-i(g_2/\Omega)\sin\Omega t \tag{2.6}$$
$$C_1(t)=\sqrt{2}(g_1g_2/\Omega^2)(\cos\Omega t-1) \tag{2.7}$$

where Ω is the effective Rabi frequency, $\Omega^2=2g_1^2+g_2^2$. If $\sqrt{2}g_1=g_2$, then the dynamics simply reflects the periodic interchange of two quanta between atom and field. After half a period, when $\cos\Omega t=-1$, the system is in the state

$$|\psi\rangle=|1\rangle\{\exp(i\phi)\sin(\theta/2)|0_f\rangle-\cos(\theta/2)|2_f\rangle\} \tag{2.8}$$

which is a product of a pure two-photon superposition state and the atomic ground state. The system is in this state periodically, and at these times the field exhibits squeezing. If we choose the phase ϕ to be zero then at these times, the variance in the a_1 quadrature is

$$(\Delta a_1)^2 = \frac{1}{4}(1+4\cos^2(\theta/2) - 2\sqrt{2}\cos(\theta/2)\sin(\theta/2))$$

which is clearly squeezed if $\sqrt{2}\cos(\theta/2)-\sin(\theta/2)$ is less than zero. Here the smallest possible variance in a_1 is $(\Delta a_1)^2=(1/4)(3-\sqrt{6})\approx 0.13763$, corresponding to about 45% squeezing below the standard quantum limit.

When the three-level atom is excited by two quantized field modes the RWA interaction picture Hamiltonian is

$$H=\hbar\{g_1(b^\dagger|1\rangle\langle 2|+|2\rangle\langle 1|b)+g_2(a^\dagger|2\rangle\langle 3|+|3\rangle\langle 2|a)\}. \qquad (2.9)$$

Two-mode squeezed states are superpositions of photon states in which both modes contain the same number of excitations. The simplest is the superposition of the two-mode vacuum state and that in which each mode contains a single photon. Such a state is generated from the two-mode vacuum by an atom prepared in the superposition state eq.(2.2):

$$|\psi(t=0)\rangle=|A(t=0)\rangle|0_a\rangle|0_b\rangle \qquad (2.10)$$

where a,b denote field modes. Now H couples $|3\rangle|0_a\rangle|0_b\rangle$ to the states $|2\rangle|1_a\rangle|0_b\rangle$ and $|1\rangle|1_a\rangle|1_b\rangle$. The zero-quantum state $|1\rangle|0_a\rangle|0_b\rangle$ does not evolve. At time t, the wavefunction is

$$\begin{aligned}|\psi(t)\rangle=&\cos(\theta/2)\{C_2(t)|2\rangle|1_a\rangle|0_b\rangle+C_3(t)|3\rangle|0_a\rangle|0_b\rangle\\&+C_1(t)|1\rangle|1_a\rangle|1_b\rangle\}\\&+\exp(i\phi)\sin(\theta/2)|1\rangle|0_a\rangle|0_b\rangle.\end{aligned} \qquad (2.11)$$

Again, solution of the Schrödinger equation for the probability amplitudes $C_i(t)$ is simple and gives

$$C_3(t)=[g_1^2+g_2^2\cos\Omega t]/\Omega^2 \qquad (2.12)$$

$$C_2(t)=-i(g_2/\Omega)\sin\Omega t \qquad (2.13)$$

$$C_1(t)=(g_1 g_2/\Omega^2)(\cos\Omega t-1) \qquad (2.14)$$

where Ω is the effective Rabi frequency $\Omega^2=g_1^2+g_2^2$. These simplify now if $g_1=g_2$ so that again two quanta are periodically exchanged, one into each of the two modes. When $\cos\Omega t=-1$, we have

$$|\psi(t=\pi/\Omega)\rangle=|1\rangle\{-\cos(\theta/2)|1_a\rangle|1_b\rangle+\exp(i\phi)\sin(\theta/2)|0_a\rangle|0_b\rangle\} \quad (2.15)$$

a product of the atomic ground state and a pure state of the two-mode field. This pure state is related to the thermofield [8] or two-mode squeezed state, generated by the squeeze operator

$$S(\zeta)=\exp(\zeta^* ab-\zeta a^\dagger b^\dagger). \quad (2.16)$$

The multimode squeezed states are superpositions only of those states in which both modes contain the same number of photons. The strong correlation between the modes generates enhanced Bose-Einstein fluctuations in each single mode but squeezing in mode superpositions.

The density matrix for the a-mode is found by tracing the field density matrix over the b-mode:

$$\rho_a=\cos^2(\theta/2)|1a\rangle\langle 1a|+\sin^2(\theta/2)|0_a\rangle\langle 0_a| \quad (2.17)$$

is a statistical mixture of one and zero photons. There is no phase-information and the quadrature operators exhibit enhanced rather than squeezed fluctuations

$$(\Delta a_1)^2=\tfrac{1}{4}(1+2\cos^2(\theta/2))=(\Delta a_2)^2. \quad (2.18)$$

A similar result holds for the b-mode.

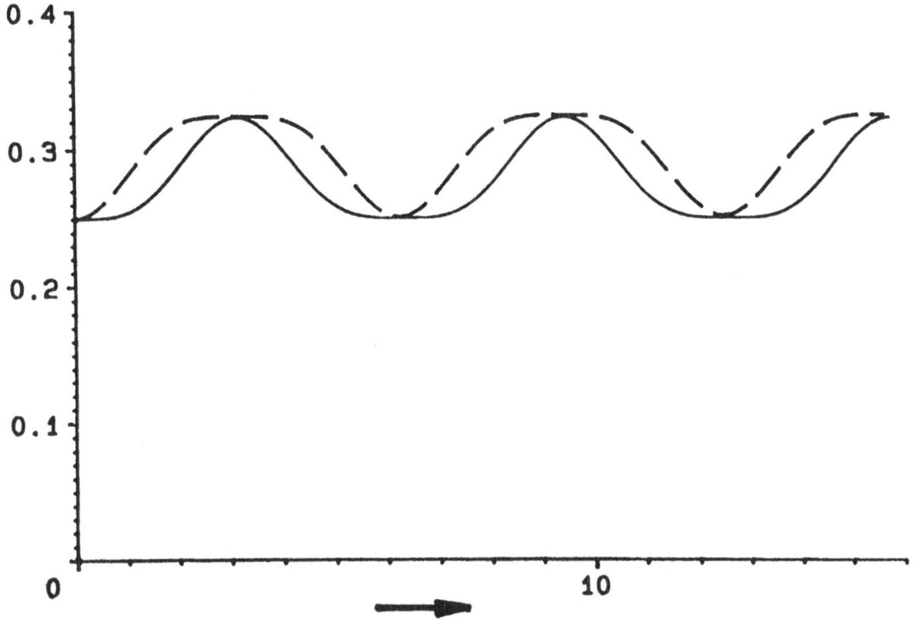

Fig. 1: The evolution of the variances in the a_1 and a_2 (– – –) and b_1 and b_2 (——) quadratures for the three-level, two-mode JCM with $\theta=3\pi/4$. The fluctuations in these modes are phase-insensitive and do not exhibit squeezing.

The two fluctuating modes are nevertheless tightly correlated. We introduce superposition modes with annihilation operators

$$\bar{a}=(a-\exp(i\epsilon)b)/\sqrt{2} \tag{2.19}$$

$$\bar{b}=(b+\exp(-i\epsilon)a)/\sqrt{2} \tag{2.20}$$

In terms of these superposition mode number states, the two-mode field wavefunction in eq. (2.15) has the form

$$|F\rangle=\exp(i\phi)\ \sin(\theta/2)\ |0\bar{a}\rangle|0\bar{b}\rangle + \frac{1}{\sqrt{2}}\cos(\theta/2)\{\exp(i\epsilon)\ |2_{\bar{a}}\rangle|0_{\bar{b}}\rangle-\exp(-i\epsilon)|0_{\bar{a}}\rangle|2_{\bar{b}}\rangle\} \tag{2.21}$$

This state would be the product of two single-mode superpositions of the two-photon type except that the contribution from $|2_{\bar{a}}\rangle|2_{\bar{b}}\rangle$ is absent. The individual superposition modes are not in pure states and

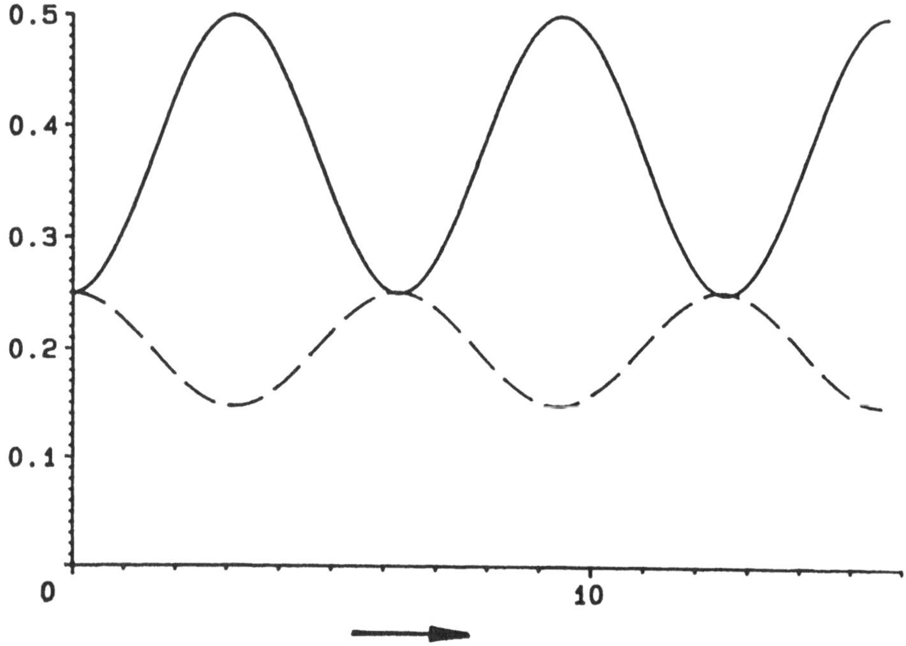

Fig.2: The evolution of the superposition mode variances in the $\bar{a}_1$ (dashed) and $\bar{a}_2$ (solid line) quadratures for the three-level, two mode JCM with $\theta=3\pi/4$. With this choice of phases the $\bar{a}_1$ quadrature exhibits squeezed fluctuations that may be reduced by up to 40% below the standard quantum limit.

mixed state effects degrade the squeezing. We define the in-phase quadrature operator

$$\bar{a}_1=(1/2)(\bar{a}+\bar{a}^\dagger) \tag{2.22}$$

to generate a variance

$$(\Delta a_1)^2=(1/4)(1+2\cos^2(\theta/2)+\sin\theta\cos(\epsilon-\phi)) \tag{2.23}$$

If $\epsilon-\phi=\pi$, the variance is

$$(\Delta\bar{a}_1)^2=(1/4)+(1/4)(1+\cos\theta-\sin\theta) \tag{2.24}$$

which is squeezed if $(\sin\theta-\cos\theta)>1$. The minimum variance in $\bar{a}_1$ is produced if the atom is prepared initially with $\cos\theta=-1/\sqrt{2}$ and $\sin\theta$

$=1/\sqrt{2}$, and is $(\Delta\bar{a}_1)^2=(1/4)(2-\sqrt{2})\approx 0.14645$, corresponding to 40% squeezing, marginally less than the squeezing associated with single-mode pure superpositions (each of the superpositions $\bar{a}$, $\bar{b}$ are put into mixed states by tracing over the other superposition mode).

3. SQUEEZING IN FREE-SPACE FLUORESCENCE FROM THREE-LEVEL ATOMS

Walls and Zoller [9] predicted squeezing in resonance fluorescence from a single two-level atom interacting with a coherent field. Three-level lambda systems were suggested as a source of squeezing because, although nonlinear interactions are present, negligible upper level population is produced on two-photon resonance because of population trapping [10]. We have analyzed the squeezing properties of lambda system fluorescence excited by two coherent fields [2]. We consider the level scheme shown in Fig.3:

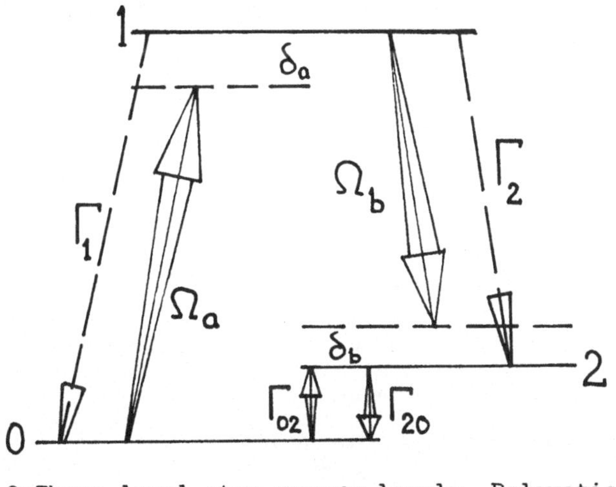

Fig.3: Three-level atom energy levels. Relaxation rates Γ_1 and Γ_2 from the excited state |1> to the ground states |0> and |2> are shown, together with ground state relaxation rates Γ_{02} and Γ_{20}

The atom, in level |0> interacts with two single-mode laser

fields a,b of angular frequencies ω_a, ω_b, polarizations $\underline{\epsilon}_a, \underline{\epsilon}_b$ which are initially in coherent states $|\alpha_a\rangle$, $|\alpha_b\rangle$, $\alpha_c(c=a,b)=(\bar{n}_c)^{1/2}$ where $\bar{n}_c$ is the mean photon number. We write the laser fields in terms of positive (E_c^+) and negative (E_c^-) frequency components

$$\underline{E}_c=(E_c^+ + E_c^-)\underline{\epsilon}_c \tag{3.1}$$

and these induce Rabi frequencies

$$\Omega_a=2(\omega_a\bar{n}_a/2\epsilon_o\hbar V)^{1/2}\langle 1|\underline{\mu}\cdot\underline{\epsilon}_a|0\rangle \;,\; \Omega_b=2(\omega_b\bar{n}_b/2\epsilon_o\hbar V)^{1/2}\langle 1|\underline{\mu}\cdot\underline{\epsilon}_b|2\rangle \tag{3.2}$$

where V is the quantization volume and $\underline{\mu}$ the dipole operator. We introduce the quadrature operators

$$E_{1c}=E_c^+\exp(i\omega_c t)+E_c^-\exp(-i\omega_c t) \tag{3.3}$$

$$E_{2c}=-i[E_c^+\exp(i\omega_c t)-E_c^-\exp(-i\omega_c t)] \tag{3.4}$$

so that the fields are

$$E_c=E_{1c}\cos\omega_c t+E_{2c}\sin\omega_c t. \tag{3.5}$$

We choose in this section to calculate normally-ordered field variances. The fluorescent dipole source fields are given in terms of the atomic dipole raising and lowering operators $|1\rangle\langle 0|$ and $|0\rangle\langle 1|$ by [11]

$$E_a^-=\phi_a|1\rangle\langle 0| \tag{3.6}$$

$$E_a^+=\phi_a|0\rangle\langle 1| \tag{3.7}$$

where

$$\phi_a=\frac{\omega_{10}^2}{4\pi\epsilon_o c^2 R}\langle 1|\underline{\mu}\cdot\underline{\epsilon}_a|0\rangle \tag{3.8}$$

and similarly for the 1-2 transition. Here R is the atom-detector distance. The normal-ordered variances are [12]

$$F_{1a}\equiv\langle:(\Delta E_{1a})^2:\rangle/\phi_a^2=[2\rho_{11}(t)-[2\mathrm{Re}\rho_{01}(t)]^2] \tag{3.9}$$

and

$$F_{2a}\equiv\langle:(\Delta E_{2a})^2:\rangle/\phi_a^2=[2\rho_{11}(t)-[2\,\mathrm{Im}\rho_{01}(t)]^2] \tag{3.10}$$

Squeezing occurs in the fluorescent field if F_{1a} or F_{2a} is negative. These variances are determined by the density-matrix elements $\rho_{11}(t)$ and $\rho_{01}(t)$ which obey the three-level Bloch equations [13,14]. We solve these by matrix methods in terms of the eigenvalues and eigenvectors [15].

The steady-state variances are found to be strongly-dependent on the ground state relaxation rates Γ_{02} and Γ_{20}. We can determine the steady-state ($t\to\infty$) quantities F_{1a}, F_{2a}, $Re\rho_{01}$ and $Im\rho_{01}$ and ρ_{11}. These are plotted in Fig.4 as a function of the two-photon detuning Δ. A negative (squeezed) variance F_{1a} is produced

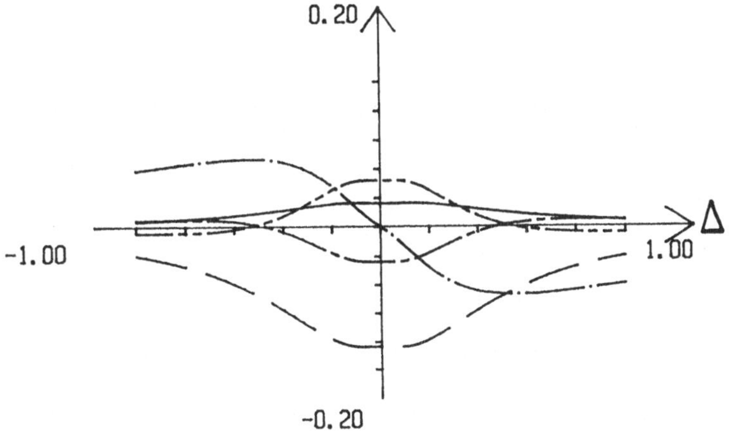

Fig.4: Dependence of the three-level atom fluorescence variables on the normalized two-photon detuning $\Delta = (1/2)(\delta_a + \delta_b)$ for $\delta_a = \delta_b$. The solid line is the excited state population; the real part of ρ_{01} is dashed, the imaginary part dash-dot; the variance F_{1a} is ___ _ ___ and F_{2a} is ___ _ _ ___. The Rabi frequencies are chosen to be $\Omega_a = 0.2(\Gamma_1 + \Gamma_2) = \Omega_b$, and $\Gamma_1 = \Gamma_2$. We set $\Gamma_{20} = (1/2)(\Gamma_1 + \Gamma_2)$ and $\Gamma_{02} = 0$.

near two-photon resonance, whereas a squeezed variance F_{2a} is produced in the wings of the lineshape. The squeezing, we show below, is entirely due to the lower-state relaxation creating essentially a two-level system from the three-level atom.

Had we set the lower state relaxations Γ_{20} and Γ_{02} equal to zero, no steady-state squeezing is produced: population trapping eliminates one-photon coherences [2]. To second-order in the two-photon detuning

Δ, we find for the purely radiatively-damped three-level atom

$$F_{1a} = [(4\Delta/\Omega)\sin 2\phi]^2 \qquad (3.11)$$

$$F_{2a} = [(4\Delta/\Omega)\sin 2\phi \cos\phi]^2 \qquad (3.12)$$

where

$$\phi = \tan^{-1}(-\Omega_b/\Omega_a) \qquad (3.13)$$

and the two-photon Rabi frequency Ω is given by

$$\Omega^2 = (\Omega_a^2 + \Omega_b^2). \qquad (3.14)$$

The lower state relaxation unbalances the symmetry between ground states to prevent complete destructive interference between the two excitation channels. We find in steady-state, that F_{1a} is negative for $\Delta/(\Gamma_1+\Gamma_2)$ between -1/2 and +1/2, and F_{2a} is negative outside this interval. The amount of squeezing increases as Γ_{20} increases. As Γ_{20} increases, the state 2 population is slaved to return to state 0 rapidly and plays no role in the dynamics, so that the three-level system reduces to a two-level system. For this effective two-level system

$$F_{1a} = 2\Omega_a^2(2\Omega_a^2+4\delta_a^2-\gamma^2)/(2\Omega_a^2+4\delta_a^2+\gamma^2)^2 \qquad (3.15)$$

$$F_{2a} = 2\Omega_a^2(2\Omega_a^2-4\delta_a^2+\gamma^2)/((2\Omega_a^2+4\delta_a^2+\gamma^2)^2 \qquad (3.16)$$

where $\gamma = \Gamma_1+\Gamma_2$ and δ_a is the 0-1 one-photon detuning. Similar results have been obtained for squeezing in two-level resonance fluorescence [9]. For $\delta_a=0$, $\Omega_a=0.2(\Gamma_1+\Gamma_2)$, they give $F_{2a} \approx 0.074$ and $F_{1a} \approx 0.063$ in agreement with Fig.4.

Transient squeezing can also be generated by the three-level lambda system. We have studied this [2] using dressed atom or uncoupled state [16] representations to calculate the time-dependent density matrix elements. In Fig.5 we plot F_{1a} and F_{2a} versus time with $\Gamma_1=\Gamma_2$ and we put $\Gamma_{02}=0$; the lasers are tuned to resonance ($\delta_a=0=\delta_b$) and we have chosen $\Omega/(\Gamma_1+\Gamma_2)=200$, $\phi=-3\pi/8$ (corresponding to $\Omega_a/(\Gamma_1+\Gamma_2)\approx 76$ and $\Omega_b/(\Gamma_1+\Gamma_2)\approx 185$). Note that F_{1a} is squeezed for almost all times

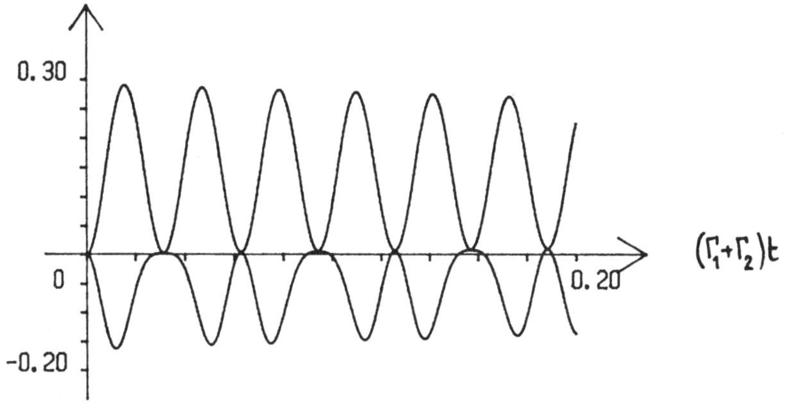

Fig.5:Transient resonant evolution of variances F_{1a} (upper) and F_{2a} (lower) plotted against $(\Gamma_1+\Gamma_2)t$ for large Rabi frequencies (see text) such that $(\Omega_a^2+\Omega_b^2)^{1/2}=200(\Gamma_1+\Gamma_2)$

in this transient regime, with a slow decay of the oscillations around a finite value. The long-time values of F_{1a} and F_{2a} decay to zero [2] as discussed above.

4.QUANTUM JUMPS AND CORRELATIONS IN THREE-LEVEL FLUORESCENCE

So far we have concentrated on the nonclassical squeezing of quadrature variances in light emitted by three-level atoms. The intensity correlations are also of interest in studies of three-level atoms. Quantum jumps in the fluorescence from three-level V-systems have attracted much theoretical [17] and experimental [18] interest. Here we examine the optical correlations in the light emitted by a V-system (Fig.6) in which a strongly-allowed transition competes with weak excitation to a metastable state $|2\rangle$. We will describe the intensity correlations for two cases:purely incoherent excitation using Einstein rate equations and coherent excitation using dressed atom methods.

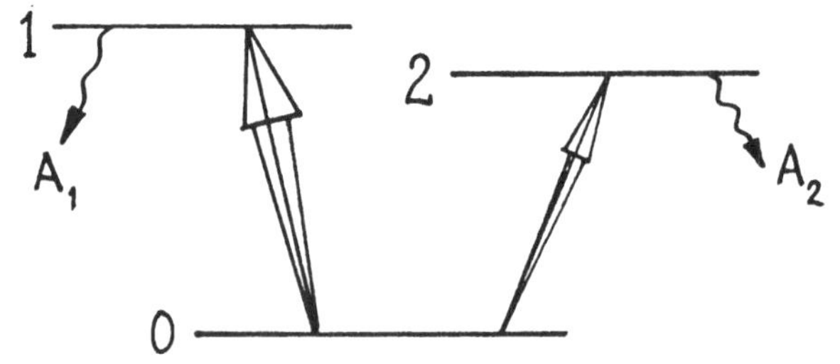

Fig.6: Three-level V-system in which a strongly-allowed (and rapidly decaying) transition competes with weak excitation to a metastable state 2 with small decay rate A_2.

The intensity correlations and the degree of second-order coherence for laser-driven atomic fluorescence is normally calculated using the quantum regression theorem [11]. Here we use instead the delay function method developed by Cohen-Tannoudji and Dalibard to describe sequential emission of photons. Let $w_2(\tau)$ be the probability density for an interval τ between one photon and the next, normalized so that

$$\int_0^{\infty} w_2(\tau)\,d\tau = 1 \qquad (4.1)$$

The probability $P(\tau)$ of no further photon to be emitted within τ of the first emission at time 0 is

$$P(\tau) = 1 - \int_0^{\tau} w_2(\tau')\,d\tau'. \qquad (4.2)$$

The probability that <u>any</u> photon is emitted between times τ and $\tau + d\tau$ after one is emitted at time 0 is $Q(\tau)d\tau$, given by

$$Q(t) = w_2(t) + \int_0^{t} Q(\tau) w_2(t-\tau)\,d\tau. \qquad (4.3)$$

The Glauber second-order coherence function $g^{(2)}(\tau)$ gives the normalized correlation function for the joint detection of a photon at time 0 followed by any other (not necessarily the next) at time τ and is given in terms of $Q(\tau)$ by

$$g^{(2)}(\tau) = Q(\tau)/Q(\infty). \qquad (4.4)$$

For incoherent excitation of the V-system, we may be interested in correlating emission from 1 to 0 and from 2 to 0 with each other. We label the 1-0 and 2-0 transition photons as 1 and 2, and the degree of second-order coherence $g^{(2)}_{ii}(\tau)$ gives the normalized correlation of photons of type i. If i=1, we are interested in the strongly-allowed fluorescence and its correlation through $g^{(2)}_{11}(\tau)$. The rate equations for the population densities in states 0,1 and 2 are

$$\frac{d}{dt}\rho_{00} = b\rho_{11} - (b+s)\rho_{00} + q\rho_{22} \tag{4.5}$$

$$\frac{d}{dt}\rho_{11} = -a\rho_{11} + b\rho_{00} \tag{4.6}$$

$$\frac{d}{dt}\rho_{22} = -q\rho_{22} + s\rho_{00} \tag{4.7}$$

where $a=A_1+B_1W_1$, $b=B_1W_1$, $q=A_2+B_2W_2$ and $s=B_2W_2$; and A and B are Einstein A and B coefficients, with W_i (i=1,2) the energy density of radiation exciting the 0-1 and 0-2 transitions. In calculating $g^{(2)}_{11}(\tau)$ we have assumed that spontaneous emission from level |1> does <u>not</u> repopulate level |0> but that spontaneous emission from |2> does repopulate |0>. This technique allows us to keep track of sequential photon emissions from the 1-0 transition and to calculate the $w_2(\tau)$ function. We can write the probability density $w_2(\tau)$ in terms of the probability of remaining in the atom-field dressed state manifold without emitting a photon:

$$w_2(\tau) = -\frac{d}{dt}(\rho_{00} + \rho_{11} + \rho_{22}) = A_1\rho_{11} \tag{4.8}$$

We write the general eq.(4.3) in Laplace-transform space

$$\tilde{Q}(z) = \tilde{w}_2(z)/(1-\tilde{w}_2(z)) \tag{4.9}$$

where $\tilde{Q}(z)=\mathcal{L}(Q(\tau))$ and $\tilde{w}_2(z)=\mathcal{L}(w_2(\tau))$. We find

$$\tilde{Q}(z) = \frac{A_1B_1W_1}{(z-\lambda_+)(z-\lambda_-)} + \frac{A_1B_1W_1(A_2+B_2W_2)}{z(z-\lambda_+)(z-\lambda_-)} \tag{4.10}$$

where $\lambda_{\pm}=T/2 \pm [T^2-4(aq+bq+as)]^{1/2}$ and $T=-(a+b+q+s)$. In the experimental studies of quantum jumps and the observation of fluorescent two-state telegraphs [18], the metastable state was exceedingly long-lived, so that $A_1 >> A_2$ and $B_1W_1 >> B_2W_2$. In this case we find from eq.(4.10) and eq.(4.4)

$$g_{11}^{(2)}(\tau) \approx 1 - \frac{2A_2+3B_2W_2}{2(A_2+B_2W_2)} e^{\lambda_+\tau} + \frac{B_2W_2}{2(A_2+B_2W_2)} e^{\lambda_-\tau} \tag{4.11}$$

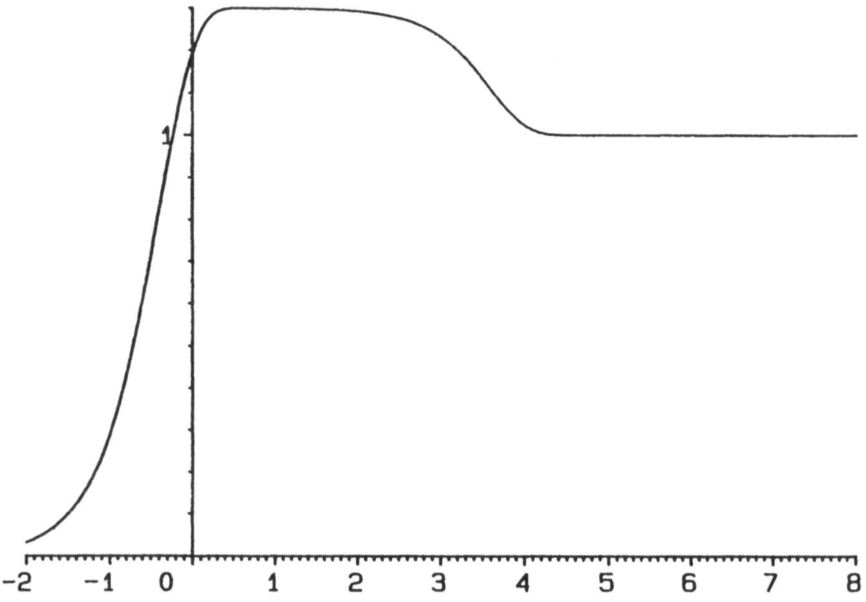

Fig.7: The degree of second-order coherence $g_{11}^{(2)}(\tau)$ versus $\ln A_1\tau$ from eq.(4.11) for incoherent excitation with $A_1=1/2$, $B_1W_1=1$, $A_2=1.0\times10^{-4}$ and $B_2W_2=1.0\times10^{-4}$.

where $\lambda_+ \approx -(2B_1W_1+A_1+\frac{1}{2}B_2W_2)$ and $\lambda_- \approx -((3B_2W_2/2)+A_2)$. This result agrees with that obtained earlier using the quantum regression theorem [3]. This is shown in Fig.7. We note that at short times the correlation function is that of

the dominant two-level part of the dynamics and exhibits antibunching but that at larger times there is a sudden drop as the metastable transitions (the "jumps") play a role. The "excess" correlations represented by the hump visible in Fig.(7) are the signature of a telegraph signal. A regular classical square wave telegraph [11] would generate a regular triangular $g^{(2)}(\tau)$. Here, a distribution of "on" and "off" times smooths the correlation and quantum effects are responsible for the initial antibunching.

When the atom is driven by two coherent laser fields, of which one is resonant to the transition frequency of the strong transition $|0\rangle \rightarrow |1\rangle$ and the other is detuned to that of the weak transition, Δ_2, $|0\rangle \rightarrow |2\rangle$, the dynamics of the atom can be described by the following equations of motion.

$$\frac{d}{dt}C_0 = -\frac{i}{2}\Omega_1 C_1 - \frac{i}{2}\Omega_2 \bar{C}_2 \tag{4.12}$$

$$\frac{d}{dt}C_1 = -\frac{i}{2}\Omega_1 C_0 - \gamma_1 C_1 \tag{4.13}$$

$$\frac{d}{dt}\bar{C}_2 = -\frac{i}{2}\Omega_2 C_0 + i(\Delta_2 + i\gamma_2)\bar{C}_2 \tag{4.14}$$

where C_0 and C_1 are the probability amplitudes of states $|0\rangle$ and $|1\rangle$ respectively and $\bar{C}_2 = C_2\exp(i\Delta_2 t)$ is used instead of the probability amplitude C_2 of state $|2\rangle$, in order to remove the fast oscillatory factor. The Rabi frequencies Ω_1 and Ω_2 drive the transitions $|0\rangle \rightarrow |1\rangle$ and $|0\rangle \rightarrow |2\rangle$ respectively. The damping factors γ_1 and γ_2 are one half of the Einstein A coefficients for the state $|1\rangle$ and $|2\rangle$. Solving the equations of motion and approximating the solutions under the conditions $\Omega_1, \gamma_1 \gg \Omega_2, \gamma_2$, Cohen-Tannoudji and Dalibard [19] have shown that

$$\begin{aligned} P(\tau) &= |C_0(\tau)|^2 + |C_1(\tau)|^2 + |C_2(\tau)|^2 \\ &= \lambda_1^{-2}(\Omega_1^2 - \gamma_1^2\cos\lambda_1\tau + \gamma_1\lambda_1\sin\lambda_1\tau)e^{-\gamma_1\tau} + |\alpha|^2 e^{-2\Gamma\tau} \end{aligned} \tag{4.15}$$

where

$$\lambda_1 = \sqrt{\Omega_1^2 - \gamma_1^2} \tag{4.16}$$

$$\alpha = \frac{i}{2}\Omega_2 \frac{(R_S - \gamma_1)}{[\frac{1}{2}(\gamma_1 + i\lambda_1) - R_S][\frac{1}{2}(\gamma_1 - i\lambda_1) - R_S]} \tag{4.17}$$

$$R_S = \gamma_2 - i\Delta_2 + \frac{\Omega_2^2}{4}\left[\tfrac{1}{4}\Omega_1^2\gamma_1 + i\Delta_2\left(\frac{\Omega_1^2}{4} - \Delta_2^2 - \gamma_1^2\right)\right] \Big/ \left[\left(\frac{\Omega_1^2}{4} - \Delta_2^2\right)^2 + \Delta_2^2\gamma_1^2\right] \tag{4.18}$$

and Γ is the real part of R_S. The first term in the probability function $P(\tau)$ is that of the two-level atom dynamics and the second term is due to the presence of the metastable state $|2\rangle$. Using the general relation eq.(4.9) and $w_2(\tau)=-dP/d\tau$ we get the probability function

$$\tilde{Q}(z) = \frac{1 - z\tilde{P}(z)}{z\tilde{P}(z)} \tag{4.19}$$

$$\approx \frac{\gamma_1\Omega_1^2 z + 2\gamma_1\Omega_1^2\Gamma}{z[z^3 + (3\gamma_1 + 2\Gamma)z^2 + (2\gamma_1^2 + \Omega_1^2)z + 2\Gamma(2\gamma_1^2 + \Omega_1^2) + |\alpha|^2\Omega_1^2\gamma_1}$$

where we have again used the assumption that the atom is in the ground state at delay time $\tau = 0$. Thus the degree of second order coherence is given by

$$\tilde{g}^{(2)}(z) = \tilde{Q}(z) \Big/ \lim_{z \to 0} z\tilde{Q}(z) \tag{4.20}$$

$$= \frac{[2\Gamma(2\gamma_1^2 + \Omega_1^2) + |\alpha|^2\Omega_1^2\gamma_1](z + 2\Gamma)/2\Gamma}{z[z^3 + (3\gamma_1 + 2\Gamma)z^2 + (2\gamma_1 + \Omega_1^2)z + 2\Gamma(2\gamma_1^2 + \Omega_1^2) + |\alpha|^2\Omega_1^2\gamma_1]} .$$

The degree of second order coherence described above differs from the normal degree of second order coherence $g_{ij}^{(2)}(\tau)$, since it gives the relation between one photon and any other photon of any kind (whether from the 1-0 or 2-0 transition) However it is very unlikely we will get a fluorescence photon due to the $|2\rangle \to |0\rangle$ transition and there is a very slim chance to have a consecutive $|0\rangle \to |2\rangle$ transition. Thus $g^{(2)}(\tau)$ will not be very different from $g_{11}^{(2)}(\tau)$.

To find the degree of second order coherence using the quantum regression theorem, involves 9th order polynomials, and 9 differential equations must be solved [20]. In contrast by using the $Q(\tau)$ function

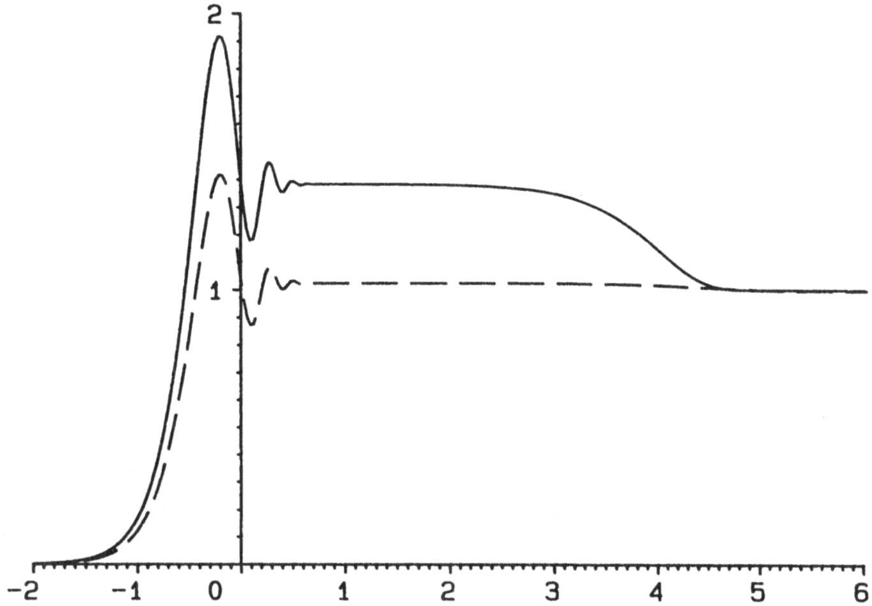

Fig. 8: The degree of second-order coherence $g_{11}^{(2)}(\tau)$ for coherent excitation versus $\ln\gamma_1\tau$, when $\Omega_1=5$, $\gamma_1=1$, $\gamma_2=1\times10^{-5}$ and $\Omega_2=1\times10^{-2}$. The dotted line is plotted for a zero detuning; the solid line is calculated for $\Delta_2=\Omega_1/2$.

we can find $g_{11}^{(2)}(\tau)$ without any major difficulties.

The degree of second order coherence is plotted in Fig. 8. When the delay time is small, we have Rabi oscillations at frequency Ω_1, and antibunching at short times. At long times we again see a plateau before the metastable state makes its contribution felt but only if the probe laser is tuned to an a.c. Stark shifted level. The presence of quantum jumps depend here upon the details of the coherent excitation.

Acknowledgements

This work was supported in part by the U.K. Science and Engineering Research Council, the Brazilian CNPq, the Royal

Society (through a Guest Research Fellowship for K.W.) and the British Council. We would like to thank Profs. R.Loudon and D.T.Pegg and Dr.J.Dalibard for discussions.

References

[1]K.Wódkiewicz, P.L.Knight, S.J.Buckle, and S.M.Barnett, Phys.Rev.A, in press(1987)

[2]F.A.M.de Oliveira, B.J.Dalton and P.L.Knight, submitted to J.Opt.Soc.Am.B (1987)

[3]D.T.Pegg, R.Loudon and P.L.Knight, Phys.Rev.A 33,4085 (1986) and references therein.

[4]P.L.Knight, Physica Scripta T12,51 (1986)

[5]D.F.Walls, Nature 306,141 (1983) and refrences therein.

[6]P.Filipowicz, P.Meystre, G.Rempe and H.Walther, Optica Acta 32,1105 (1985)

[7]S.Haroche, "New Trends In Atomic Physics" ed.G.Grynberg and R.Stora (Elsevier Science Publishers BV,1984)

[8]S.M.Barnett and P.L.Knight, J.Opt.Soc.Am. B2,467 (1985)

[9]D.F.Walls and P.Zoller, Phys.Rev.Lett 47,709 (1981)

[10]M.D.Reid, D.F.Walls and B.J.Dalton, Phys.Rev.Lett 55,1288 (1985)

[11]R.Loudon, "The Quantum Theory Of Light" second edition (Clarendon Press,1983)

[12]B.J.Dalton, Physica Scripta, T12,43 (1986)

[13]B.J.Dalton and P.L.Knight, Opt.Comm 42,411 (1982)

[14]B.J.Dalton and P.L.Knight, J.Phys.B 15,399 (1982)

[15]R.M.Whitley and C.R.Stroud Jr. Phys.Rev A14,1498 (1976)

[16]D.T.Pegg and W.R.MacGillivray, Opt.Comm. 59,113 (1986)

[17]R.J.Cook and H.J.Kimble, Phys.Rev.Lett. 54,1023 (1985)

[18]W.Nagourney, J.Sandberg and H.Dehmelt, Phys.Rev.Lett. 56,2727 (1986); J.C.Bergquist, R.G.Hulet, W.M.Itano and D.J.Wineland, Phys.Rev.Lett. 57,1699 (1986); T.Sauter, W.Neuhauser, R.Blatt and

P.E.Toschek Phys.Rev.Lett. 57,1696 (1986).

[19]C.Cohen-Tannoudji and J.Dalibard, Europhys.Lett. 1,441 (1986).

[20]A.Al-Hilfy and R.Loudon, J.Phys.B 18,3697 (1985)

INTERFEROMETRIC DETECTION OF GRAVITATIONAL RADIATION AND NONCLASSICAL LIGHT

Walter Winkler[1], Gerhard Wagner[1] and Gerd Leuchs[1,2]
[1]Max-Planck-Institut für Quantenoptik,
Postfach 1513, D-8046 Garching
[2]Sektion Physik der Universität München

I. Introduction

Prototypes of laser interferometric gravitational wave detectors have been developed for more than a decade and have now reached a stage where large, km-long interferometers are planned in several countries. In the first four sections the predictions for sources of gravitational radiation and the present detector performance is briefly reviewed. The prototype interferometers have essentially reached the shot noise limit[1)]. However, this fundamental limit may be overcome using techniques being developed in quantum optics. A graphical approach to the noise analysis of the interferometer is presented in the last two sections, showing that squeezed states of the radiation field may improve the interferometer sensitivity for measuring gravitational waves.

II. Characteristics of Gravitational Waves

Every mass is associated with a corresponding gravitational field surrounding it. Intuitively one would expect, that accelerated masses would give rise to the emission of some sort of radiation - analogous to the emission of electromagnetic waves by accelerated electrical charges. Einstein predicted the existence of gravitational radiation already in 1916. Corresponding to his Theory of General Relativity gravitational waves are expected to be transversal and to propagate with the speed of light like electromagnetic waves. However, there is an important difference between electomagnetism and gravitation: the sign of electrical charges may be positive or negative, whereas there is only one type of mass. As a consequence, there is no gravitational dipole-radiation. This can also be seen from the analogy to the expression for the power emitted by an oscillating electro-magnetic dipole:

$$\left.\frac{dE}{dt}\right|_{\text{el.mag.dipole}} \propto \ddot{p}^2 = \left(\frac{d}{dt}\int v\,dq\right)^2 \qquad (1)$$

It is proportional to the square of the second time derivative of the dipole-moment p. The time dependent velocity of the charge q is denoted by v. The analogous expression for any moving mass distribution is the square of the time derivative of the total momentum:

$$\left.\frac{dE}{dt}\right|_{\text{grav. dipole}} \propto \left(\frac{d}{dt}\int v\,dm\right)^2 = 0$$

Owing to conservation of momentum in a closed system this expression vanishes. Therefore, in the case of gravitation, the lowest nonvanishing order multipole radiation is the quadrupole radiation. The emitted power is proportional to the square of the third time derivative of the mass-quadrupole-moment:

$$\left.\frac{dE}{dt}\right|_{\text{grav.quadrupole}} \propto \dddot{Q}^2 \qquad (2)$$

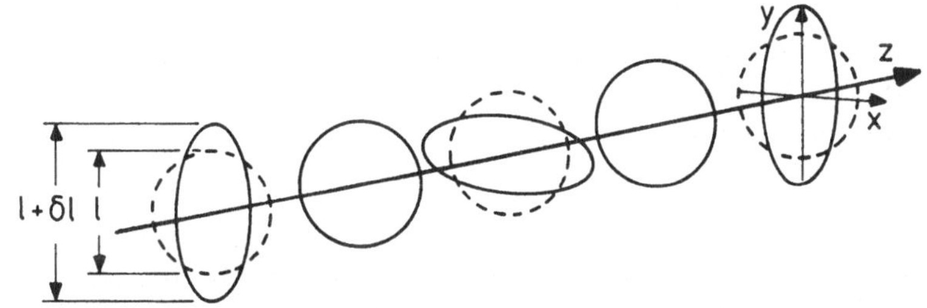

Fig. 1: Spatial strain pattern produced by a gravitational wave.

A gravitational wave manifests itself in a variation of the metric of spacetime - e.g. in a change of the optical distance between free testmasses, which can be monitored by registering the light travel-time between these testmasses. The effect of a gravitational wave can be described as a time dependent index of refraction of the lightpath; another equivalent description introduces a time-dependent strain of space, as indicated in Fig. 1.

The quadrupolar characteristics of the gravitational wave can be seen from the elliptic deformation of a circular arrangement of free testmasses, if the plane of the circle is perpendicular to the direction of propagation of the wave. A positive strain in one direction is accompanied by a negative strain in the perpendicular direction; half a period later the signs of deformation have changed. Fig. 1 shows one of the two independent polarizations of the wave; for the second one the pattern is rotated by 45°. The amplitude h of a gravitational wave is given by a dimensionless quantity, the strain of space it introduces:

$$h = 2\delta \ell / \ell$$

$\delta\ell/\ell$ is the relative change of distance between testparticles, e.g. of the axes of the ellipses indicated in Fig. 1. Unfortunately the quantity h is usually discouragingly small for any measurement in the vicinity of the earth. In order to get a strong wave $\dddot{Q}$ has to be as large as possible according to Eq. (2). This implies, that the masses involved have to be accelerated as fast as possible. As simple calculations show, gravitational waves of measurable strength cannot be produced in a laboratory. Consider for example as a source a metallic cylinder with a length of 1m and a mass of 1 ton. This cylinder emits gravitational radiation, if it rotates around an axis perpendicular to its axis of symmetry. The emitted power increases strongly with increasing frequency of rotation. Just below the limit of rupture the amplitude h produced at a distance of one wavelength of the gravitational wave is only on the order of 10^{-40} - much too small to be seen by any presently conceivable detection technique.

Fortunately astrophysics tells us about scenarios, where huge masses are so strongly accelerated, that despite their big distance the amplitudes of gravitational waves produced are not totally out of reach for experimental access[2)].

A favourable type of source for gravitational waves are compact binaries; these are stellar objects consisting of two neutron stars or even black holes. Such objects emit gravitational radiation while rotating around their common center of mass.

The related energy loss leads to a shrinking of the relative distance and a corresponding increase in angular frequency. Finally the two objects - with masses in the order of a solar mass - are accelerated close to the speed of light, and during the following collapse a huge amount of energy is emitted in form of gravitational radiation - most of it in a time interval shorter than one second. An expected signal of this type is drawn in Fig. 2 indicated as line (a)[3].

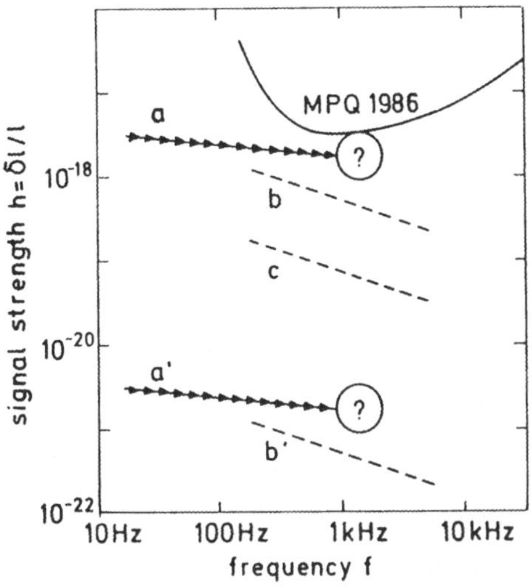

Fig. 2: Signal strength of some sources of Gravitational Radiation as a function of frequency f. The bandwidth for evaluation of the detector-signal was chosen to be equal to f for supernovae, and f/n for the n cycles to be observed for compact binaries.

a: compact binary in the center of our galaxy
a': compact binary in the Virgo-cluster
b: Supernova in the center of our galaxy
b': Supernova in the Virgo-cluster
c: Supernova in the Magellan cloud

Another conceivable source are supernovae of type II. Under particular circumstances they occur at the end of the life of a normal star, when the nuclear fuel is used up. The thermal motion of the atoms decreases and can no longer balance the gravitational pressure; the inner part of the star starts to collapse. The huge amount of energy released shoots off the outer shell leading to the well known optical spectacle. In the center, however, - not to be seen optically from outside - a very dense core of strongly accelerating masses develops. Again a short, intense pulse of gravitational radiation is likely to be emitted, if the collapse takes place asymmetrically. Such an asymmetry will usually develop owing to the initial rotation of the stars. A signal expected for this type of source is drawn in Fig. 2 as line (b). The supernova which occured at the beginning of this conference in the

Magellan cloud about 55 kiloparsec away would have produced a pulse with Fourier components along line (c), if we assume that the emitted energy was 10^{-3} of a solar mass. The signals a and b are expected to be produced by events in our galaxy - but most likely the event rate will be low, only a few per century. The Virgo cluster, however, contains a few thousand galaxies, leading to a correspondingly higher event rate - but the signal strength will be much lower due to the one thousand times larger distance (signals a'and b'). From Fig. 2 one can deduce, that it would be most desirable to get strain sensitivities on the order of 10^{-21} for the detectors of gravitational radiation.

III. Detectors for Gravitational Radiation.

When Einstein first introduced the concept of gravitational waves, nobody dared to attempt an experimental verification for decades because the effect is discouragingly small. Nevertheless, in the sixties Joseph Weber of the University of Maryland started his pioneering work developing resonance detectors. The basic element of his antenna was an aluminum bar with dimensions on the order of one meter and a mass of more than a thousand kilograms. The spatial strain introduced by gravitational waves should couple to the longitudinal fundamental mode of the bar and change the state of oscillation of the bar, if the assumed shortpulse excitation had Fourier components at the fundamental eigenfrequency. The motion of the bar was sensed with piezocrystals attached to its surface. Based on Weber's work the Munich-Frascati coincidence experiment[4)] was optimized to give a strain sensitivity of several times 10^{-17}. Only the thermal motion of the antenna and no correlated signals have been observed.

In the meantime work went on to cool the bars down to liquid helium-temperature and to use electromechanical transducers less noisy than the original piezo crystals. Presently the best strain sensitivity reached[5)] is 10^{-18}, an impressive figure but still almost three orders of magnitude above the level of the signals expected to occur a few times per month.

A completely different approach towards detecting gravitational waves is the concept of a broadband antenna. The basic idea is to sense optically the gravitational wave induced change of the metric of space-time, e.g. to measure the variations of the distance betweeen testmasses interferometrically. The simplest approach is a Michelson-interferometer, where the beamsplitter and the two mirrors serve as testmasses. For a gravitational wave of optimum polarization and direction of propagation the two arms experience strains of opposite sign as can be seen from Fig. 1. The created armlength-difference can be sensitively detected. The observed phase-difference between the two interfering beams follows the influence of the gravitational wave directly and therefore the setup is inherently broadband. This allows the exact reconstruction of the signal in contrast to a resonant system, which shows only the spectral density of the signal at the resonance frequency.

Another advantage of the broadband antenna is the possibility to increase the lightpath up to half a wavelength of a gravitational wave $L \leq c\, T_{grav}/2$, thereby increasing the quantity to be measured, namely the path difference between the two arms. In contrast to this the dimension of a resonance detector for a given frequency is limited by the speed of sound: $L \leq v\, T_{grav}/2$; therefore the Gravity-wave induced displacement in a broadband detector is larger by a factor of c/v.

Certainly it is expensive to realize very long optical paths, and therefore, in several laboratories prototype interferometers have been constructed first. In order to get a long optical path the light beams are reflected back and forth in the interferometer arms many times before they are recombined at the beamsplitter. The beams in each arm can either be put on top of each other as in the case of a Fabry-Perot or be more or less well separated as in an optical delay line. The latter approach was chosen in the presently most sensitive setup which is run by a group at the Max-Planck-Institut für Quantenoptik (Fig.3). The mirror distance l is 30 m, and for the latest measurements 90 beams have been used to give a total pathlength of $L = 2.7$km.

The beamsplitter is adjusted to sit in the symmetry plane between the two near mirrors. Consequently there are no spatial fringes to be seen at the output; the output power varies sinusoidally as a function of pathdifference between the two interfering beams as indicated in Fig.4.

The information about the time dependent pathdifference is provided by a fast servo loop, which compensates deviations from a given point of operation by a voltage across Pockels cells sitting in the lightpath. This voltage is taken as the output signal. In practice only one output port is used, and the point of operation is one of the minima in Fig. 4. This implies, that a modulation technique has to be used for stabilizing the interferometer.

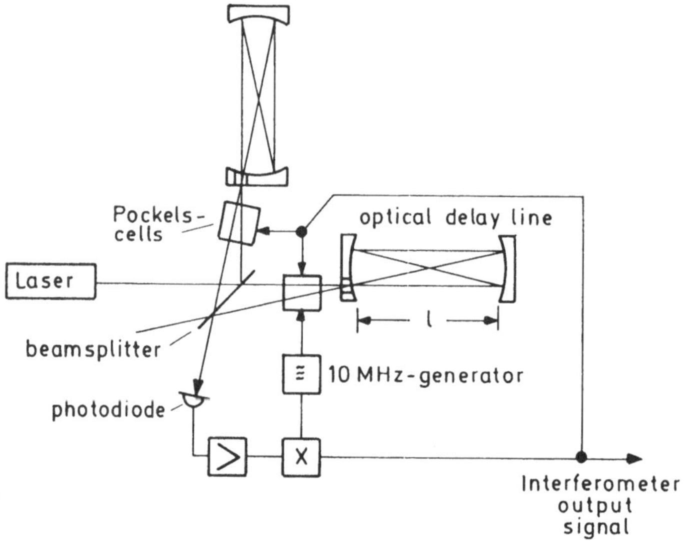

Fig. 3: Schematic diagram of the prototype interferometer showing the servo-loop used for stabilization.

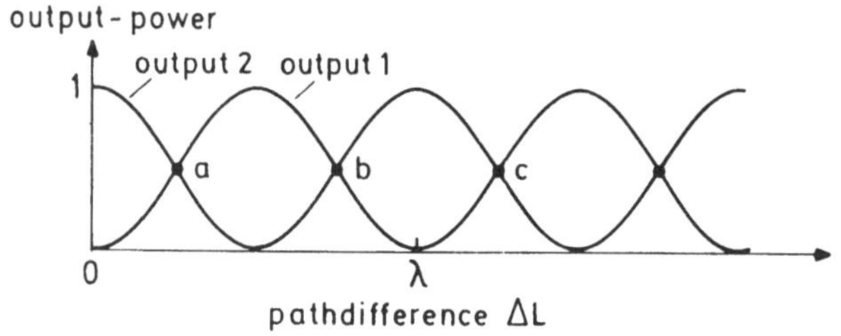

Fig. 4: Light power at the two output ports of the interferometer as a function of optical path difference in the two arms.

More easily to understand is another nulling method, where the difference between the photocurrents at the two output ports is taken. The point of operation is one of the points a, b, c, in Fig. 4 of equal output power. Again a servo loop keeps the interferometer at the chosen point of operation. If now the input power fluctuates, then the two photocurrents fluctuate correspondingly, and the difference remains zero. But if a mirror is shifted, the power in output 1 e.g. increases, in output 2 it decreases, leading to a compensating voltage across the Pockels cells.

IV. Noise sources

The most fundamental limitation of the sensitivity inherent to the detection technique arises because of the quantum nature of light. In a coherent state the photon number fluctuates according to Poisson-statistics. This shot noise will be dealt with in the next two sections. Here we discuss a few additional noise sources which have to be identified and controlled. The most prominent ones up to now have been the following:

1) Mechanical perturbations from outside.
Motions of the ground either of seismical or technical origin as well as acoustical noise are transmitted more or less to the interferometer and thus give rise to signals. All relevant optical components are therefore suspended as pendulums in vacuum. The suspension point is formed by a mass, which in turn is suspended by springs. There are several servo loops to damp the motion at the resonance frequencies of the pendulums and to keep the interferometer at the desired position. The two stage pendulum suspension turned out to be sufficient for the present setup at frequencies above several hundred Hz. If necessary, further stages may be added.

2) Thermal excitations of eigenmodes.
If all mechanical excitation from outside is reduced sufficiently, the thermally excited eigenmodes of the optical components become visible. The corresponding amplitudes are orders of magnitude above the acceptable level in an antenna for gravitational waves. Unfortunately in a mechanical setup resonances are very likely to show up just in the interesting frequency range around one kHz. By a very careful design of the mechanical structure one has to arrange all relevant eigenfrequencies above the frequency window of observation. In addition, the internal mechanical damping of the optical components has to be as low as possible. In this case the thermally induced motion is mostly concentrated around the resonance frequencies and the wings of the resonances, i.e. the contribution at other frequencies, are kept low.

3) Frequency fluctuations of the laserlight.
Frequency fluctuations give rise to a signal in the interferometer, when the pathdifference between interfering beams is not zero. In a setup with delay lines such a pathdifference cannot be

avoided easily, because of the difficulty of producing large radii of curvature with high accuracy. For a delay line a particular ratio between mirror distance and radius of curvature has to be chosen in order to bring the output beam to a given position. Therefore, a difference in the radii of curvature of the mirrors in the two arms implies also a different mirror separation and thus a different optical path.

Another mechanism to convert frequency fluctuations into interferometer signals is caused by scattered light, which finds its way to interfere with the main beam and which may have a huge path difference with respect to it. Up to now a careful two-stage stabilization of the laser frequency could eliminate spurious signals of this kind of noise.

4) Time dependent beam geometry.
The geometry of the laser beam like position, orientation and diameter, fluctuates by tiny amounts; e.g. the position of the beam may vary by about 10^{-10} m on a timescale of milliseconds. Usually nobody cares about such a small effect. However, in a not ideally symmetric interferometer these fluctuations are converted into signals; e.g. in connection with a small angle misalignment of the beamsplitter. It is not possible to align an interferometer perfectly and keep it in this condition. Therefore a beam cleaning device was introduced, i.e. either a single mode Fabry-Perot-resonator or a single mode glass fiber. The laser beam has to be matched to this particular mode. Variations in beam geometry can be described as admixture of other modes, for which the beam cleaning device is not resonant. These other modes are therefore not transmitted they are rather reflected, and only a small fluctuation of the transmitted light power is produced. This does not harm if as usual a nulling method is used in the measurement.

In the Garching set-up it was possible to reduce the influence of all these noise sources below the shot noise limit discussed in the following two sections. In the frequency window of interest between 500 Hz and 5 kHz the sensitivity is close to the shot noise limit of about 0.3 W, which is determined from the Poisson like photon counting statistics at the output of the interferometer. The strain sensitivity in a bandwidth of 1 Hz is 10^{-19}, and correspondingly $3x10^{-18}$ in a bandwidth of 1kHz. To increase the sensitivity one has to increase the pathlength and the light-power.

The pathlength can still be enhanced by a factor of 30, before half a wave length of the gravitational radiation is reached; to get the remaining factor of 100, a lightpower on the order of 1kW will be necessary. This might be possible by adding up coherently the beams of several strong cw-lasers and by making use of the output power of the interferometer. In this so called recycling scheme the interferometer is run at minimum output power in one output port; almost all of the light therefore leaves through the other output port and can be added coherently to the input beam. If the overall losses can be kept small, a considerable lightpower can build up inside the interferometer.

On the other hand, a very high power density at the optical components may introduce new problems. To increase the sensitivity of the detectors, it would at any rate be very helpful, if the new approach to reduce the photon counting error by using squeezed states would prove to be feasible.

V. Noise introduced by a beam splitter.

The optical beam splitter is an important part of any optical interferometer and it is responsible for the Poisson-type fluctuations in the photon flux at the detector, also referred to as shot noise. The understanding of the beamsplitter is essential for the proposed application of nonclassical light fields towards improving the interferometer sensitivity[6].

Consider a semitransparent mirror which splits the incoming laser radiation into two beams of equal intensity (Fig. 5). The number of photons impinging onto the beam splitter per sampling time interval is N. If the laser light is in a perfectly coherent state than the photon noise is described by Poisson-statistics, $\sqrt{<\Delta N_{in}^2>} = \sqrt{<N_{in}>}$. In each output port of the beam splitter one finds half the number of photons $<N_{out}> = <N_{in}>/2$. As is well known the root mean squared fluctuations do not reduce by a factor of 2 but by $\sqrt{2}$, $\sqrt{<\Delta N_{out}^2>} = \sqrt{<N_{in}>/2}$. This can be derived by a statistical analysis taking into account that each photon can exit through either one of the two output ports but not through both[7].

The result of this statistical picture, however, does not readily comply with the naive picture of the beam splitter, where both the mean value and the fluctuations are cut down by a factor of 2. At first sight the naive picture is not unreasonable since one deals with

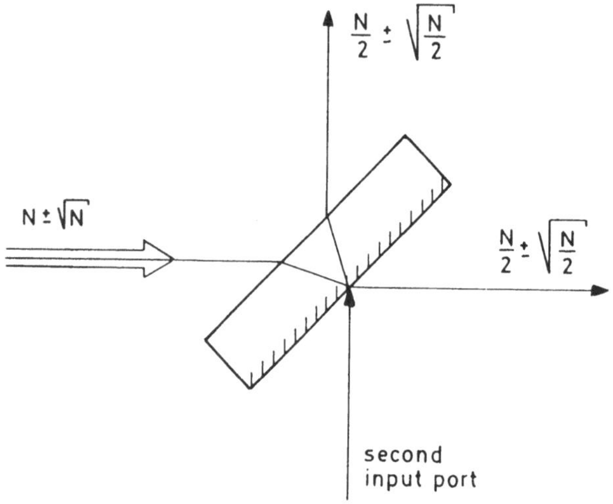

Fig. 5: Photon number fluctuations at the input and output ports of a beamsplitter.

quite macroscopic fluctuations. For 1 Watt visible laser power and a sampling time of 1 ms the root mean squared fluctuations are of the order of $\sqrt{<\Delta N^2>} \approx 10^8$ photons!

As Carlton M Caves[8)] has shown, the factor of $\sqrt{2}$ missing in the naive picture can be accounted for by recognizing that there is a second normally not used input port to the beam splitter. Through this second port at least the zero point fluctuations of the electromagnetic field are coupled in if nothing else. These amplitude fluctuations are not negligible since they are of the same size for a coherent and the vacuum state!

Both pictures of the beam splitter give the same result for the noise and in both cases it is the particle nature of light which ensures that the intensity fluctuations stay at the Poisson level. In the first picture the beamsplitter adds noise, because it statistically sorts out photons and sends them one way or the other. The second picture at first sight looks like a wave-type picture. However, the zero point fluctuations entering through the second port are a direct consequence of field quantization. Therefore, also in this case the added noise is a result of the particle nature of light. The second picture although leading to the same result has a huge advantage over the statistical approach. It tells the experimenter which knob to turn in order to modify the shot noise limit.

VI. The shot noise limit of an interferometer.

The second normally not used input port of the beamsplitter plays a dominant role in the noise analysis of a Michelson-interferometer which we show graphically[9)]. Based on these results it will be straight forward to see why squeezed light may improve the interferometer sensitivity.

Fig. 6 shows a schematic diagram of a Michelson-interferometer, the endmirrors are tilted so that the second input port is spatially separated from the output ports. The fields entering the main (laser) and the second input port are denoted $\mathcal{E}_L = E_L \cos(\omega t + \Phi)$ and $\mathcal{E}_S = E_S \cos(\omega t + \phi)$ respectively. It is now assumed that the phases are near zero, $\Phi \approx \phi \approx 0$, and that the amplitude of the laser is dominant, $E_L >> E_S$. For an optical phase difference of $\Delta \approx \pi/2$ in the two arms the two output beams have nearly equal power $I_1 \approx I_2$. Operating the interferometer near this point, the difference of the two output powers $I_2 - I_1$ may serve as the signal that contains information about the applied strain. When calculating the fields in the two output ports one has to take into account the phase shift at the beamsplitter. Independent of

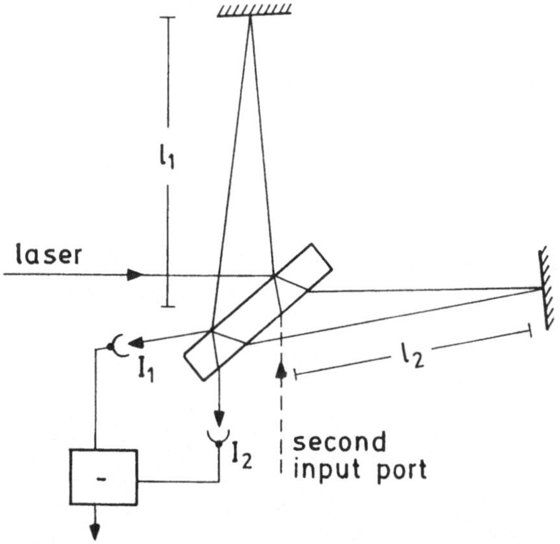

Fig. 6: Schematic diagram of a Michelson-interferometer showing the second input port of the beamsplitter.

the type of beamsplitter these phase shifts can be obtained from a time reversal or energy conservation argument. If two beams combined by a beamsplitter interfere constructively in one output port they must interfere destructively in the other output port. In other words, if the phase difference of the two interfering beams is β in one of the two exits then it must be $\beta+\pi$ in the other one (see Fig. 4).

With the above definitions the fields in the two output ports are

$$E_1=\tfrac{1}{2}[E_L\sin(\omega t+\Phi+\Delta)+E_L\sin(\omega t+\Phi)+E_S\sin(\omega t+\phi+\Delta)+E_S\sin(\omega t+\phi+\pi)] \qquad (3a)$$

$$E_2=\tfrac{1}{2}[E_L\sin(\omega t+\Phi+\Delta)+E_L\sin(\omega t+\Phi+\pi)+E_S\sin(\omega t+\phi+\Delta)+E_S\sin(\omega t+\phi)] \qquad (3b)$$

The beam splitter phase difference π appears in one output port for the E_S and in the other one for the E_L - beams owing to the symmetry in the geometrical arrangement.

We will now use the graphical representation in a phase diagram to study the influence of phase and amplitude fluctuations on the signal $I_2 - I_1$. In these diagrams the vectors representing the field amplitudes coming back from both arms of the interferometer are shown as well as their vectorial sum. Fluctuations in either of the two amplitudes and phases are discussed separately. The results shown hold of course only as long as the optical path difference in the two arms is less than the coherence length of the laser light (see Sec. IV).

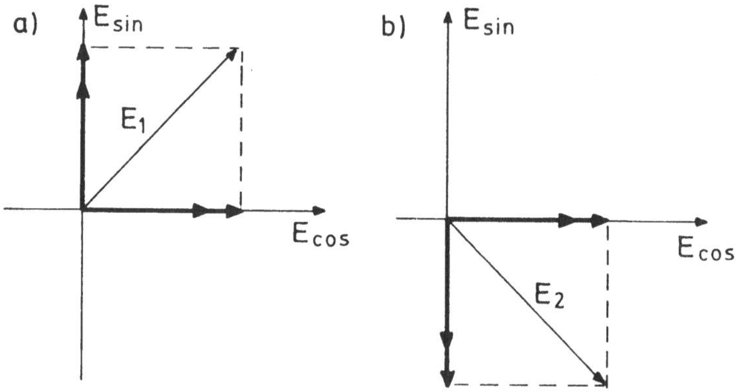

Fig. 7: Phase diagram for the graphical determination of the electric field amplitudes E_1 and E_2 at the two outputs of the interferometer. Each field is decomposed into two parts, E_{sin} and E_{cos}, oscillating in phase with sin t and cos t respectively. The figure shows the influence of amplitude fluctuations of the laserlight.

1) Fluctuations in the amplitude E_L of the laser field.
Fig. 7a shows the field vectors for the case E_S, Φ, and $\phi = 0$ at the point of operation $\Delta = \pi/2$. If there are amplitude fluctuations of E_L, here represented by a positive excursion (small added vectors) then also the resulting field amplitude E_1 will fluctuate. In the other output port the fields which have to be added are nearly the same, the only difference being that the phase of one of the recombined fields is shifted by π (Fig. 7b). Here the field vectors span the same rectangle as in Fig. 7a. The magnitude of the resulting field vector E_2 fluctuates like the one of E_1. In fact E_1 and E_2 correspond to the two diagonals of the parallelogram they span. Since for a rectangle the length of the two diagonals is the same, the difference in the photo currents $I_2 - I_1 = E_2^2 - E_1^2$ stays zero irrespective of the amplitude fluctuation of the laser.

2) Fluctuations in the phase Φ of the laser field.
A phase fluctuation is represented by a small vector of length $E_L \Phi$ which is added at right angles to the main fields (Fig. 8a,b). The sign of these small vectors follows from Eqs. 3a, b. Again one finds that the field amplitudes at the two outputs correspond to the diagonals of the same rectangle and $E_1 = E_2$. Consequently, the signal is also indepent of phase fluctuations in the laser field.

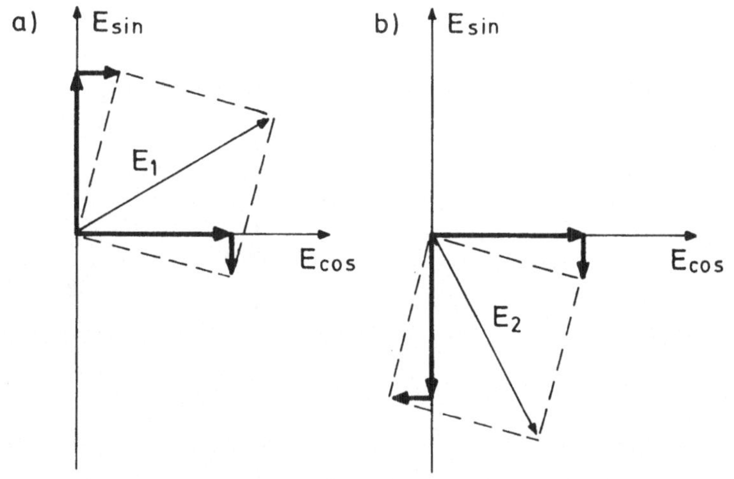

Fig. 8: same as Fig. 7, but influence of phase fluctuations of the laserlight.

3) Fluctuations in the field amplitude E_s at the second input port. The sign of the small vectors representing the amplitude fluctuations of E_s have to be traced carefully. Fig.9a and b show that, as before, the output fields E_1 and E_2 fluctuate in correlation as to cancel out the corresponding photocurrent fluctuations, $I_1-I_2 = 0$.

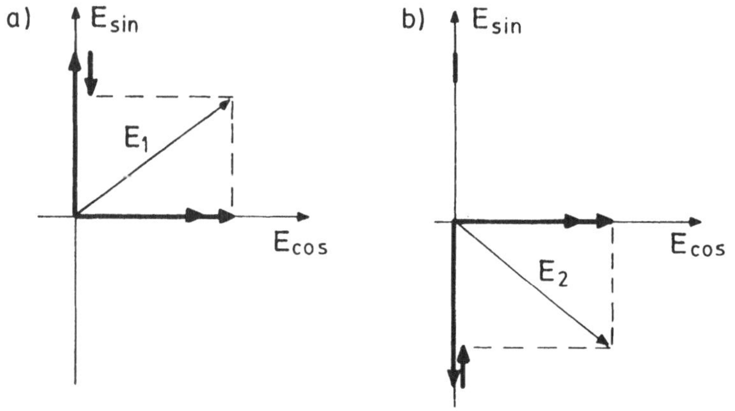

Fig. 9: same as Fig. 7, but influence of amplitude fluctuations at the second input port.

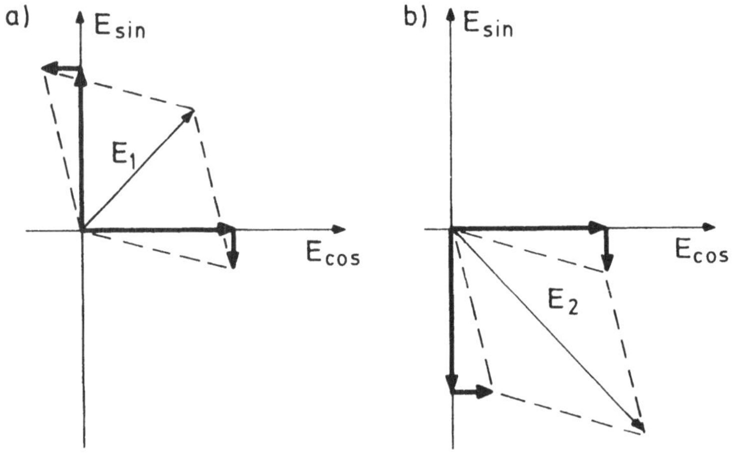

Fig. 10: same as Fig. 7, but influence of phase fluctuations at the second input port.

4) Fluctuations in the phase ϕ at the second input port. Finally there is a type of fluctuation which produces noise in the signal $I_2 - I_1$. The resulting field vectors span a parallelogram which is not a rectangle, $I_2 - I_1 \neq 0$(Fig. 10a,b). A quantitative evaluation of Eqs. 3a,b yields the noise in the signal

$$(I_2 - I_1)_{noise} \propto 2\phi E_S E_L, \quad \text{for } \phi \ll 1 \text{ and } E_S \ll E_L$$

The term $\phi \cdot E_s$ represents the amplitude fluctuations in the cosine part of the field at the second input port

$$(I_2 - I_1)_{noise} \propto 2E_L\sqrt{\langle \Delta E_{s,cos}^2 \rangle}$$

In units where the field amplitude is given by the square root of the photon number, the quantum mechanical rms-fluctuations of a coherent field are $\sqrt{\langle \Delta E_{s,cos}^2 \rangle} = 1/2$. The size of these rms - amplitude fluctuations are independent of the meanfield and are the same as for the vacuum field.

The noise in $I_2 - I_1$ determines the sensitivity with which the optical phase difference $\Delta = \pi/2 + \psi$ and hence the optical path difference in the two interferometer arms can be measured. It has to be compared to the signal

$$(I_2 - I_1)_{signal} \propto \psi E_L^2$$

For $E_L^2 = n$ photons in the laser mode one finds the quantum limit for the phase sensitivity to be

$$(\delta\psi)_{min} = 2\sqrt{\langle \Delta E_{s,cos}^2 \rangle} / \sqrt{n}$$

This corresponds to a smallest detectable mirror displacement of

$$(\delta l)_{min} = \lambda(\delta\psi)_{min}/(4\pi)$$

In the case where the field at the second input port is the vacuum or a coherent state, the fluctuations are $\sqrt{\langle \Delta E_{s,cos}^2 \rangle} = 1/2$, which yields the usual shot noise limit

$$(\delta l)_{min} = \lambda/(4\pi\sqrt{n})$$

VII. Improving the interferometer sensitivity using squeezed states.

The discussion in Sec. VI illustrates that the noise in the signal is due to only one out of four possible noise sources i.e. the fluctuations at the second input port out of phase with the main laser field. This explains why light in a squeezed state with $\sqrt{\langle \Delta E_{s,cos}^2 \rangle} < 1/2$ coupled into the second input port should lead to a displacement sensitivity better than the shot noise limit.

The fact that Heisenberg's uncertainty relation requires the in-phase amplitude fluctuations at the second input port to be correspondingly larger, $\sqrt{\langle \Delta E_{s,\sin}^2 \rangle} > 1/2$, does not harm since the signal is not sensitive to these fluctuations as discussed under point 3) in Sec. VI.

The sensitivity increase that can be hoped for is of course limited. Ultimately there is a limitation caused by fluctuating photon pressure on the mirrors since the latter is enhanced when using squeezed states. Apart from that there are some technical considerations. Unlike coherent or thermal light fields, squeezed states change their characteristics under linear attenuation towards that of a coherent state. Therefore, it is essential to have a photodetector quantum efficiency as close to 100% as possible and to avoid losses. It also turned out that a visibility less than one critically limits the sensitivity gain to be expected when using squeezed states[10]. Nevertheless, it is reasonable to hope for a sensitivity which can otherwise be reached only for a 5 to 10 times higher laser power. In view of the high laser powers which are aimed at, the squeezed state technology looks promising and viable.

References

1. D. Shoemaker, W. Winkler, K. Maischberger, A. Rüdiger, R. Schilling, L. Schnupp: Proc. 4th Marcel Grossmann Meeting on General Relativity and Gravitation, R. Ruffini (ed.), pp. 605-614. Elsevier Science Publishers 1986.
2. K.S. Thorne: Rev. Mod. Phys. 52, 285 (1980).
3. P. Kafka: Naturwissenschaften 73, 248 (1986).
4. H. Billing, P. Kafka, K. Maischberger, F. Meyer, W. Winkler: Results of the Munich-Frascati gravitational-wave experiment, Lett. Nuovo Cimento 12, 111-116 (1975).
5. P.F. Michelson: The low temperature gravitational wave detector at Stanford University, T. Piran. North Holland Publ. Co., 465-474 (1983).
6. C.M. Caves: Phys. Rev. D23, (1981).
7. R. Loudon: The Quantum Theory of Light, Clarendon Press, Oxford 1983.
8. C.M. Caves: Phys. Rev. Lett. 45, 75 (1980).
9. G. Wagner and G. Leuchs: Laser und Optoelektronik 1, 37-44 und 52 (1987).
10. J. Gea-Banacloche and G. Leuchs: "Applying Squeezed States to Non-Ideal Interferometers" (to be published).

PART III: Quantum Jumps in Three-Level Systems

"QUANTUM JUMPS" OBSERVED IN SINGLE-ION FLUORESCENCE

Th. Sauter, R. Blatt, W. Neuhauser,
and P. E. Toschek*
Universität Hamburg, D-2000 Hamburg, Fed.Rep.Germany

Abstract

We have demonstrated that interruptions of macroscopic duration appear in the 493-nm fluorescence of a single trapped and cooled Ba^+ ion. They are caused by sudden transitions of the ion into the "dark" $^2D_{5/2}$ state. - Multiple simultaneous jumps of three ions indicate cooperative interaction with the light.

Introduction

The canonical interpretation of quantum mechanics attributes calculable expectation values to statistical mean values of variables. With conventional measurements in atomic physics, large ensembles contribute to the measured quantity, and averaging is inherent to the act of measurement. If a *single* particle is observed and evolves on a time scale fast compared with the time of resolution of the measuring device, we invoke ergodicity to expect time averages as the result of the measurement. However, if time resolution is good enough, we have no safe prediction for the evolution of the atom to be observed in "real time". According to Bohr's model of the hydrogen atom |1|, atoms undergo instantaneous transitions from one eigenstate of energy to another one upon interactions with the radiation field. It is certainly conceivable, on the other hand, that even with fast and repeated single-atom detection, only time-averaged quantities are meaningful results of the measurement. Consequently, quantum-mechanical superposition states might depict individual atoms even *on the small time scale set by the sequence of measurements*, i.e. they might be "real".

This alternative is open to decision by experiment. Recently, the preparation of single, cold atomic particles - ions - in specified in-

* Also: JILA, University of Colorado and NBS, Boulder, Colo. 80309, USA

ternal and external states has been demonstrated |2-4|. Ensemble averaging, unavoidable in conventional measurements, is absent in experiments with single atomic particles: these experiments allow, for the first time, direct proof of one of the most basic concepts in quantum mechanics by repeated identical preparation and observation of an individual atomic system. Intuitive arguments predict that transitions on a very weak line are detectable via the excitation of resonance fluorescence on a strong transition coupled by a common level |5|. When such a rare transition, say, an absorption event, occurs on the weak signal line, the fluorescence on the neighbouring line is supposed to be quenched since the atom - or ion - in the upper, metastable level of the weak line is no longer available for the fast excitation and fluorescence cycles.

These conclusions have been confirmed by rigorous quantum-statistical calculations |6-11|. They show that interruptions in the fluorescence are expected which are on the order of the lifetime of the metastable level. These "dark" or "off" periods are the signature of quantum jumps; they have been observed recently |12-15|. In our experiments |12,15|, a single Ba^+ ion is localized in an electrodynamic ion trap and optically cooled to less than 10 mK |16,17|. The relevant energy levels of Ba^+ are shown in Fig. 1. Resonance fluorescence is excited at the 493-nm line with a cw laser, and a second laser beam at 650 nm couples the $^2D_{3/2}$ level to the continuously excited $^2P_{1/2}$ level. When the ion, once in a while, drops into the 5 $^2D_{5/2}$ level - upon laser-pumped electronic Raman-Stokes transitions -, the fluorescence becomes suppressed. Thus, the transitions into and out of the "dark" 5 $^2D_{5/2}$ state are observed with 100% detection efficiency, and with time resolution which corresponds to the mean time separation of the photoelectron counts of the fluorescence signal.

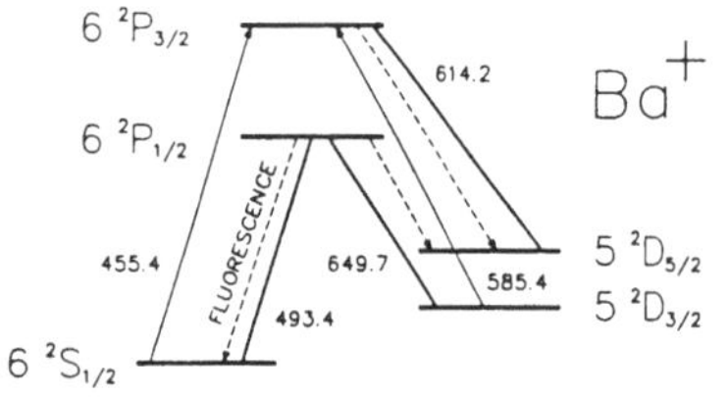

Fig. 1: Simplified energy level scheme of Ba^+. Wavelength values in nm.

Experimental

A thermal beam of barium atoms is ionized by impact with 1-s pulses of a very weak electron beam. After several unsuccessful attempts to generate an ion, eventually green fluorescence signalsthe apprearance of a Ba^+ ion in the center of the 1-mm trap, where the foci of the two coaxial laser beams are located. One beam is generated by a cw Coumarine-102 laser, the other one by a cw DCM laser. The green laser is down-tuned from resonance by 150 MHz for optimum cooling the ion, whereas the red laser is kept at the center of the $6\ ^2P_{1/2} - 5\ ^2D_{3/2}$ line.

The fluorescence signal is detected by a microscope, cooled photomultiplier selected for low dark current, and a photon counter.

An additional Ba hollow cathode lamp permits us to weakly excite the $6\ ^2P_{3/2}$ level from the ground or $5\ ^2D_{3/2}$ metastable states |13|, and a third laser at 614 nm serves for the selective release of the ion from the "dark" level.

Interrupted Fluorescence

Fig. 2 shows a recording of the ion's fluorescence at 493 nm. The mean "on" time τ_+ is 136 s $\pm$ 13 s, determined by the probability for off-resonant Raman-Stokes transitions (via $6\ ^2P_{3/2}$) excited by the green

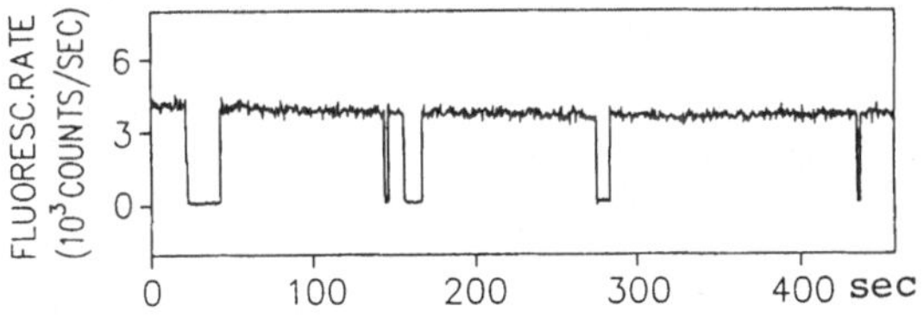

Fig. 2: Recording of the laser-excited fluorescence, at 493 nm, of a single Ba^+ ion.

and red laser light at 60 and 10^3 fold saturation of the respective transitions $^2S_{1/2} - {}^2P_{1/2}$ and $^2D_{3/2} - {}^2P_{1/2}$ (Fig. 3). The mean "off" time, $\tau_- = 8$ s, is dominated by Raman anti-Stokes transitions. There is also a small contribution from collisionally quenching the "dark" state $5\ ^2D_{5/2}$.

Irradiation with the light of the Ba hollow cathode lamp excites the ion to the real $^2P_{3/2}$ level with a chance to decay into the "dark" state that is higher than for the far off-resonant Raman excitation. Thus, this irradiation reduces the mean "on" time to 24 $\pm$ 4 s.

For unambiguous identification of the "dark" state, we have irradiated the ion by additional cw laser light at 614 nm corresponding to the $5\ ^2D_{5/2} - 6\ ^2P_{3/2}$ transition. When a 0.4-s pulse is applied after the observed fluorescence went off, the jump is immediately undone by re-excitation, and fluorescence reappears (Fig. 4). With continuous

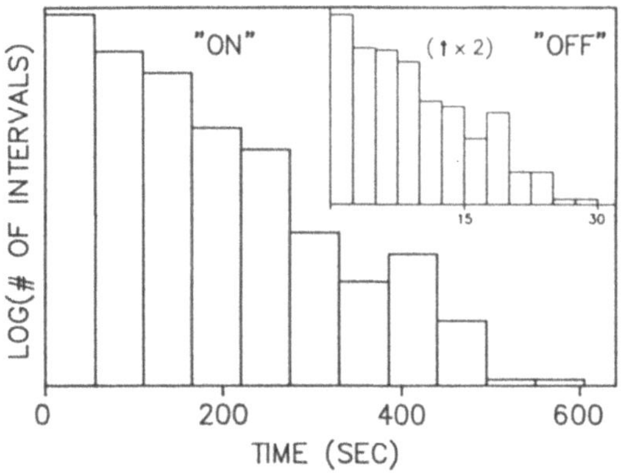

Fig. 3: Distributions of the lengths of "on" and "off" times.

irradiation, no jumps appear at all, since the $^2D_{5/2}$ level is coupled to the superposition of levels which forms the "on" state. This kind of manipulation or "shelving prevention" represents the active control of an internal degree of freedom of the ion.

From recordings of the fluorescence signal as in Figs. 2 and 4, the two-time intensity correlation has been calculated. There are predictions of this quantity |9,10|: a superposition of two exponentials, one fast and one slowly decaying, which correspond to the strong and weak

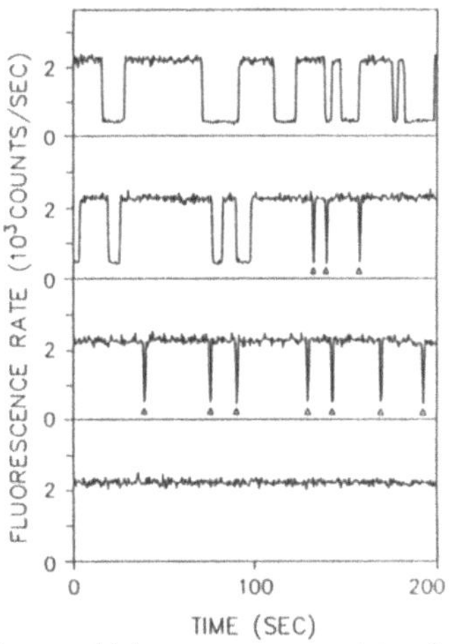

Fig. 4: Single-ion "on" and "off" intervals of fluorescence (top). Removing the ion from the "off" state ($^2D_{5/2}$) by manually pulsed laser light at 614 nm (Δ, center). Coupling the $^2D_{5/2}$ level to the "on" state by continuous 614 nm laser light, which results in "shelving prevention" (bottom). Full length of uninterrupted fluorescence recording: 20 min. (From Ref. 15)

transitions, respectively. Only the latter one is observable with the 0.4-s experimental sampling time, and the corresponding modified expression becomes

$$\langle I(t)I(t+\tau)\rangle / \langle I(t)\rangle^2 = 1 + \frac{\tau_-}{\tau_+} \exp\left[-(\tau_-^{-1}+\tau_+^{-1})\tau\right]$$

Fig. 5 shows two-time intensity correlations with and without excitation of $^2P_{3/2}$ by coherent light. They are in agreement with the val-

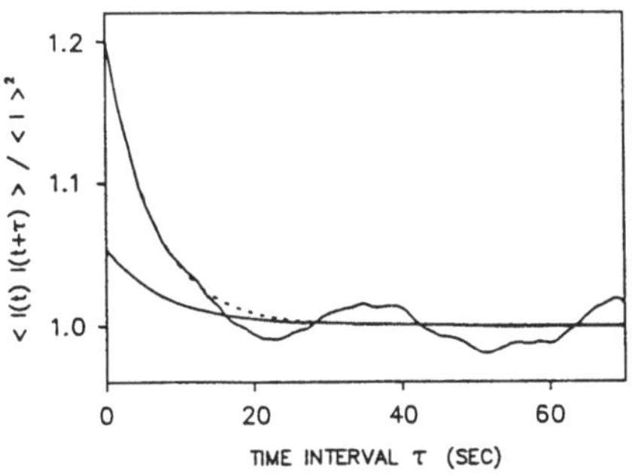

Fig. 5: Two-time intensity correlation function calculated from fluorescence recordings with laser excitation only (lower trace), and with additional lamp excitation of the $^2P_{3/2}$ level (upper trace). The oscillatory feature indicates the existence of population pulsations. (From Ref. 15)

ues of τ_+ and τ_- derived from the distributions of "on" and "off" times. The total rate $\tau_0^{-1} = \tau_+^{-1} + \tau_-^{-1}$ is dominated by τ_-^{-1}. The combination of $^2P_{3/2}$ excitation and subsequent decay into the $^2D_{5/2}$ level with anti-Stokes back pumping forms a cyclic process which shows up as modulation of the intensity correlation function (s. Fig. 5), the signature of population pulsations. Note that without excitation of the $^2P_{3/2}$ state by incoherent light the generalized Rabi frequencies at the pump and Stokes transitions, and also the effective two-photon Rabe frequency, are very large due to the lasers being detuned far off the transitions involving the $^2P_{3/2}$ relay level |18|. Thus, the population cannot adiabatically follow, and it does not pulsate.

Single-Ion Spectra

Excitation spectra of single-ion fluorescence have been recorded by scanning the 650-nm light across the $^2P_{1/2}$ - $^2D_{3/2}$ resonance and detecting the green scattered resonance light |19|. With a well-cooled ion, these spectra show sudden reduction of the signal, when the up-scanned light has crossed the line center. This phenomenon is caused by optically heating the ion, i.e. making its orbit grow |20|. With a rather *hot* ion, this effect is negligible, and occasional transitions of the ion into the dark $^2D_{5/2}$ state unambiguously reveal themselves as breaks in the spectrum (see Fig. 6). Since the ion moves most of the time far off the trap centre, it feels the rf electric drive field, at 35 MHz, which considerably modulates its speed. The well-known narrow resonance on the low-frequency wing of the spectrum which is generated by electronic Raman transitions, $^2S_{1/2}$ - $^2D_{3/2}$, develops conspicuous motional sidebands.

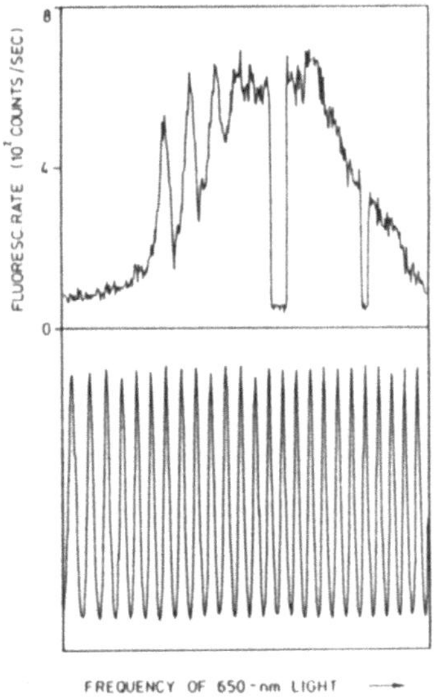

Fig. 6: Fluorescence signal of single hot ion *vs*. frequency of 650-nm laser. Interruptions (centre and right) mark quantum jumps. Raman resonance (left) shows motional sidebands (top). Frequency marker: 20.8 MHz/fringe (bottom).

Multiple Jumps

The fluorescence of a small cloud of three ions shows four discrete intensity levels corresponding to three, two, one, or no ions in the "on" state (Fig. 7). Upon inspection of the recorded traces it is obvious that simultaneous jumps of two or even three ions happen much more often than expected as random coincidences. This phenomenon has been substantiated by quantitative evaluation of the rates of random multiple jumps |15|. It turns out that the observed rates exceed the random rates by more than two orders of magnitude. This observation indicates that the ions interact collectively with the light fields. Moreover, this collective action does not require ensemble averaging in order to become detectable as in conventional experiments on super-radiance |21|. It involves real coupling of individual particles, as is proved by the huge excess of simultaneous jumps. Macroscopic collective phenomena, on

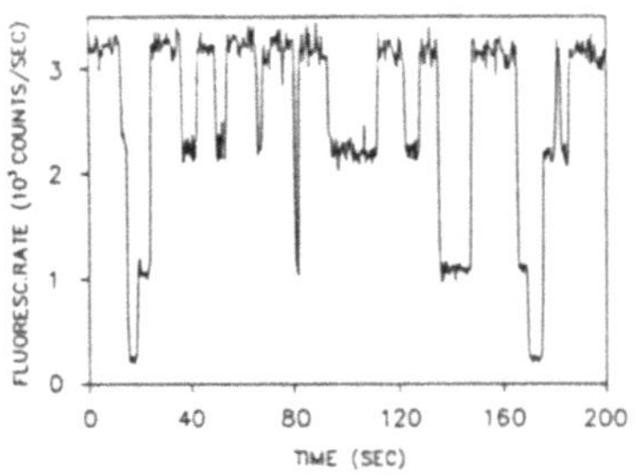

Fig. 7: Multiple jumps documented in the laser-excited 493-nm fluorescence of *three* Ba^+ ions. (From Ref. 15)

the other hand, are described in terms of enhanced probability for a microscopic process as caused by the presence of the entire ensemble.

Conclusions

We have observed interruptions of random time duration in the fluorescence of a single Ba^+ ion stored and optically cooled in an electrodynamic trap. The mean duration of the bright and dark intervals has been evaluated for the ion interacting with two resonant laser beams, and also for additional excitation to the $^2P_{3/2}$ state by incoherent light. The $^2D_{5/2}$ level has been unambiguously identified as the "off" state by re-exciting the dark ion to the $^2P_{3/2}$ state in order to make it join the "on" state - a superposition of $^2S_{1/2}$, $^2P_{1/2}$, and $^2D_{3/2}$ - again. This procedure establishes "quantum-manipulation" of an internal degree of freedom of a single atomic particle. The observed two-time intensity correlation function agrees with the time distributions, and indicates the existence of population pulsations. Three trapped ions show simultaneous multiple jumps at a rate vastly exceeding random coincidence. The cloud interacts collectively with the light by coupling the individual particles.

The novel type of measurements exercised in these experiments does not rely on ensemble averaging. Instead, one particle is prepared under specified conditions over and over again, and a very large number of individual measurements is carried out - essentially one measurement for each photoelectron counted.

This approach has allowed us, by proving the existence of macroscopic pauses in the single-ion fluorescence, to verify that the dynamics of single atomic particles is governed by sudden transitions. They are not artifacts of the temporal discreteness of the measuring process. They do not establish, on the mircroscopic time scale, statistically averaged populations in states. Rather, they make an atom indeed occupy a particular state for a time interval on the order of its lifetime.

This work was supported by the Deutsche Forschungsgemeinschaft. - One of us (P.E.T.) thanks the JILA Visiting Fellows Program for support.

References

1. N. Bohr, Phil. Mag. 26 (1913) 1, 476.
2. W. Neuhauser, M. Hohenstatt, P.E. Toschek and H. Dehmelt, Phys. Rev. A 22 (1980) 1137.
3. D.J. Wineland and W.M. Itano, Phys. Lett. 82A (1981) 75.
4. G. Janik, W. Nagourney and H. Dehmelt, J. Opt. Soc. Am. B2 (1985) 1251.
5. H. Dehmelt, Bull. Am. Phys. Soc. 20 (1975) 60.
6. R.J. Cook and H.J. Kimble, Phys. Rev. Lett. 54 (1985) 1023.
7. C. Cohen-Tannoudji and J. Dalibard, Europhys. Lett. 1 (1986) 441.
8. J. Javanainen, Phys. Rev. A 33 (1986) 2121.
9. A. Schenzle, R.G. DeVoe and R.G. Brewer, Phys. Rev. A 33 (1986) 2127, A. Schenzle and R.G. Brewer, Phys.Rev. A 34 (1986) 3127.
10. P. Zoller, M. Marthe and D.F. Walls, Phys.Rev. A 35 (1987) 198.
11. D.T. Pegg, R. Loudon, and P.L. Knight, Phys. Rev. A33, 4085 (1986)
12. Th. Sauter, R. Blatt, W. Neuhauser, and P.E. Toschek, IQEC '86 post-deadline paper, San Francisco, June 1986, and Phys. Rev. Lett. 57, 1696 (1986).
13. W. Nagourney, J. Sandberg and H.G. Dehmelt, loc. cit., and Phys. Rev. Lett. 56 (1986) 2797.
14. J.C. Bergquist, R.G. Hulet, M.W. Itano and D.J. Wineland, IQEC '86, loc. cit., and Phys. Rev. Lett. 57, 1699 (1986).
15. Th. Sauter, R. Blatt, W. Neuhauser, and P.E. Toschek, Opt. Communic. 60, 287 (1986).
16. W. Neuhauser, M. Hohenstatt, P.E. Toschek and H. Dehmelt, Phys. Rev. Lett. 41 (1978) 233.
17. D.J. Wineland, R.E. Drullinger and F.L. Walls, Phys. Rev. Lett. 40 (1978) 1639.
18. R.G. Brewer and E.L. Hahn, Phys. Rev. A 11, 1641 (1975).
19. W. Neuhauser, M. Hohenstatt, P.E. Toschek and H. Dehmelt, in *Spectral Line Shapes*, B. Wende, ed., de Gruyter, Berlin, 1981, p. 1045.
20. Th. Sauter, W. Neuhauser, and P.E. Toschek, in *Fundamentals of Laser Interactions*, F. Ehlotzky, ed., Vol. 229, Lecture Notes in Physics, Springer, Heidelberg, 1985.
21. See, e.g., "Cooperative Effects in Matter and Radiation", C.M. Bowden, D.W. Howgate, and H.R. Robl, eds., Plenum, New York 1977 - Q.H.F. Vrehen and H.M. Gibbs, in "Topics in Current Physics 27, Dissipative Systems in Quantum Optics", R. Bonifacio, ed., Springer, New York 1981, p. 111.

MACROSCOPIC QUANTUM JUMPS

Axel Schenzle
Fachbereich Physik, Universität Essen
4300 Essen, West Germany

Abstract

When an electron is prepared in a metastable state, it will remain there for some period of time, until it eventually jumps back to the ground state by the emission of a resonant photon. This event can be monitored conveniently by coupling the ground state to a dipole allowed state through a resonant laser field. Then the return of the electron from the forbidden state manifests itself in a sudden appearance of fluorescence from the allowed transition. When the forbidden transition is being driven resonantly as well, the fluorescence will be quenched again, while the electron returns to the forbidden level. If this intuitive picture is basically correct, then it is possible to monitor the individual quantum jumps, which are accompanied by a single photon event only, by the random appearance and extinction of a strong fluorescence signal - by a macroscopic signal. This process is a unique single atom phenomenon which is washed out gradually, when more and more particles participate in the scattering process.

1. Introduction

Quantum mechanics without any doubt is one of the corner stones of our present day understanding of fundamental processes in nature. But even after more than 60 years since the formulation of the concept, we still have - now and then - our difficulties with quantum theory. It is not the application of the formal apparatus to some specific problem which is bothering us, it is the lack of an intuitive view which can leave us puzzled when considering a new quantum phenomenon. One may argue that intuitivity is not a relevant ingredience of theoretical physics - all that is necessary is a consistent theory, which allows one to predict the outcome of any conceivable experiment in a satisfactory way. This view is correct as long as a well defined physical question is formulated and waits to be answered theoretically. However, in order to come up with new ideas, and to suggest new experiments, experiments where the quantum nature of the processes dominates the dynamics, we have to rely on intuition first, in order to motivate a subsequent analytical treatment of the quantum problem.

Intuition has to do with experience and therefore can be acquired to a certain extent. In the special case here, it seems that experience can be obtained by considering a number of simple and transparent physical examples, where the quantum nature of the process is essential for its understanding. Experiments on a small number of atoms, or even on single particles are the least likely to disguise the underlying quantum mechanical nature and its subtleties that may otherwise be washed out by the average over a large ensemble. In the field of quantum optics e.g. single particle experiments have become available recently either through weak atomic beams (1,2), or by the use of ion traps (3,4), and the Jaynes-Cummings model (5,6,7), as well as Anti-Bunching (1,2,8,9,10), two fundamental single particle quantum effects, have already been realized experimentally.

The early quantum mechanics of Bohr was able to predict the stationary properties of the hydrogen atom i.e. the energy levels with great accuracy but the dynamics had to be introduced artificially, by introducing the concept of sudden jumps among these eigenstates, the Quantum Jumps. This ad hoc assumption, which had no counterpart in the dynamics of the Schrödinger equation caused a vivid controversy in the early days of quantum mechanics. The discussion finally subsided long ago, since the adoption or rejection of this concept was felt to be more a matter of taste since there had been no conceivable experiment that could demonstrate the discontinuous fashion of these transitions. The main reason being that all experiments until now dealt with an enormous number of particles, and all individual jumps would have been smeared out by the average over the ensemble. Spontaneous emission of a collection of atoms results in a continuous signal on the photon detector, even if one insists that the signal is made up from a discrete number of individual events.

The progress in experimental techniques over the last years, which could certainly not to be anticipated in the early days of quantum mechanics, now makes it possible to perform experiments with just a single particle. In which way would fluorescence from a small number or even a single atom be substantially different from fluorescence at an entire ensemble of atoms. Obviously, whenever large fluctuations are expected to occur in the single particle experiment, they will be averaged out completely in the experiment with a fluorescing gas, and noise is minimized due to the presence of many atoms.

In a single atom experiment this averaging does not occur, and we can expect to see a signature of the quantum fluctuations which are responsible for the spontaneous decay. When a dipole allowed transition is driven in saturation, up to 10^8 photons per second might be scattered and the photon multiplier will produce a continuous signal and no traces of the individual counting events are visible. A dipole forbidden transition on the other hand may typically produce a single scattering event per second, a signal that is unquestionably way below the noise level of the detector. The only way to combine the high intensity of the allowed transition with the convenient time scale of the forbidden one is to consider a three level configuration. This shelving idea that was first suggested by H. Dehmelt (11) as a means of performing high resolution spectroscopy, would have gone almost unnoticed if Cook and Kimble (12) would not have pointed out the fundamental character of this effect and its relation to the quantum measurement process.

Dehmelt suggested a three level system in V-configuration, where the ground state is coupled to an allowed as well as a forbidden excited state which are both driven resonantly by two individual laser fields as indicated in Fig. 1.

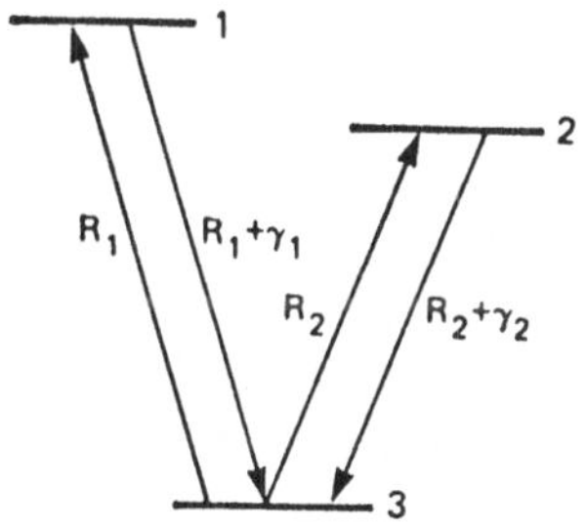

Fig. 1

The ground state is connected with two excited states via an allowed transition $|3\rangle$ -- $|1\rangle$ and a forbidden transition $|2\rangle$ -- $|1\rangle$. The corresponding life times are γ_1^{-1}, and γ_2^{-1} . The two transitions are driven by their respective laser fields at rates R_1 and R_2.

The transition frequencies are supposed to be well separated, such that the fluorescence signals can be spectrally resolved and the competition of these two transitions will result in a unique feature of the fluorescence signal. At first glance it seems to be quite evident that the electron, when driven along the allowed transition, will emit intense fluorescence of frequency $\omega 1$, which, is quenched eventually, when the electron is being shelved in the forbidden state. The resulting period of darkness is expected to last for a life time of this state, and the spontaneous return of the electron to the ground state will trigger the strong fluorescence again. Thus, also the atom is driven in a purely continuous fashion, it is expected not to respond continuously, but the strong fluorescence from the allowed transition is interrupted at random time instances by random periods of darkness.

This is just one of those examples mentioned above, where we have been relying entirely our intuitive understanding of quantum mechanics in predicting the outcome of a new experiment. The tacit assumption made, however, was that in any instance of time, the atom can only occupy a single energy eigenstate. A mixed quantum mechanical state described by a statistical operator with vanishing off - diagonal elements, might well be described intuitively by atoms that only exist in eigenstates at a time, however, quantum mechanical superposition states cannot be viewed in this way. Since the atom is driven coherently, one may argue that the coherent superposition state is closer to physical reality than the incoherent mixture. Loosely speaking, in a superposition state all the levels are occupied simultaneously and the prediction of random jumps just on intuitive grounds is much less obvious. Since we are dealing with an interesting and fundamental problem, a detailed quantum statistical treatment is in place in order to resolve the question such that no doubts about the nature of the process remain. This can only be done when we do not assume a priori that the quantum jumps exist (12,13), and only describe their statistical properties, but prove their existence in a first principle calculation (14,15).

When we try to devise a theoretical description of this process, we tend to be a little irritated by the fact that quantum mechanics is a statistical theory based on the concept of ensembles and ensemble averages, a concept which at first sight cannot be applied to a single particle experiment. An individual trajectory like the temporal fluctuations of the fluorescence intensity I(t) cannot be predicted theoretically in its actual time dependence. Only the average over many repeated and identically prepared experiments represent an ensemble and can be compared with theory. In the present case, an ensemble average over the single particle trajectories inevitably averages over the random dark periods, and leaves us with a constant mean intensity. To be a little more specific, let us consider a three level system driven in saturation by both resonant laser fields, then the average fluorescence intensity of the strong transition is given by:

$$\langle I(t) \rangle = 1/3\ \gamma_1$$

where γ_1 is the corresponding spontaneuous life time. This is to be compared with the intensity obtained in a single run, which is expected to be:

$$I(t) = 1/2\ \gamma_1$$

during the emission period and zero in the dark times. When we assume that the emission period lasts twice as long as the dark time, then a time average over the stochastic signal is identical with the ensemble average above. Needless to say that the ensemble average of the intensity alone does not say anything about the existence of the quantum jumps.

This situation, however, is not at all different from what we know from classical statistical mechanics. The Brownian particle e.g. traces out complicated and entangled trajectories in phase space, which cannot be predicted theoretically in any detail. The average coordinate e.g. remains at its initial value and does not display any sign of the random evolution. But in classical statistics we can do better and calculate higher order moments that contain more statistical information. Instead of calculating the entire hierarchy of moments of the coordinate x we can also calculate the probability density in space, as it evolves in time:

$$P(x,t)$$

This function is obtained as the solution of a Fokker Planck equation and contains the entire statistical information of the process, provided the Brownian motion is suitably described by a Markov process - and there is nothing more that one can say theoretically about the process.

Also in quantum mechanics we can do better than merely calculating the average over the basic variables, like the intensity in our case here. We can also calculate variances and cummulants of higher order, and describe the statistics in increasing details. Instead of the moments of the intensity we could also calculate those of the photon counts detected in a given collection period T:

$$\langle n \rangle \, , \, \langle n^2 \rangle , \ldots$$

The entire hierarchy is contained in the photon counting probability W(n,T) i.e. the probability of observing n events in a collection time T:

$$\langle n^\ell \rangle = \sum_{n=0}^{\infty} n^\ell \, W(n,T)$$

In case that we can show that this probability allows us to distinguish between smooth or intermittent fluorescence, a quantum mechanical calculation of W(n,T) would provide the required first principle evidence of the existence of quantum jumps.

2. Correlation Functions

The electromagnetic field, as created in any elementary process is subject to fluctuations, which may result from thermal noise in the source. The quantum nature of the source and the field is an inevitable source of fluctuations, which makes it necessary to use a quantum statistical description. We will restrict ourselves to the mode picture, and disregard all spacial properties of the field. $E_i(t)$, $E_i^+(t)$ is the quantized form of the field amplitude of frequency ω_i, and we will consider here only its stationary properties. One way to characterize the statistical fluctuations of the light field in increasing detail is to set up the hierarchy of normally ordered correlation functions (16,17).

The most elementary correlation function that does not already vanish due to the phase symmetry of the field is given by:

$$G_i^{(1)}(t) = \langle E_i(t) E_i^+(0) \rangle = tr\, \rho\, E_i(t) E_i^+(0)$$

which for t=0 represents the average intensity, ρ is the statistical operator of the ensemble. The temporal correlations characterized by $G_i^{(1)}(t)$ display the loss of coherence due to phase fluctuations. The amount of randomness in the field intensity is measured by the following correlation function (18,19) which is not sensitive to phase noise:

$$G_{ij}^{(2)}(t) = \langle E_j(0) E_i(t) E_i^+(t) E_j^+(0) \rangle$$

Intuitively speaking, $G_{ij}^{(2)}(t)$ is a measure for the probability of observing a photon of frequency ω_j at t=0 and another photon of frequency ω_i a time t later. The correlation of these events characterizes the fluctuations of the field intensity. From the definition of $G_{ij}^{(2)}(t)$ it is convenient to construct a conditional probability in the following way:

$$P_{ij} = G_{ij}^{(2)}(t) / G_j^{(1)}(0)$$

where $P_{ij}(t)$ is the probability of observing a photon of frequency ω_i at time t, after a photon of frequency ω_j had been detected with certainty at t=0. Since we are only looking at two chosen instants in time and not at what has been happening in between, $P_{ij}(t)$ provides only an incomplete picture of the process, and it is natural to proceed to correlation functions of higher order. $P_{ij}(t)$ only provides the information that there are photons of frequency ω_j and ω_i at t=0 and t respectively, but not whether there had also been emission events in between, i.e. P does not guarantee that the observed photon ω_i at t is the first to be detected after the observation of the photon at ω_j.

Before we procede to discuss the hierarchy of photon correlations in the next chapter, we want to demonstrate here that the intensity correlation function already contains enough information to give us a hint, whether it is realistic to expect the quantum jumps or not. For this purpose we compare the two photon correlation function of the driven three level system, with the corresponding and well-known two level result.

In order to keep the calculations simple and for most of the time in analytical form, we assume for the moment that the dynamics of the atomic system can reasonably be characterized by rate equations.

The electromagnetic field is created by the oscillating dipoles:

$$E^{+}(t) \sim P(t') \quad , \quad E(t) \sim P^{+}(t)$$

where P, P^{+} is the operator of the atomic polarisation, and t and t' differ by the time of propagation from the source to the detector (20,21). The dynamics of the field is thereby traced back to the Bloch dynamics of the atomic source. With the help of the quantum regression theorem (22) we are able to calculate all desired correlation functions provided that we know the general solution of the corresponding Bloch equations for arbitrary initial conditions. Using the notation explained in Fig. 1 we find for the tree level system the following auto-correlation functions:

$$P_{11}(t) = \frac{1}{3}\left(1 + \frac{1}{2}\left(e^{-\frac{3}{2}\gamma_2 t} - 3\,e^{-2\gamma_1 t}\right)\right)$$

and

$$P_{22}(t) = \frac{1}{3}\left(1 - e^{-\frac{3}{2}\gamma_2 t}\right)$$

For the cross correlations we find the same results as above, depending on the sequence of observation.

$$P_{12}(t) = P_{11}(t)$$

$$P_{21}(t) = P_{22}(t)$$

In case the weak transition is not driven, we are left with a two level system which is characterized by:

$$P_{11}'(t) = \frac{1}{2}\left(1 - e^{-2\gamma_1 t}\right)$$

P_{11} (t) is compared with P_{11}' (t). At a first look there is already a significant difference between the correlation of the strong fluorescence in the presence and the absence of the weak transition. While P_{11}'(t) is a monotonously increasing function, P_{11} (t) follows the same functionality only over the short time scale and finally decays towards the uncorrelated result on the long time scale of the metastable state. We will see that this hump in Fig. 2 is a typical indication of the appearance of random dark times.

The conclusions that can be drawn from these four correlation functions are summarized in Fig. 3. Assuming that the fluorescence from the tree level system is intermittent, we have sketched in the upper part the strong emission from the allowed transition and have indicated schematically the individual emission events - equidistant for convenience. The lower part symbolizes the weak emission from the forbidden state. We expect the dark time to last for a life time of the metastable state γ_2^{-1}. In case of saturation, the period of emission must then last for about

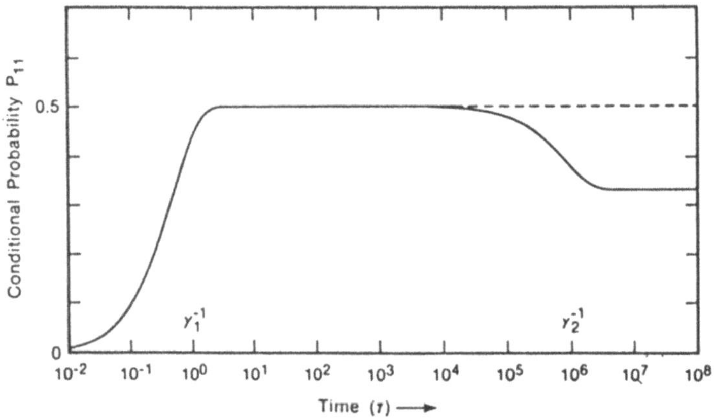

Fig. 2

Conditional Probability of observing two photons of frequency ω_1 a time interval t apart. The solid line indicates the saturated three level system, while the dashed curve represents the saturated two level case. The rates had been chosen γ_1/γ_2=10^6.

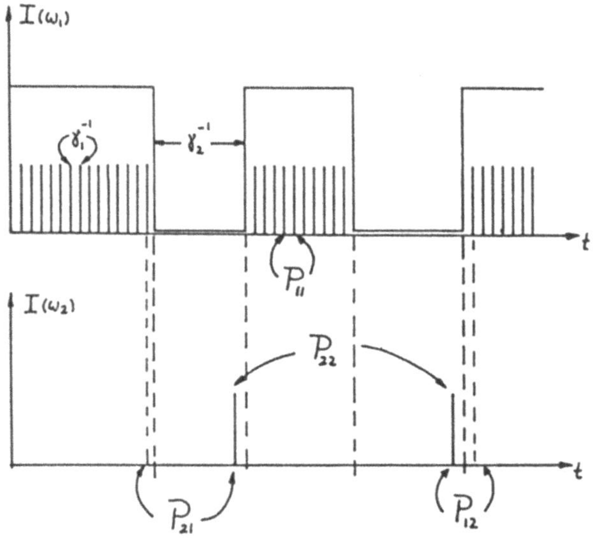

Fig. 3

Intermittent fluorescence from the strong transition is shown in the upper part, where the small lines symbolize the individual photon events in the bright periods. The lower half correlates the spontaneous emission events from the forbidden state with the periods of emission and darkness.

twice that period in order to yield the correct average intensity. The length of the individual fluorescence and the dark periods, however, will fluctuate randomly about these average values.

$P_{11}(t)$ rises from zero on a time scale γ_1^{-1}, the average time between the emission events, indicating that two photons are not likely to be emitted arbitrarily close to each other - this is well-known under the name of anti-bunching. In order to see the first photon with certainty, t = 0 must lie inside an emission period, which itself lasts roughly for the life time of the metastable state. It is very probable to find another photon a time t later as long as $\gamma_1^{-1} < t < \gamma_2^{-1}$. On this time scale the emission is expected to be identical with the two level fluorescence. Only when we search for the second photon, at times $t > \gamma_2^{-1}$ after the first one, it is more and more likely that the instant t falls into a dark period and no signal is registered. This is indicated by the deviation of the correlation function from the two level result, and the drop in probability. The humped correlation function can therefore be understood qualitatively using the picture of intermittent fluorescence.

$P_{22}(t)$: When the electron returns from the metastable to the ground state it is most likely driven up to the allowed level which marks the end of a dark period, and the strong fluorescence signal reappears. Therefore, shortly before the beginning of this signal, a photon of the weak transition must have been emitted, roughly a time γ_1^{-1} earlier. Since the times of reappearance of fluorescence are approximately γ_2^{-1} seconds apart, it is very unlikely to see a second photon of frequency ω_2 earlier than $t \sim \gamma_2^{-1}$.

$P_{12}(t)$ characterizes the conditional probability of observing a photon of frequency ω_1 a time t after a photon of frequency ω_2 had been detected. Since the emission from the forbidden state triggers the strong fluorescence, this probability rises on a time scale γ_1^{-1} and will fall off again on the long time scale γ_2^{-1}. Since the photon at ω_2 is emitted very close to the beginning of the fluorescence period it doesn't really matter which of the photons had been detected first, it is only the last one which characterizes the correlation and therefore it is rather obvious that $P_{12}(t) = P_{11}(t)$.

$P_{21}(t)$: A photon of frequency ω_1 is emitted in the bright periods, while the photon from the metastable state is emitted at the end of the dark period, therefore we have to wait at least a time γ_2^{-1}, until the conditional probability $P_{21}(t)$ rises appreciably from zero.

The correlation functions so far do not prove, but strongly support our intuitive picture from an intermittent fluorescence signal. In order to show that the statistical properties of the emission are uniquely related with a signal that displays random sequences of emission and darkness, we have to procede to higher order correlation functions or to an equivalent statistical description. This will be done in the next chapter, by deriving the photon counting statistics of the process.

Before proceding that way, this might be the place to clarify the use of rate equations in describing the quantum jumps, since earlier we had been somewhat puzzled by the role of coherent superposition states. It is correct that an ideal coherent laser field will drive an atom into a state of superposition, and off - diagonal elements of the statistical operator don't vanish. But this superposition or phase memory will only last on the a time scale of the relaxation time, since the event of spontaneous emission prepares the atom in its ground state and the phase information is completely lost. It will be established again, without relation to the previous coherence, in a Rabi cycle and will be lost again in a subsequent spontaneous emission act. For an ensemble of atoms, as described by the Bloch equations, we always prepare a macroscopic number of atoms in a superposition state, which is responsible for the macroscopic coherent polarisation. Individual atoms drop out from this collective coherent state in a spontaneous emission event, and are brought back in a Rabi period by the action of the laser field.

Since we are dealing with a stationary ensemble, dissipation is inevitable in order to establish this state. Stationarity is only reached after many relaxation times, and the phase memory has decayed long before. We conclude from here that the distinction between mixed states and pure superposition states, intuitive to some extent, was somewhat artificial and misleading in this context, because it did not account for dissipation, something that is vital for the observation of fluorescence. Therefore we expect that this distinction is not really essential, and the presence of coherent Rabi oscillations will only alter the picture quantitatively, but the basic conclusions will simply carry over from the picture of a rate process. To support this view, we have calculated the two-photon correlation function for coherent driving, by diagonalizing the 9 by 9 matrix of the three level Bloch equations and find typically the behaviour sketched in Fig. 4, where the two level and the three level correlations are compared. As expected, the two traces coincide over the short time period where anti-bunching and Rabi oscillations occur but deviate later. The "hump" again is clearly visible and indicates the appearance of dark periods.

3. Photon Counting Statistics

One way to determine the statistical properties of light is to count the number of photons that fall on a detector in a given time interval T. Due to the discreteness of the photon counting events, the results will fluctuate randomly and display Poissonian statistics. This is the randomness introduced solely by the detection scheme. The fluctuations of the field intensity itself will be superimposed

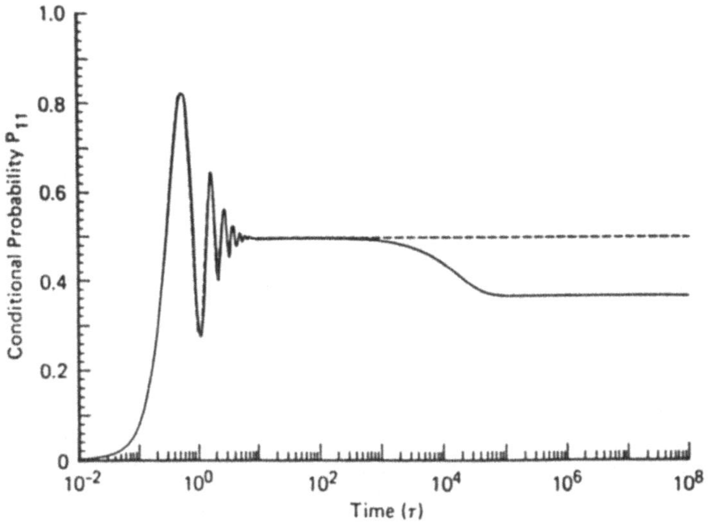

Fig. 4

Conditional probability as in Fig. 2 but with coherent driving fields (solid curve) in comparison with the corresponding two level case (dashed curve).

and cause an additional broadening of the counting distribution. The intrinsic statistical properties of the light field have to be derived therefore from the deviation of the actual counting distribution from the ideal Poissonian one. A coherent field with its stabilized intensity is Poissonian, and one might expect that additional fluctuations of the intensity can only result in further uncertainty and therefore in broadening. This is correct when we describe the field in terms of classical electro - dynamics. However, the fluctuations characteristic of quantum fields can either broaden or squeeze the statistical distribution from the ideal Poissonian result, a feature that is not easily understood on intuitive grounds. The Sub - Poissonian statistics of resonance fluorescence is a unique quantum effect (23,24) which has been observed experimentally (25,10).

The quantum mechanical photon counting distribution has been derived by Glauber (26) and by Kelley and Kleiner (27) from a first principle consideration of the source, the emitted field, the detection process and their mutual interaction. This result generalizes naturally the classical photon counting distribution, derived previously by Mandel (28,29):

$$W(n,T) = \mathrm{tr}\,\rho\,\hat{T}\,\frac{1}{n!}\,(IT)^n_T\,\exp(-IT)$$

where we introduced the operator $I(T) = \gamma\eta/T \cdot \int_0^T b^+(t)\,b(t)\,dt$; γ^{-1} is the spontaneous radiative life time of the excited state, η the quantum efficiency, and $\hat{T}$ guarantees time and normal ordering of the operator products. In the present form, this result closely resembles the classical one, where I(t) is the time averaged intensity, and the ensemble average is carried out over the classical fluctuations of the field. The compact form of the quantum result is lost instantly when we attempt to calculate the distribution for any practical problem, since in order to comply with the ordering prescription, the exponential must be expanded and the multiple time integration must be turned into iterated convolutions. This demonstrates clearly that the photon counting distribution comprises the entire hierarchy of multiple photon correlation functions.

As in the previous chapter, the properties of the field will be traced back to the properties of the emitting dipoles. In this way we can determine normally ordered correlation functions of arbitrary order from the general solution of the corresponding Bloch equations: (14)

$$\rho_{ij}(t) = \sum_{\ell m} K_{ij}^{\ell m}(t-t')\,\rho_{\ell m}(t')$$

The correlation function of order n+l e.g. assume the following form:

$$G^{(n+\ell)}(\{t_i\}) = \prod_{s=2}^{n+\ell} K_{11}^{33}(t_s - t_{s-1}) \cdot \rho_{11}^{ss}$$

where ρ_{11}^{ss} is the stationary population of the fluorescing excited state. For convenience, we will use the abbreviation:

$$K_{11}^{33}(t) = h(t)$$

The convolutions over the multiple product of h(t) is most conveniently carried out by using the tool of Laplace transformations, and we find

$$W(n,T) = \int_0^T dt\, \mathcal{L}_t^{-1}\, \Gamma(n,z)$$

where $\Gamma(n,z)$ is given by:

$$\Gamma(n,z) = \rho_{11}^{ss} \cdot \lambda \cdot \frac{1}{z} \cdot \frac{h^{(n-1)}(z)}{(1+\lambda h(z))^{n+1}}$$

and

$$\lambda = \gamma\eta$$

The structure of $\Gamma(n,z)$ suggests that the general photon distribution $W(n,T)$ can be obtained by simple differentiation of the zero count probability:

$$W(n,T) = (-1)^n \frac{\lambda^n}{n!} \frac{\partial^n}{\partial \lambda^n} W(n=0,T)$$

It is not difficult to show that the photon counting distribution in the general definition above has the following properties:

i) $W(n,T)$ is normalized

$$\sum_{n=0}^{\infty} W(n,T) = 1$$

and determines the average counting properties.

ii) In the limit of a vanishing counting interval $T \to 0$, the probability of no event approaches unity while the n count probability must vanish:

$$\lim_{T \to 0} W(n,T) = \delta_{n,0}$$

iii) The probability of observing precisely n events in an ever increasing interval must vanish:

$$\lim_{T \to \infty} W(n,T) = 0$$

The goal of our present considerations is to show on the basis of a first principle calculation that the fluorescence from Dehmelt's shelving scheme, while driven in steady state, is intermittent and displays long periods of darkness. This is to be distinguished from a continuous fluorescence signal with only Poissonian fluctuations about the average value. If the fluorescence is interrupted by dark-periods of the order of the lifetime of the metastable level, say 1 second, then for a saturated three level system we expect that in every third experiment we observe no counting events, as long as the collection time T is kept shorter than a second. On the other hand, if fluorescence would be continuous with an average counting rate of typically 10^8 per second it would be extraordinarily improbable to see no event during a period of one second. This follows directly from the derived counting statistics, when we assume for a moment that I(T) is a constant c-number. For a Poissonian process with average $\langle n \rangle$ = 10^8 the probability of seeing no event is:

$$W(n=0, T=1sec) = \exp(-10^8)$$

which for all practical purposes is identical to zero. In order to present an unbiased approach, we must set up the calculation in a way that the result may lie anywhere in this wide range of probabilities. This requires that we do not rely on a perturbative approach around the average intensity or zero intensity by an appropriate expansion of the exponential. The series in n - photon correlation functions that would be generated thereby cannot be truncated at any low order, since the correlations tend to become negligible only when its order becomes large compared with the average number of counts. For the present case this would require the calculation of the first 10^8 correlation functions. For this reason we will simplify the model to be used to such an extent that we can derive the statistics in an analytical and nonperturbative way.

Under the assumption - as above - that the system can reasonably be described by rate equations, and in the limit of strong driving of the allowed transition, it is possible to derive the counting probability in closed analytical form. For the probability of darkness over an interval T we find: (14)

$$W(n=0,T) = \frac{\gamma_2 + R_2}{2\gamma_2 + 3R_2}\left(\frac{R_2}{\gamma_2 + R_2} e^{-(\gamma_2+R_2)T} + 2\, e^{-\frac{1}{2}\gamma_1 \eta T}\right)$$

and we can immediately decide whether this result predicts quantum jumps or not. For an intermediate time interval T_0 :

$$R_1^{-1}, \gamma_1^{-1} \ll T_0 \ll \gamma_2^{-1}, R_2^{-1}$$

the probability of no events is found to be:

$$W(n=0,T_0) = \frac{R_2}{2\gamma_2 + 3R_2}$$

which in case that we saturate also the forbidden transition, leads to the expected probability:

$$W(n=0,T_0) = 1/3$$

This result is only consistent with the picture of an intermittent fluorescence signal and not with the Poissonian statistics of continuous fluorescence. The complete time dependence of $W(n=0,T)$ is plotted in Fig. 5. The curve starts at the expected value of one for vanishing counting intervals and then drops rapidly over a period of the lifetime of the dipole allowed transition. This drop would continue if it would not be for the metastable state. The shelving becomes evident in this plot through the plateau in the intermediate time regime. The height and the width of the plateau indicates the probability of occurence and the length of the dark periods. As the saturation parameter $S = R_2 / (R_2 + \gamma_2)$ is increased in the plot, the dark periods become more and more frequent and the plateau rises, while their duration decreases, due to the induced downward transitions that reduce the survival time in the forbidden state. Above a time interval of the order of spontaneous life time of the metastable state, the probability drops rapidly towards zero, indicating that longer dark periods become excessively improbable. The probability for observing a given finite number of counts is easily derived by mere differentiation. Since only the second term in $W(n=0,T)$ depends on $\gamma_1\eta$, only the rapid time constant γ_1 enter the probability W (n, T) for n $>$ 0:

$$W(n,T) = 2\,\frac{\gamma_2 + R_2}{2\gamma_2 + 3R_2}\cdot\frac{1}{n!}\left(\tfrac{1}{2}\gamma_1\eta T\right)^n\cdot e^{-\frac{1}{2}\gamma_1\eta T}$$

i.e. a Poissonian distribution with the exception of the zero count rate. This means that during the emission periods, the fluorescence is well described by Poissonian statistics. A qualitative plot of the distribution is shown in Fig. 6.

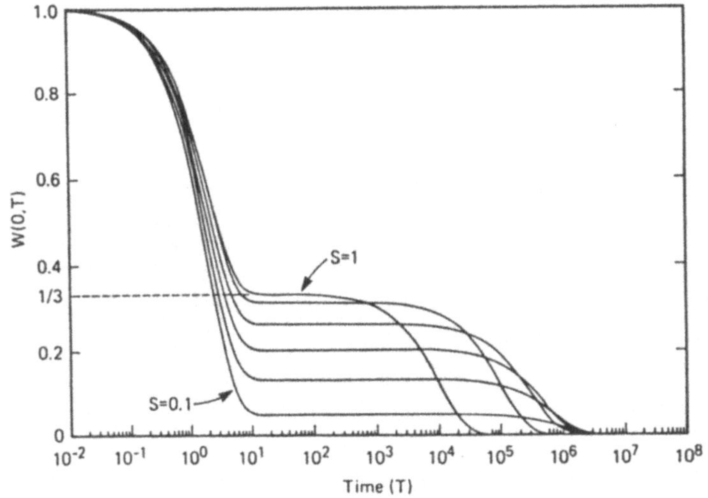

Fig. 5

Probability of observing no events in a time interval T. The parameter is the saturation rate of the forbidden transition. For strong saturation the plateau rises up to the expected value of 1/3.

It may be illustrative to note that the probability of observing a single or a few counts in a period T_0 is extremely small, since this would correspond to a marginal event, where precisely the last photon of the previous period, or the first, but only the first of the next period falls into the counting interval. The derived distribution demonstrates this clearly:

$$W(n=1, T=1\,\text{sec}) = 10^8\cdot\exp(-10^8)$$

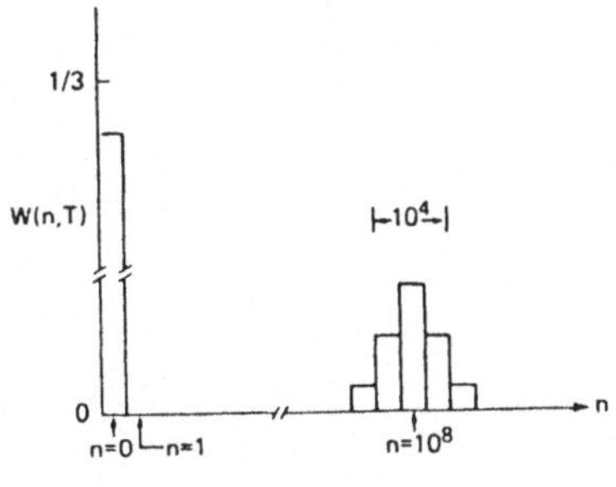

Fig. 6

Photon counting distribution W(n,T), schematically for γ_1^{-1} = 10^8 sec, γ_2^{-1} = T = 1 sec.

which again can be identified with zero for any practical purposes.

The dark time is easily determined experimentally. Since it varies largely about its average value, it may be interesting to derive also its statistics and compare it to the experimentally determined histograms. Obviously, the zero count probability must be associated in some way with the probability P (T) of observing a break in the strong emission of length T. In Fig. 7 we derive in an intuitive way P (T) by the following arguments:

i) All events where the dark time is larger than the chosen interval T contribute to W (n,T), $n=0$.

ii) When we vary $T \rightarrow T + dT$, the change of probability:

$$W(n,T+dT) - W(n,T) \sim \frac{dW}{dT} \cdot dT$$

is a measure for the probability of observing a last or first event in connection with a following or previous period no events lasting for T seconds. This is indicated in the middle of Fig. 7. This is obviously not yet the quantity that we want to compare with experiment, since the associated dark period is still shorter than the actual one.

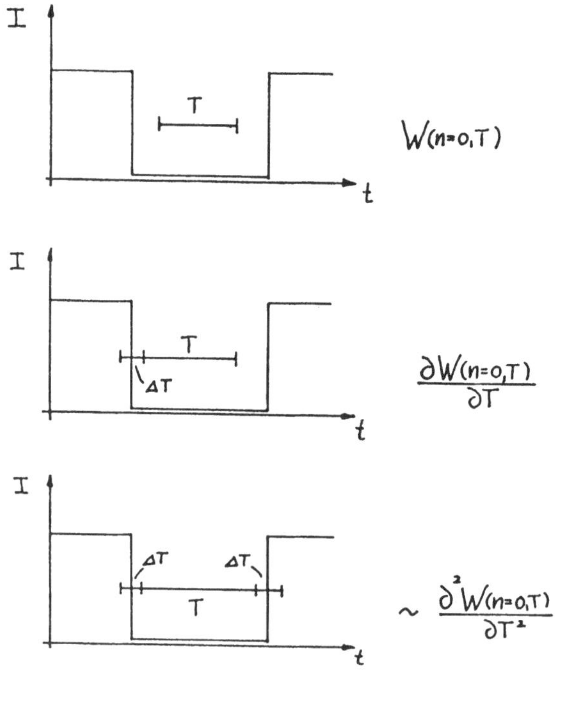

Fig. 7

Dark time probability P(T)dT is related to W(n=0,T) through differentiation as sketched in this plot.

iii) A second variation in the time interval, properly normalized, however, leads us to the desired probability density of observing a period of darkness sandwiched between two emission events:

$$\mathcal{P}(T) = -\frac{1}{W'(n=0,T=0)} \cdot \frac{d^2W(n=0,T)}{dT^2}$$

This quantity is positive, due to the monotonous property of W(n=0,T). It contains the probability of darkness between individual photon emission events during the bright periods, as well as the probability of long dark intervals; i.e. it contains the complete information that could only be obtained with a photo multiplier of arbitrary time resolution. This probability allows us to illustrate the intermittent statistical fluorescence signal, by drawing random numbers from a computer, which are distributed according to the calculated density. Then for any given number we draw a line of unit length that is separated from the previous line by a distance proportional to the last random number. So most frequently small numbers are drawn, and the lines lie densely together, until a large number creates a gap, the dark period, see Fig. 8. This illustrative plot has first been devised by Zoller et.al.

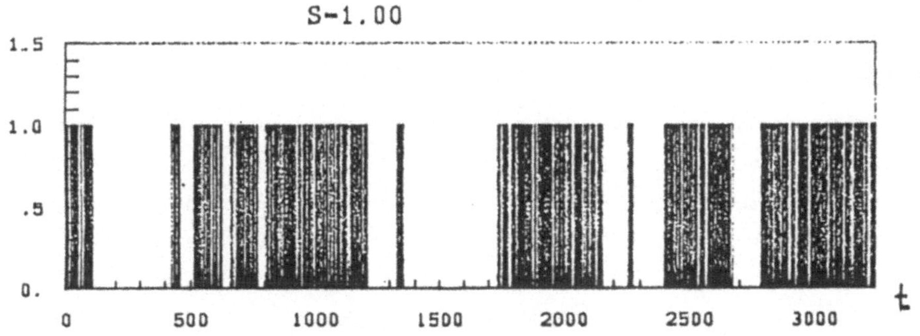

Fig. 8

Simulation of the fluorescence from a three level system by drawing random numbers according to the dark time distribution P(T). This simulates an experiment with arbitrary time resolution.

using a different but equivalent theoretical basis. The plot would represent a typical experiment if the detector wouldn't have a dark time of itself and wouldn't average over a finite time interval. The averaging can be done also numerically and a more realistic signal results as shown in Fig. 9. In the meanwhile, experiments at different places have observed the quantum jumps independently (30-33). The first observation has been made by Dehmelt, who almost 10 years ago has first suggested

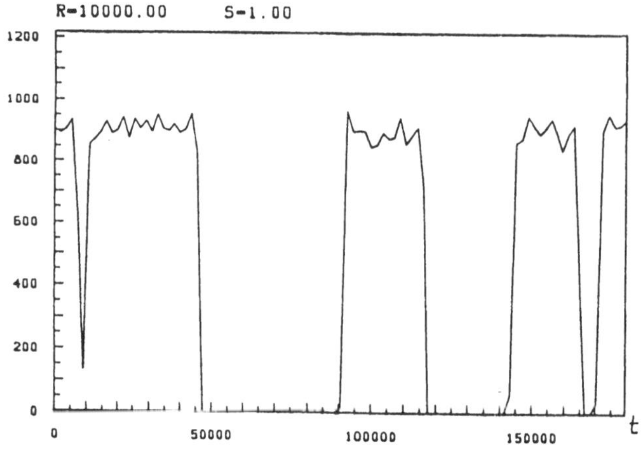

Fig. 9

The same simulation as in Fig. 8 but with a finite time resolution.

this experiment. Qualitatively this result looks very much like the simulated plot, quantitatively a comparison is not that easily available, since the experiment had been made on a more complicated level structure.

4. Conclusions

In classical mechanics it does not cause any intrinsic contradictions when we describe states of a mechanical system and its time evolution, without considering explicitly the way we interact with that system through a detection device. For a quantum mechanical system this is fundamentally different. It does not mean anything to say that an electron is in any specific state, or has jumped to a different one at a given time, without considering the measuring process that provides this information. As in a single atom fluorescence experiment, we have no way to determine the dynamics of an isolated electron, what it is doing when we do not look at it. It is the fluorescence signal, the counting event, the click in the detector that indicates that the electron has returned to its ground state. This event is discrete, and occurs at a given time. It is meaningless to consider the question whether the electron dynamics by itself determines the discreteness or the photon emission process or eventually the detection mechanism in the photon multiplier. It is the entire combined physical system that displays this feature, and there is no way to dissect the quantum system and separate it into the basic process and the measurement. In this sense, the discreteness of the quantum jumps and the discreteness of the photo effect has the same origin.

What has revived this idea again after all these years is the progress in experimental technology which now has made it possible to perform the essential step and to do experiments on single atoms or molecules. This allows one to look at individual events emitted from a single particle and not only on the average behaviour of a large ensemble. In finally detecting these single events, the shelving idea of Dehmelt has been essential, since it provides a simple way of amplifying the weak microscopic signal that indicates the quantum jump up to a macroscopic scale. In a way this is a very transparent and simple example of a quantum mechanical measurement, where a microscopic signal i.e. the photon from the weak transition is amplified in order to move a macroscopic pointer, the position of which we can read off, without interfering any more with the microscopic system. It is well-known that this quantum mechanical amplification process inevitably introduces noise, which limits the precision of our observation, in agreement with the uncertainty principle. This is rather obvious in this example.

It was not the aim of this paper to get involved again in the old controversy, or to say anything new about the electronic quantum jumps. The aim was to describe the statistics of the light field, emitted from a continuously driven three level system, and to determine whether the intuitive picture, based on the simple quantum jump concept, leads to the correct prediction for the experiment or not. In order to do this, the theoretical approach had to be free from any ad hoc assumptions that would introduce the quantum jump concept through the back door.

In our calculation of the photon statistics, we have demonstrated that the discontinuous jumps in the fluorescence are directly related to the angular momentum statistics of the n level atom, and are a basic feature of the quantum system. It may be worthwhile to stress that the description of the jumps does not require to introduce an additional time constant that would not already be present in the atomic dynamics and the detector response. Even with a detector of arbitrary time resolution, the occurence of a jump can never be determined to a higher accuracy than the time between individual emission events. And even this is only an average number. If the time after the last event gets longer than the average time, then this can indicate an excessively improbable event, and we are still in the bright period. It is only when the darkness lasts for many spontaneous life times that we have to conclude finally that the electron is shelved and darkness will last for much longer, i.e. the lifetime of the metastable state.

The quantum jumps in the fluorescence signal are a unique quantum phenomenon, which is characteristic for a single particle system. Its satisfactory theoretical description in a simple and transparent model, together with the beautiful experimental verification of the intermittent fluorescence makes this phenomenon a key problem in quantum mechanics, which is also of great pedagogical value for the understanding of fundamental processes, and the quantum mechanical measurement.

References

1. H.J. Kimble, M. Dagenais and L. Mandel, Phys. Rev. Lett. 39,691 (1977)
2. M. Dagenais and L. Mandel, Phys. Rev. A 18,2217 (1978)
3. W. Neuhauser, M. Hohenstatt, P.E. Toschek and H.G. Dehmelt, Phys. Rev. Lett. 41, 233 (1978)
4. D.J. Wineland, Science 226,395 (1984)
5. E.T. Jaynes and F.W. Cummings, Prog. IEEE 51,89 (1963)
6. J.H. Eberly, N.B. Narozhny and J.J. Sanchez-Mondragon, Phys. Rev. Lett. 44,1323 (1980)
7. G. Rempe, H. Walther, and N. Klein, Phys. Rev. Lett. 58,353 (1987)
8. C. Cohen Tannoudji, in Frontiers in Laser Spectroscopy, ed. by R. Balian, S. Haroche and S. Liberman (North Holland, Amsterdam, 1977)
9. H.J. Carmichael and D.F. Walls, J. Phys. B9,1199 (1976)
10. F. Diedrich and H. Walther, Phys. Rev. Lett. 58,203 (1987)
11. H.G. Dehmelt, Bull. Amer. Soc. 20,60 (1975)
12. R.J. Cook and H.J. Kimble, Phys. Rev. Lett. 54,1023 (1985)
13. C. Cohen-Tannodji and J. Dalibard, Eur. Lett. 1, 441 (1986)
14. A. Schenzle and R.G. Brewer, Phys. Rev. A34,3127 (1986)
15. P. Zoller, M. Marthe and D.F. Walls, Phys. Rev. A
16. A. Schenzle, R.G. DeVoe and R.G. Brewer, Phys. Rev. A33,2127 (1986)
17. D.T. Pegg, R. Loudon and P.L. Knight, Phys. Rev. A33,4085 (1986)
18. R.J. Glauber, Phys. Rev. 130,2529 (1963)
19. R.J. Glauber, Phys. Rev. 131,2766 (1963)
20. G.S. Agarwal, Quantum Optics, Vol. 70 of Springer Tracts in Modern Physics (Springer Verlag, berlin 1974) p. 39
21. J.R. Ackerhalt and J.H. Eberly, Phys. Rev. D10,3350 (1974)
22. M. Lax, Phys. Rev. 157,231 (1967)
23. L. Mandel, Opt. Lett. 4,205 (1979)
24. R.J. Cook, Phys. Rev. A23,1243 (1981)
25. R. Short and L. Mandel, Phys. Rev. Lett. 51,384 (1983)
26. R.J. Glauber, in Quantum Optics and Electronics, ed. by C. DeWitt, A. Balandin and C. Cohen-Tannoudji (Gordon and Breach, New York, 1965), p. 178
27. P.L. Kelley and W.H. Kleiner, Phys. Rev. 136,316 (1964)
28. L. Mandel, Proc. Phys. Soc. London 72,1037 (1958)
29. L. Mandel, Prog. Phys. Soc. London 74,233 (1959)
30. W. Nagourney, J. Sandberg and H.G. Dehmelt, Phys. Rev. Lett. 56,2727 (1986)
31. J.C. Berquist, R.G. Hulet, W.M. Itano and D.J. Wineland, Phys. Rev. Lett. 57,1699 (1986)
32. T. Sauter, W. Neuhauser, R. Blatt and P.E. Toschek, Phys. Rev. Lett. 17,1469 (1986)
33. T. Sauter, R. Blatt, W. Neuhauser and P.E. Toschek, Opt. Commun. 60,287 (1486)

PART IV: Quantum Electrodynamics in a Cavity

SPONTANEOUS EMISSION IN CONFINED SPACE

by

S. HAROCHE

Ecole Normale Supérieure (Paris) and Yale University (New Haven)

The quantum noise of the electromagnetic field usually sets an intrinsic limitation to the precision of any optical experiment. Interferometric methods are essentially based on the determination of the phase of a quasi-monochromatic field, whose ultimate fluctuations are of quantum nature. Similarly, spectroscopic measurements of atomic energy intervals have their precision ultimately limited by the natural width of the excited electronic states. This width reflects -through the Heisenberg uncertainties- the spontaneous decay of these states due to their coupling to the quantum fluctuations of the vacuum field... It has recently been recognized however that the effects of the vacuum field noise can be greatly reduced in some physical observations and during the last two years, several experiments have demonstrated the reduction or even the nearly complete cancellation of photon noise effects. These experiments fall into two categories : squeezed states generation experiments [1] make use of non-linear optical processes in order to decrease the quantum noise on one phase of the field -at the expense of the quadrature component on which the noise is increased. In Cavity Quantum Electrodynamics experiments [2-5], atomic systems are confined in small cavities in which the mode distribution of the vacuum field is strongly modified with respect to its free space value, entailing important alterations of the radiative properties of the atoms. The quantum noise resonant with the atomic transition can be either suppressed or increased, leading to either inhibition or enhancement of the excited states spontaneous decay.

Recent squeezed state generation experiments are discussed in other contributions to these proceedings. In this paper, I will discuss spontaneous emission modifications induced by a cavity, describe a recent experiment in which the suppression of spontaneous decay has been observed for the first time on an optical transition and discuss some implications of this effect.

Spontaneous emission in a cavity : theoretical background

The change of spontaneous emission in a cavity is a quite simple effect, which can be understood by an analysis of the mode density of the vacuum radiation field surrounding the atom [6-7]. An atomic excited state $|e\rangle$ undergoes spontaneous decay towards a final state $|f\rangle$ because the system can radiate a photon in any mode of the vacuum field surrounding the atom. The probability Γ of photonemission per unit time is simply given by the Fermi-Golden rule :

$$\Gamma \sim \frac{2\pi}{\hbar} |\langle g|\vec{D}|e\rangle|^2 . \hbar\omega . \rho(\omega) \qquad (1)$$

In Equ. (1), $\langle g|\vec{D}|e\rangle$ is the matrix element of the atomic dipole operator $\vec{D}$ between the initial and final states, $\hbar\omega$ measures the vacuum field fluctuations per mode at the frequency ω of the atomic transition and $\rho(\omega)$ is the density per unit volume and frequency of final photon states with the polarization of the atomic transition. In free space, $\rho(\omega)$ is isotropic : $(\rho_0(\omega) = \omega^2/\pi^2c^3)$. If the atom happens to be confined in an electromagnetic cavity, the boundary conditions at the walls modify the mode density and accordingly change the emission rate. Of particular interest is the case of an atom in a cavity so small that $\rho^{(cav)}(\omega) = 0$ (cavity beyond cut-off). Then the field quantum noise is totally suppressed at the relevant frequency and the excited atomic state survives for ever [7] -at least in principle...

To be more specific, we now discuss the simple situation of an atom radiating between two plane parallel mirrors separated by a gap d. The calculation of the mode density in such a structure is a text book problem of classical electromagnetism. Figure 1a) represents the mode density $\rho_\sigma^{(cav)}(\omega)$ and $\rho_\pi^{(cav)}(\omega)$ corresponding to fields having at the midplane position z=d/2 an electric field respectively parallel and normal to the cavity mirrors (σ and π polarizations respectively). The variations of $\rho_\sigma^{(cav)}$ and $\rho_\pi^{(cav)}$ versus ω are compared to the one of the free space mode density ρ_0, shown by a dotted line on the same figure. The most striking feature is the cancellation of $\rho_\sigma^{(cav)}(\omega)$ for ω below the cut-off frequency $\omega_0=\pi c/d$, whereas $\rho_\pi^{(cav)}(\omega)$ remains non-zero down to $\omega=0$ and is actually larger than $\rho_0(\omega)$ for $\omega \leqslant 1.5\ \omega_0$. These features have a very simple explanation. The modes with an electric field parallel to the mirrors must present a vanishing tangential electric component at the metallic boundaries, which requires the existence of at least one standing wave

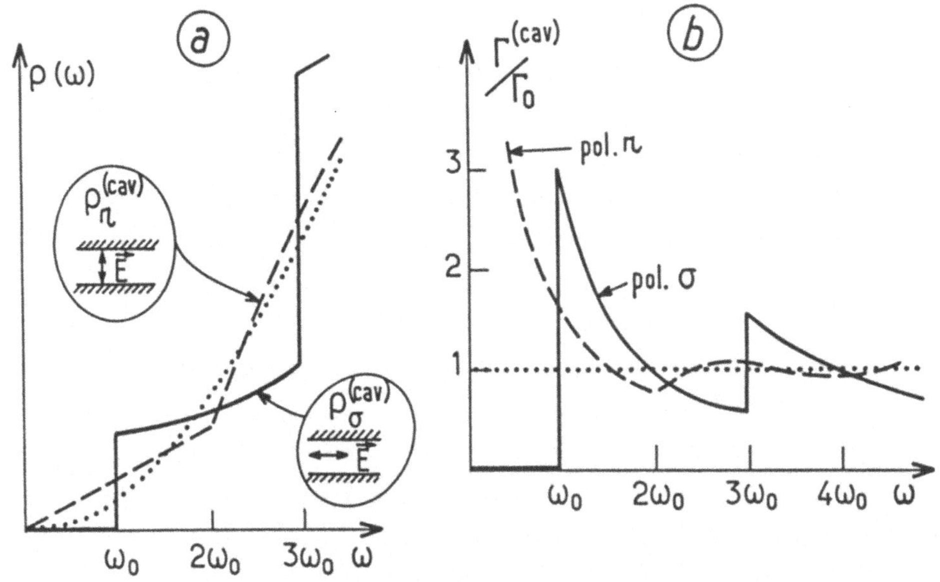

Figure 1 : a) Mode density $\rho(\omega)$ versus ω in a cavity made of two plane parallel mirrors separated by a gap d. Density evaluated at midgap position z=d/2; full line : σ polarization; dashed line : π polarization. For comparison, the free space mode density is represented in dotted line.
b) Ratio $\Gamma^{(cav)}/\Gamma_0$ versus frequency ω; full line : spontaneous emission in polarization σ; dashed line : spontaneous emission in polarization π

between z=0 and z=d, i.e. $d > \lambda/2$ or equivalently $\omega > \omega_0$. The modes with their electric field normal to the mirrors, on the other hand, correspond for $\omega=0$ to the electrostatic configuration of a parallel plate capacitor. These modes thus exist down to zero frequency. The linear variation of $\rho_\pi^{(cav)}(\omega)$ versus ω for ω small has also a simple interpretation. The only modes surviving in the cavity below $\omega = \omega_0$ have their wave vector $\vec{k}$ parallel to the mirrors. The $\vec{k}$ vector associated to a frequency ω have a length ω/c and, in phase space, their tips belong to a circle of radius ω/c, whose length is proportional to ω/c. In free space on the other hand, the $\vec{k}$ vectors corresponding to frequency ω have their tips on the surface of a sphere, whose area is proportional to ω^2/c^2. The mode density $\rho_\pi^{(cav)}(\omega)$ is thus larger than $\rho_0(\omega)$ by a ratio proportional to c/ω, i.e. to λ. Actually, this dimensionless ratio is $3\lambda/4d$ ($\rho_\pi^{(cav)}(\omega)/\rho_0(\omega) = 3\lambda_0/4d = 1.5$ for the cut-off wavelength $\lambda_0 = d/2$).

According to the above discussion, the spontaneous emission rate $\Gamma^{(cav)}$ of an atom located at midplane between the two mirrors is equal to the free space rate Γ_0 multiplied by the factor $\rho_\sigma^{(cav)}(\omega)/\rho_0(\omega)$ if the atomic transition is polarized parallel to the conducting surfaces and by the factor $\rho_\pi^{(cav)}(\omega)/\rho_0(\omega)$ if it is polarized perpendicular to them. Figure 1b shows the variations of these ratios versus ω. For σ polarized transition, the spontaneous emission rate is totally suppressed when $\omega < \omega_0$. It undergoes a sharp increase at $\omega=\omega_0$ and other resonances occur for $\omega=3\omega_0$, $5\omega_0$... For π-polarization, the emission rate is enhanced with respect to Γ_0 for $\omega < \omega_0$, being equal to $1.5\ \Gamma_0$ for $\omega=\omega_0$.

The spontaneous emission rate alterations are important only for ω smaller than or of the order of a few ω_0. For larger atomic frequencies, or equivalently for larger cavity sizes, $\Gamma_\sigma^{(cav)}$ and $\Gamma_\pi^{(cav)}$ rapidly become very close to Γ_0. These cavity induced effects thus require cavities whose sizes are of the order of the atomic wavelength.

For the sake of simplicity, we have restricted this analysis to atoms located at $z=d/2$. At other locations in the gap between the mirrors, the mode density also undergoes resonances for $\omega=2\omega_0$, $4\omega_0$... (the corresponding modes have a node at the mid plane position and their contribution thus does not appear on Figure 1). The important point for our forthcoming discussion is that the field amplitude in the cavity does not depend upon z for $\omega < \omega_0$, so that all the conclusions derived above remain true even if $z\neq d/2$, provided $\omega < \omega_0$: the dramatic effects of spontaneous emission inhibition (for σ polarization) and enhancement (for π polarization) are position independent in the cavity when $\omega < \omega_0$. The above discussion has assumed perfect mirror conductivity. Small cavity losses have the effet to smooth the sharp resonances of $\Gamma^{(cav)}/\Gamma_0$ and result in a small non-zero radiation rate in σ-polarization for $\omega\leq\omega_0$.

I have chosen here to discuss cavity Q.E.D. effects according to a field mode expansion analysis, following the point of view of other theoretical studies [8], from which most of the above discussion can be derived, although not always very directly. This approach has the advantage of emphasizing the photon noise point of view and clearly demonstrates that the cavity effectively cuts down or amplifies the vacuum field noise components resonantly coupled to the atom. Another equivalent point of view chooses an electric image approach [9] : the atom in the cavity is described as a dipole interacting with its own self-radiation field and with the field reflected from the mirror which is viewed as being radiated by image dipoles induced in the cavity walls. This alternative model describes the atom + cavity radiation process as an interference effect between the fields radiated by the atom and its images. It yields of course exactly the same conclusions as the mode expansion model. We will not discuss it any more in this paper.

Cavity Q.E.D. experiments in the microwave domain

The first evidence of radiative rate modifications near metallic boundaries came from fluorescence measurements performed on complex molecules radiating near a surface [10]. I will not discuss these early experiments here and I will restrict my analysis to the more quantitative and more precise experiments performed recently on isolated and simple atomic systems in cavities. These experiments have to overcome the difficulty of confining excited atoms in metallic structures whose dimension is of the order of the atomic transition wavelength. One solution consists in studying Rydberg atomic states radiating on long wavelength centimetric or millimetric transitions. The cavities are then fairly large and atoms moving at thermal velocities stay in these cavities during a time varying from a few microseconds to a fraction of millisecond, depending upon cavity geometry.

The first experiment of this kind [2] has been carried out at Ecole Normale Superieure in 1983 : we have observed the enhancement of the spontaneous emission rate on the 23S to 22P transition of Sodium atoms at a frequency $\omega/2\pi$ = 340GHz ($\lambda \sim$ 0.88mm transition). Atoms crossing the cavity were excited in a cm-size cavity operating in a high-order mode, which was resonant with the atomic transition. The cavity was of Fabry-Perot type, with one plane and one spherical mirror. This configuration -slightly different from the simple biplanar geometry discussed above- is much more convenient for the observation of resonant enhancement of spontaneous emission rates. The semi-cofocal Fabry-Perot structure has the advantage of considerably increasing the density for transverse electric field modes resonant with the cavity. In such a configuration, it is more appropriate to analyze the mode density in terms of the cavity quality factor Q. If the cavity has a volume $\mathcal{V}$, $\rho_\sigma^{(cav)}(\omega)$ is at resonance equal to $Q/\omega\mathcal{V}$ (one mode per frequency interval ω/Q and volume $\mathcal{V}$). It is thus enhanced with respect to free space by the factor $\sim \omega^3 Q/\mathcal{V}c^3 \sim \lambda^3 Q/\mathcal{V}$. Thus, for high Q's and even if the cavity does not operate in its lowest orders ($\mathcal{V} \gg \lambda^3$), large enhancement factors can be obtained. In our experiment, Q was of the order of 10^6 and an enhancement of about 500 was achieved, much larger than the maximum factor 3 in the biplanar configuration (see Figure 1b). The free space spontaneous emission rate $\Gamma_0 \sim 150s^{-1}$ was increased to $\Gamma_\sigma^{(cav)} \sim 8.10^4 s^{-1}$.

The first observation of spontaneous emission inhibition in the plane mirror cavity configuration was carried out at M.I.T. by Hulet, Hilfer and Kleppner [4] in 1985. The cavity was made of two plane aluminized mirrors separated by a d=0.23mm gap. Cesium atoms were prepared in a high angular momentum "circular" Rydberg state [11] with principal quantum number n=22, before entering in the cavity. The Rydberg electron was then in a plane, oriented parallel to the mirrors so that the electric dipole of the atomic transition was σ polarized. The atomic frequency (Rydberg n=22 → n=21 transition at ~ 663GHz) was tuned just below the cavity cut-off frequency $\omega_0 = c\pi/d$ by Stark effect induced in an external static electric field. The atoms were crossing at thermal velocity the 12cm long cavity in a time of about 0.5ms, i.e. of the order of the natural life time of this transition in free space (corresponding to $\Gamma_0 \sim 2000s^{-1}$). The absence of excited state decay during that time (monitored by detecting the atomic state after the cavity crossing) was the evidence for a nearly complete suppression of spontaneous emission in the cavity.

Rydberg atoms are very convenient for these cavity Q.E.D. experiments because their very weakly bound electron spontaneously emits long wavelength radiation. The "geonium" system, weakly bound state of an isolated electron in a Penning trap constitutes another choice system for such studies. By measuring the damping of the cyclotron motion of an electron in such a trap, Gabrielse and Dehmelt at the University of Washington [3] have observed that for some values of the magnetic field the cyclotron damping occured slower than in free space, whereas it was faster for other magnetic field values... In this experiment, the trap electrodes themselves made a cavity with eigenfrequencies close to the cyclotron frequency (~ 164GHz corresponding to $\lambda \sim 2mm$). By slightly changing the field strength, the electron frequency was swept across the cavity frequencies, inducing either cavity enhancement or inhibition of the spontaneous emission rate. In this experiment, the free space rate $\Gamma_0 \sim 12s^{-1}$ was increased to about $30s^{-1}$ or decreased down to $3s^{-1}$. This observation, made in 1985, preceeded the MIT group demonstration by a few months, but its quantitative analysis is somewhat more difficult because of the complex cavity geometry.

Suppression of spontaneous emission at optical frequencies

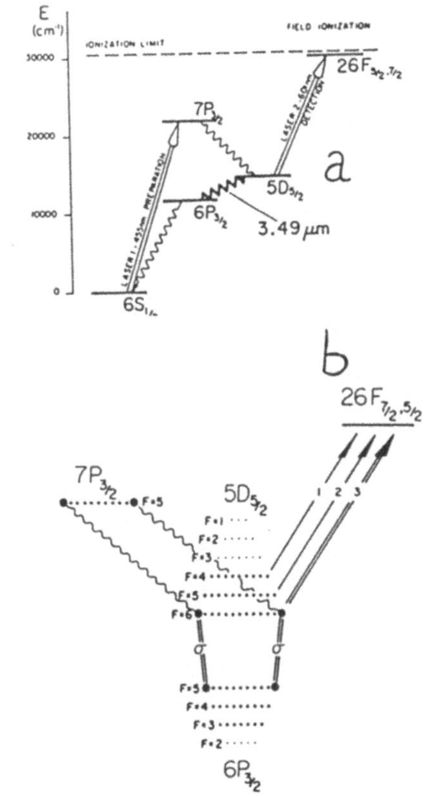

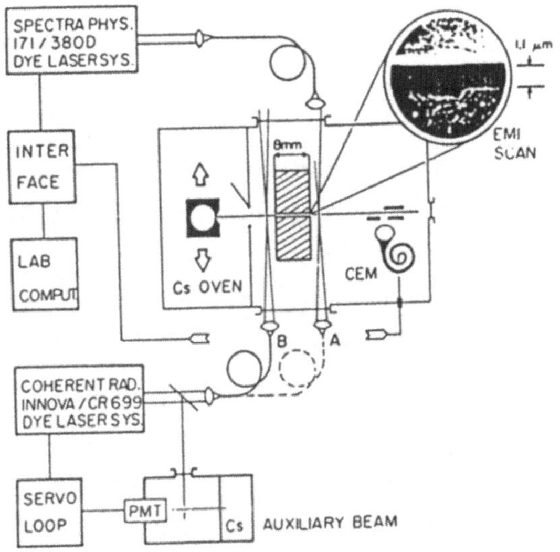

Fig. 2 : a) Cesium energy levels relevant to the optical spontaneous emission inhibition experiment.
b) Close-up showing the hyperfine structure of the $5D_{5/2}$ and $6P_{3/2}$ states

Fig. 3 : Experimental set-up of the optical spontaneous emission inhibition experiment. The insert shows a scanning electron microscope picture of the 1.1μm mirror cavity exit.

In all these microwave Q.E.D. experiments, the spontaneous rates to be modified by the cavity were quite small (10 to $10^3 s^{-1}$ range), the quantum noise being indeed very weak in this frequency domain. In order to demonstrate the prospects of cavity Q.E.D. for an effective quantum noise suppression, it was important to extend these experiments to much higher frequencies, up to the optical domain where spontaneous decay is a much stronger effect, becoming a real limitation to spectroscopic resolution. The problem is then to realize micron-sized cavities and to confine excited atoms in these structures during a time exceeding several natural life times. Such an experiment has very recently been carried out at Yale University [5] on excited atoms sent between two plane parallel mirrors spaced by a 1.1μm gap.

The experiment has been performed on the $5D_{5/2}$ state of Cesium, which has a natural life time $\tau_0 = \Gamma_0^{-1} = 1.6\mu s$. The energy levels relevant for the experiment are shown on Figure 2a. The $5D_{5/2}$ level decays with a branching ratio 1 to the $6P_{3/2}$ level, emitting 3.49μm wavelength radiation which is cut-off in the micron-wide cavity. For the experiment analysis, it is important to describe the hyperfine structure of the levels (shown in Figure 2b). Due to the coupling between the electrons and the Cs nuclear spin I=7/2, the $5D_{5/2}$ level is split into 6 hyperfine levels (F = 1 to 6) and the final $6P_{3/2}$ state into 4 levels (F' = 2 to 5). Each of these levels is made of 2F+1 magnetic sublevels $|F,M_F\rangle$, $|F',M_{F'}\rangle$, which are eigenstates of the total angular momentum projection F_z along the quantization axis (chosen as the normal to the mirrors). The radiation rate Γ_{F,M_F} of each $|F,M_F\rangle$ sublevel of the $5D_{5/2}$ state can be divided into a σ and a π contribution, corresponding respectively to the emission of photons polarized parallel and perpendicular to the mirrors surface :

$$\Gamma_{F,M_F} = \Gamma^{(\sigma)}_{F,M_F} + \Gamma^{(\pi)}_{F,M_F} \qquad (2)$$

These contributions are associated respectively to $\Delta M_F = \pm 1$ and $\Delta M_F = 0$ transitions to the final $6P_{3/2}$ $F'M_{F'}$ states (σ and π photons carry respectively 1 and 0 units of angular momentum along the quantization axis). In free space (or in a large sized cavity), the total emission rate is the same for all substates (the vacuum fluctuations are then isotropic : $\Gamma_{FM_F}^{(\text{free space})} = \Gamma_0$). Between two ideal mirrors with a spacing $d < \lambda/2$, we expect on the other hand :

$$\Gamma^{(cav,\sigma)}_{F,M_F} = 0 \ ; \quad \Gamma^{(cav,\pi)}_{F,M_F} = \frac{3\lambda}{4d}\,\Gamma^{(\pi)}_{F,M_F} \qquad (3)$$

(In our d=1.1μm wide cavity, $3\lambda/4d = 2.38$). Each $|F,M_F\rangle$ substate is thus expected to have a modified emission rate :

$$\Gamma^{(cav)}_{F,M_F} = \frac{3\lambda}{4d}\,\Gamma^{(\pi)}_{F,M_F} \qquad (4)$$

which depending upon the ratio of the π to σ emission channels for this particular state can be either smaller or larger than Γ_0. Of special interest are the F = 6, M_F = ±6 maximum angular momentum states which decay only via σ transition (see Figure 2b). The radiation rate for these states is $\Gamma_{6,\pm6}^{(cav)}$ = 0. In general, the radiation rates can be readily determined from equation (4) and standard angular momentum algebra yielding $\Gamma_{FM_F}^{(\pi)}$ for each sublevel. We find :

$$\Gamma^{(cav)}_{6,M_F} = \frac{3\lambda}{4d}\cdot\frac{36-M_F^2}{66}\cdot\Gamma_o\ ;\quad \Gamma^{(cav)}_{5,M_F} = \frac{3\lambda}{4d}\cdot\frac{60-M_F^2}{150}\,\Gamma_o$$

$$\Gamma^{(cav)}_{4,M_F} = \frac{3\lambda}{4d}\,\frac{1090+M_F^2}{3850}\,\Gamma_o \qquad (5)$$

The experiment being described in details elsewhere [5], I will recall here only its main features. The set-up is sketched on Figure 3. Cesium atoms are produced in an atomic beam by an oven. They are sent through a 1.1μm wide 8mm long cavity made by two flat gold coated blocks stacked against each other with thin Nickel foil spacers between them. The $5D_{5/2}$ level preparation is achieved by a c.w. laser beam (laser n°1) tuned to the $6S_{1/2}$ (F=4) - $7P_{3/2}$ (F=5) transition. About 13% of the atoms excited in this way are transferred by spontaneous cascade into the $5D_{5/2}$ level just before entering into the mirror gap. The crossing time of the gap lasts about 20μs, i.e. ~ 13 natural life times of the $5D_{5/2}$ state. In free space, only ~ 2 atoms in 10^6 would survive in their excited state this crossing. At the mirror exit, a second c.w. laser beam (laser n°2) excites the atoms remaining in the $5D_{5/2}$ level up to the 26F Rydberg state, which is subsequently field ionized, the resulting electrons being detected by a channeltron electron multiplier (CEM). The laser 2 frequency can be tuned across the hyperfine structure of the $5D_{5/2}$ level and the resulting Rydberg electron signal is recorded versus its frequency. The obtained spectrum thus yields the relative populations of the various $5D_{5/2}$ levels as they have survived the cavity crossing. This spectrum-recorded with laser 1 in position B upstream the mirror cavity is compared with the same spectrum obtained when laser 1 is moved in position A downstream, i.e. in superposition with laser 2. In this configuration, we detect Cs atoms which have crossed in the ground state the cavity and which have been excited into the $5D_{5/2}$ level and immediately detected. We thus get a normalization signal which provides the relative populations of the $5D_{5/2}$ F hyperfine levels as they are actually prepared by the $6S_{1/2} \rightarrow 7P_{3/2} \rightarrow 5D_{5/2}$ stepwise process.

Spectra with laser 2 respectively in position A and B are shown in Figure 4. Figure 4a shows that the excitation stage prepares atoms in hyperfine levels F = 4, 5 and 6 (arrows # 1, 2, 3). Of these only atoms in hyperfine level F = 6 have their spontaneous emission inhibited enough to survive the gap crossing and to yield a large absorption signal with laser 2 in position B (arrow # 3 in Figure 4b). From an analysis of these data, we conclude that the spontaneous emission process is essentially suppressed in the gap for the F=6, $M_F=\pm6$ substates (we have to correct for excited state decay before entering and after exiting the gap; more over only ~ 68% of the atoms in the F=6 level are aligned in the $M_F=\pm6$ sublevels). For all the other substates (F=6, $M_F \leq 5$ or F<6), the cavity modified life times remain short enough so that the atoms do not survive the gap crossing in their excited state.

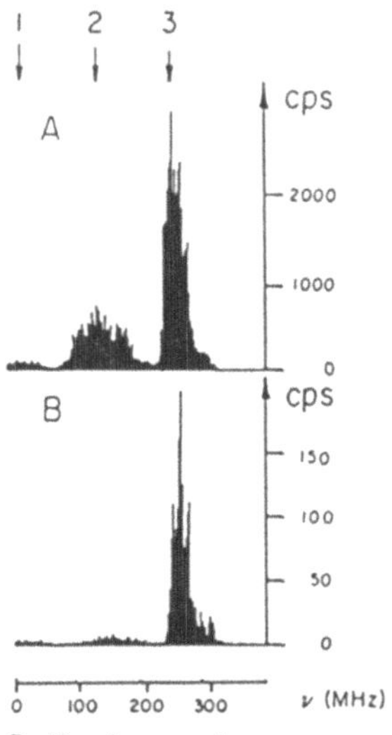

Fig. 4 : Optical spontaneous emission inhibition experiment : spectra of the $5D_{5/2} \rightarrow 26F$ transition recorded with laser 1 respectively in positions A and B (see Fig. 3). Recording b is evidence for suppression of spontaneous decay from the $5D_{5/2}$ level in the cavity.

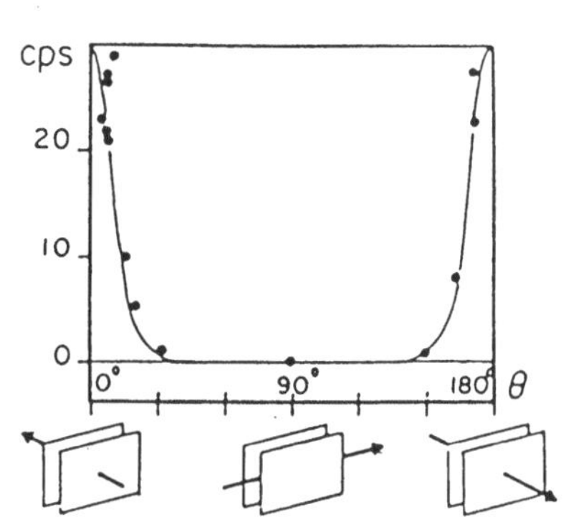

Fig. 5 : Excited state transmission through the tunnel versus the angle θ between the magnetic field and the normal to the mirrors

The magnetic dependence of the signal provides a clear evidence that we are actually observing an anisotropic vacuum field effect in the cavity. In order to observe the absorption peaks of Figure 4b, it was necessary to apply a magnetic field of about 2Gauss along the direction normal to the mirrors. This field prevented the $M_F = \pm 6$ states from being mixed with shorter lived $M_F < 6$ levels by the stray laboratory field having a non zero component along the cavity mirrors. The magnetic mixing effect is illustrated in Figure 5 which shows the excited atom transmission signal as a function of the angle θ between the directions of the applied magnetic field $\vec{B}$ and the normal to the mirror (the applied field being much larger than the stray lab-field). When $\vec{B}$ is along the mirror axis (θ=0 or $\theta \sim 180°$), a large transmission is observed. For $\vec{B}$ at an angle with the normal to the mirrors, the magnetic mixing becomes important and all the magnetic sublevels acquire a life time short enough that they do not survive anymore the mirror gap crossing. The solid line in Figure 5 corresponds to a calculation of the effective life time of the magnetically mixed levels as a function of θ (we use for this calculation the $\Gamma_{FM_F}{}^{(cav)}$ rates given by Equ.(5)). The good agreement between theory and experiment indicates that we are in effect measuring here the $\Gamma_{FM_F}{}^{(cav)}$ emission rates (for F=6, $|M_F| \leqslant 5$).

This experiment demonstrates that the quantum noise responsible for spontaneous emission can be suppressed up to optical frequencies provided the excited atom can be confined in micronsized metallic structure. It also shows that this suppression effect is anisotropic, reflecting the breaking of the vacuum isotropy by the plane parallel geometry of the mirrors. The main difficulty of this kind of experiment comes of course from the extreme collimation required for the atomic beam. The angle subtended by the mirror gap is only $\sim 10^{-4}$ radian and the Cesium oven has to be moved with precision until the atomic beam is perfectly aligned on the mirror gap axis [12]. Contamination of surfaces by the atomic beam and pressure build-up inside the gap are also sources of noise and experimental problems (the small residual contributions of atoms in levels F = 4 and 5 in the spectrum of Figure 4b probably comes from collisional transfers of atoms from the long lived $M_F = \pm 6$ states in the last mm before detection).

<u>Limitation to excited atomic state survival in a cavity : the Van der Waals interaction with the walls</u>

An interesting question we should ask at this stage is "how long can an excited atom be kept disconnected from vacuum fluctuations ?". The Yale experiment has achieved the largest life time lengthening ratio so far. Could this ratio be further increased and excited atoms propagated during hundreds of natural life times in micronsized structures ? First of all, the spontaneous emission inhibition is limited by finite cavity losses, allowing a small "residual" vacuum field to leak into the cavity beyond cut-off. In our experiment (gold surface reflexion coefficient ~ 96%), this effect corresponds to a minimum inhibited rate ~ $0.04\Gamma_0$. This limit could be improved by using better conducting cavities in which we could try to propagate excited atoms along longer pathes... Apart from pure geometrical collimation problems, we would then rapidly run into a fundamental limitation related to the Van der Waals interaction [13] of the atoms with the cavity walls. Excited as well as ground state Cesium atoms are pulled to a metallic surface at distance z by a Van der Waals force which, at close atom metal range, is proportional to z^{-4}. Only atoms exactly at mid-gap would in theory survive without falling on the mirrors, but this is a very unstable equilibrium situation... Fundamentally, the Van der Waals atom-metal attraction cannot be separated from the cavity induced radiative decay modification. We have considered above the resonant coupling of the atom with field modes having the frequency of the atomic transition (dissipative part of the atom-field interaction). The atomic system is also coupled to the non-resonant modes of the vacuum field (dispersive contribution). This coupling can be analyzed in terms of virtual photon emission and reabsorption processes and is responsible for radiative energy shifts. The cavity induced changes of the mode density also modify these processes (virtual photons -as real ones- can only be emitted in modes compatible with the cavity geometry). As a result, the atomic energy levels are altered in the cavity. These modifications depend upon the position of the atom since it can be coupled only to these modes which have a non-zero electric field amplitude at its location. The derivative of these energy level shifts with respect to z is nothing but the Van der Waals force mentioned above. The analysis of the Van der Waals interaction in term of Q.E.D. virtual processes in a cavity has been carried out in many theoretical papers [8][14], the first of which being a pioneering article by Casimir and Polder [15].

We present now a qualitative argument showing that the Van der Waals force usually restricts the excited state survival time in the cavity to a few tens of natural life times. In a biplanar cavity structure at cut-off ($d=\lambda/2$), the maximum distance an atom can be from mirrors is $z=\lambda/4$. The Van der Waals energy shift of an excited state in configuration $n\iota m$ is then of the order of

$$\Delta E\ (z \sim \frac{\lambda}{4}) \sim \frac{8q^2}{4\pi\epsilon_o} \frac{1}{\lambda^3} < x^2 + y^2 + 2z^2 >_{n\iota m} \qquad (6)$$

where q is the electron charge; x, y, z are the valence electron coordinates in the atom frame and $< \quad >_{n\iota m}$ denotes an average in the excited atomic state. Let us compare this shift to the spontaneous emission rate in free space for an atomic transition of wavelength λ between levels $|n\iota m>$ and $|n'\iota'm'>$:

$$\Gamma_o = \frac{q^2}{3\pi\epsilon_o\hbar} \frac{8\pi^3}{\lambda^3} |< n\iota m |\vec{r}| n'\iota'm' >|^2 \qquad (7)$$

The squared matrix element in Equ.(7) involves an overlap integral between different states and is usually about an order of magnitude smaller than the diagonal matrix element in Equ.(6). Keeping track of the various π factors, we thus get as a general order of magnitude :

$$\Delta E\ (z \sim \frac{\lambda}{4}) \sim \frac{\hbar\Gamma_o}{4} \qquad (8)$$

This result, independent of λ, shows that the cavity induced dispersive corrections are -around the midgap point- of the same order as the dissipative corrections (which completely cancel Γ_o). The Van der Waals force F_{VW} attracting the atom to the mirrors is the derivative of ΔE versus z and of the order of $3\Delta E/(\lambda/4) \sim 12\Delta E/\lambda$. Thus, around midgap :

$$F_{VW} \sim \frac{3\hbar\Gamma_o}{\lambda} \qquad (9)$$

This force corresponds to an acceleration $\gamma_{VW} \sim 3\hbar\Gamma_o/M\lambda$ (M = atom mass). The distance an atom falls towards the mirrors during a short time τ is :

$$\delta z \sim \frac{3\hbar\Gamma_o \tau^2}{2M\lambda} \qquad (10)$$

Since F_{VW} increases very quickly when z diminishes, it is legitimate to assume that the maximum time τ_{max} an atom will survive before colliding with a mirror is such that $\delta z\ (\tau_{max}) \sim \lambda/10$ or :

$$\tau_{max} \sim \left[\frac{M\lambda^2}{15\hbar\Gamma_o} \right]^{1/2} \qquad (11)$$

The maximum life time enhancement ratio is thus :

$$\frac{\tau_{max}}{\tau_o} \sim \left[\frac{M\lambda^2\Gamma_o}{15\hbar} \right]^{1/2} \qquad (12)$$

which can be rewritten in the more transparently homogeneous form as :

$$\frac{\tau_{max}}{\tau_o} \sim \left[\frac{Mc^2 . \hbar\Gamma_o}{\hbar^2\omega^2} \right]^{1/2} \qquad (13)$$

in which we have introduced the atom rest-mass energy (Mc^2), the atomic transition energy ($\hbar\omega$) and the excited state Heisenberg energy uncertainty ($\hbar\Gamma_o$). In our experiment [$Mc^2 \sim 1.2\ 10^{11}$eV; $\hbar\omega = 0.35$eV; $\hbar\Gamma_o \sim 3.9\ 10^{-10}$eV], we find $\tau_{max}/\tau_o \sim 20$ which is of the same order of magnitude than our observed enhancement factor (~ 13). Actually a more precise computation of the atomic Van der Waals trajectories in our 1.1μm wide mirror gap shows that 20% of the atoms only survive colliding with the walls. The excited state atomic transmission would completely vanish if our cavity length was extended beyond $\iota \sim 1$cm. Equ.(13) shows that τ_{max}/τ_o is of the order of the square root of the ratio between the natural width ($\hbar\Gamma_o$) and the photon recoil shift ($\hbar\omega^2/2Mc^2$) of the atomic transition. This ratio cannot be varied too much in the optical domain. (It slightly increases with M, which justifies the choice of a heavy element in this experiment).

The last question which remains to be addressed is whether these life time enhancement experiments would be interesting for pure spectroscopic or metrology applications. Would it be possible to make use of spontaneous emission inhibition to reduce the spectral line widths below their natural limit and achieve better ultimate resolution ? The answer is here again unfortunately given by the Van der Waals interaction. The long lived excited states are indeed energy shifted by an amount at least of the order of the natural width (or a non-negligible fraction of it: see Equ. (8)). This shift is furthermore quite inhomogeneous in the cavity so that it seems that any spectroscopic measurement will suffer a perturbation at least as large as the natural width suppressed in the cavity ! This is of course only a first order answer and one can imagine schemes in which one selects atoms travelling close to mid-gap, along trajectories where the cavity shifts could be minimized and controlled. It is worth noting that similar frequency shifts are expected in the Penning trap electron experiment. The precision of this experiment aiming at remeasuring g−2 will soon be such that a good understanding of these shifts will become essential for its data analysis [16].

In conclusion, we can say that by suppressing photon noise at the atomic frequency, we have been unable to avoid perturbing the atom by the dispersive part of the atom-metal interaction. The system we are then studying is no longer the atom in vacuum (atom dressed by the vacuum modes), but the "atom + cavity" entity (atom dressed by the cavity perturbed vacuum). The impossible dream of decoupling the atom from the quantum field is only partially fulfilled...

References

[1] R.E. Slusher, L.W. Hollberg, B. Yurke, J.C. Mertz and J.F. Valley, Phys. Rev. Lett. 55, 2409 (1985); L.A. Wu, H.J. Kimble, J.L. Hall and H. Wu, Phys. Rev. Lett. 57, 2520 (1986)

[2] P. Goy, J.M. Raimond, M. Gross and S. Haroche, Phys. Rev. Lett. 50, 1903 (1983)

[3] G. Gabrielse and H. Dehmelt, Phys. Rev. Lett. 55, 67 (1985)

[4] R.G. Hulet, E.S. Hilfer and D. Kleppner, Phys. Rev. Lett. 55, 2137 (1985)

[5] W. Jhe, A. Anderson, E.A. Hinds, D. Meschede, L. Moi and S. Haroche, Phys. Rev. Lett. 58, 666 (1987)

[6] E.M. Purcell, Phys. Rev. 69, 681 (1946)

[7] D. Kleppner, Phys. Rev. Lett. 47, 233 (1981)

[8] G. Barton, Proc. Roy. Soc. London A320, 251 (1970); M.R. Philpott, Chem. Phys. Lett. 19, 435 (1973)

[9] P.W. Milonni and P.L. Knight, Opt. Comm. 9, 119 (1973)

[10] K.H. Drexhage, in Progress in Optics XII, E. Wolf ed., North Holland 1974; see also F. DeMartini and G. Innocenti (in Quantum Optics IV, J.D. Harvey and D.F. Walls ed., Springer Verlag 1986) who have recently performed experiments on liquid dyes excited between mirrors.

[11] R.G. Hulet and D. Kleppner, Phys. Rev. Lett. 51, 1430 (1983)

[12] A. Anderson, S. Haroche, E.A. Hinds, W. Jhe, D. Meschede and L. Moi, Phys. Rev. A 34, 3513 (1986)

[13] J.E. Lennard-Jones, Trans. Farad. Soc. 28, 336 (1932)

[14] C. Lütken and M. Raindal, Phys. Rev. A31, 2082 (1985); Phys. Scripta, 28, 209 (1983)

[15] H.B. Casimir and D. Polder, Phys. Rev. 73, 360 (1948)

[16] L.S. Brown, G. Gabrielse, K. Helmerson and J. Tan, Phys. Rev. Lett. 55, 44 (1985).

THE MICROMASER AS A PROBLEM IN "QUANTUM CHAOLOGY"

T. A. B. Kennedy, P. Meystre and E. M. Wright
Optical Sciences Center
University of Arizona
Tucson, AZ 85721

1. INTRODUCTION

We consider a micromaser[1,2] consisting of a single mode, high-Q microwave cavity in which two-level atoms are injected at such a low rate that at most one atom at a time is present inside the cavity. In recent work, we have shown that its *semi-classical* version typically can lead to instabilities and chaos.[3] In contrast, general theorems indicate that its *fully quantized* version always evolves towards a unique steady-state.[4] These results confront us with an interesting paradox, since one would like to believe that quantum mechanics contains classical and semi-classical physics as limits. This, of course is the central question of quantum chaos.

The general approach to "quantum chaology" involves the detailed analysis of model systems which are believed to be representative of the "generic" case, yet are simple enough to be handled theoretically with some rigor. Here, as in classical chaos, one distinguishes between conservative and dissipative systems. Classically, the former are described by the Kolmogorov-Arnold-Moser theorem, and the central idea is that of invariant tori, while strange attractors are the signature of chaotic dissipative systems. Most theoretical work on quantum chaos deals with conservatives systems, a notable exception being offered by the work of Graham and coworkers.[5] Because in any experiment some element of loss is always present, we feel that the analysis of weakly dissipative systems deserves much more attention than is presently the case.

What makes the micromaser attractive for chaos studies is that it can actually be built in the laboratory, and that the experimental parameters are under exceedingly good control. In particular, the atom-field interaction time and cavity damping rate can be varied almost at will. Also, despite the fact that the intracavity field can not (yet) be measured directly, the dynamics of the system can be monitored by studying the state of the successive atoms as they escape the resonator, and this with almost unit quantum efficiency. We thus feel that this system is ideally suited to study the elusive "quantum chaos" both theoretically and experimentally.

The rest of this paper is organized as follows. Section 2 briefly reviews the quantum mechanical description of the micromaser. Section 3 presents a semi-classical version of the system and gives the evolution of the intracavity field in terms of a return map. In the presence of cavity damping the semi-classical micromaser exhibits a number of coexisting fixed points, whose basins of attraction are intricately entwined. We use simple pre-image arguments[6] to show that there are domains of initial conditions where these basins of attraction are fractals. Graham *et al.* have studied quantized versions of dissipative maps whose classical versions exhibit chaos[5] and found that quantum mechanical fluctuations and tunneling between different attractors lead, at least in the semi-classical limit $\hbar \to 0$, to the reduction of the quantum mechanical maps to classical maps with additional noise. Because we are dealing with spin-1/2 systems, the semiclassical limit $\hbar \to 0$ is meaningless in the micromaser. Still, these results suggest that something interesting might happen when noise is added to the semi-classical micromaser. Section 4 discusses recent preliminary results along these lines. They show that in the micromaser, an average of the semiclassical trajectories over a large random set of initial conditions reproduces at least some of the quantum mechanical features of the micromaser. This indicates that somehow, the quantum-mechanical dynamics forces the micromaser to visit all classical attractors. Finally, Section 5 is a summary and conclusion.

2. REVIEW OF THE QUANTUM-MECHANICAL DESCRIPTION

In this Section, we briefly review the quantum-mechanical description of the micromaser. Details can be found in Ref. 2. We consider a single-mode, low-loss resonator into which excited two-level atoms are injected at a rate low enough that at most one atom at a time is inside the resonator. The atom-field interaction time t_{int} is much shorter than the cavity damping time γ^{-1}, so that the relaxation of the resonator field mode can be ignored while an atom is inside the cavity, the coupled field-atom system being simply described by the Jaynes-Cummings Hamiltonian.[6] During the intervals between successive atoms the evolution of the field is governed by the master equation for a harmonic oscillator interacting with a thermal bath.

Under these conditions, the reduced density matrix ρ_f for the cavity field alone at the time t_{i+1} when the $(i+1)$th atom is injected inside the cavity is given by the return map

$$\rho_f(t_{i+1}) = \exp(Lt_p)\, F(t_{int})\rho_f(t_i)\ , \tag{1}$$

where $t_p = t_{i+1}-t_i-t_{int} \cong t_{i+1}-t_i$ is the time interval between atom i leaving the resonator and atom $i+1$ entering it, t_{int} is the (constant) interaction time between an atom and the cavity field, and $F(t_{int})$ is defined through

$$\rho_f(t_i+t_{int}) = Tr_a[U(t_{int})\rho(t_i)U^{\dagger}(t)] \equiv F(t_{int})\rho_f(t_i)\ \ , \tag{2}$$

where $\rho(t)$ is the atom-field density matrix, $U(t)$ is the Jaynes-Cummings unitary evolution matrix, $L(t)$ the Liouvillian of a damped harmonic oscillator, and Tr_a stands for trace over the atomic variables. Successive iterations of the return *map* (2) eventually yield a steady-state field density matrix $\rho_{f,st}$ which is the solution of this equation with $\rho_f(t_{i+1}) = \rho_f(t_i)$.

We consider the case where the field density matrix is initially diagonal in the number state representation, and atoms without initial coherence are injected inside the resonator. We also assume that the atoms enter the cavity according to a Poisson process with mean spacing $1/R$ between events, where R is the atomic flux. This allows us to obtain[2] a closed form solution for the steady-state photon statisics $\bar{p}_n$

$$\bar{p}_n = C\left[\frac{n_b}{1+n_b}\right]^n \prod_{k=1}^{n}\left[1+\frac{N_{ex}}{n_b}\frac{\beta_k}{k}\right], \tag{3}$$

where at resonance $\beta_n = \sin^2\left[\kappa/2\sqrt{n\, t_{int}}\right]$ and $N_{ex} = R/\gamma$, γ being the cavity damping rate, κ the atom-field coupling constant and n_b a measure of the resonator temperature. N_{ex} can be interpreted as the mean number of atoms transiting the cavity during its damping time γ^{-1}. We have checked the resulting distribution against an exact numerical solution *of* (2), and found no significant difference in the context of the present paper. We will return to the photon *statistics* (3) in Section 4.

3. REVIEW OF THE SEMICLASSICAL DESCRIPTION

There is no unique way to define a semiclassical limit. It is often interpreted as the limit $\hbar \to 0$. But this limit has no meaning in spin-1/2 systems, and we use instead the conventional[9] quantum optics semiclassical limit obtained by factorizing the full quantum mechanical Heisenberg equations of motion. As shown in Ref. 8 for a lossless case and generalized in Ref. 3 for the case of a micromaser with weak cavity damping, this procedure leads to the return map for the field $\mathcal{E}_{n+1}$ at the time of injection of atom $(n+1)$:

$$\mathcal{E}_{n+1} = \alpha\mathcal{F}(\mathcal{E}_n) \equiv \mathcal{D}(\mathcal{E}_n)\,, \tag{4}$$

where the attenuation coefficient α is

$$\alpha = \exp(-\gamma\tau_0/2)\,, \tag{5}$$

τ_0 is the (constant) time between injection of successive atoms, and $\mathcal{F}$ is given implicitely by

$$\int_{\mathcal{E}_n}^{\mathcal{F}(\mathcal{E}_n)} d\mathcal{E}\left\{1 - \left[\frac{\mathcal{E}_n^{\,2}-\mathcal{E}^2}{2} + \eta\right]^2\right\}^{-1/2} = \eta\tau_{int}\,. \tag{6}$$

Here η is a parameter indicating the state of the injected atoms, $\eta = 1$ for inverted atoms and $\eta = -1$ for ground state atoms, and τ_{int} is the atom-field interaction time, all times being in dimensionless units.[3]

In some of the numerical work, we have found it useful to reexpress (6) explicitly in terms of elliptic functions as

$$\mathcal{E}_{n+1} = \alpha \mathcal{E}_n \sqrt{\frac{1 + \mathcal{E}_n^2/4}{1 + \mathcal{E}_n^2/4 - sn^2(\tau_n, K_n)}} \,, \tag{7}$$

where

$$\tau_n = \tau_{int} \sqrt{1 + \mathcal{E}_n^2/4} \tag{8}$$

and

$$K_n = 1/\sqrt{1 + \mathcal{E}_n^2/4} \tag{9}$$

An example of the return map $\mathcal{D}$ is shown in Fig. 1.

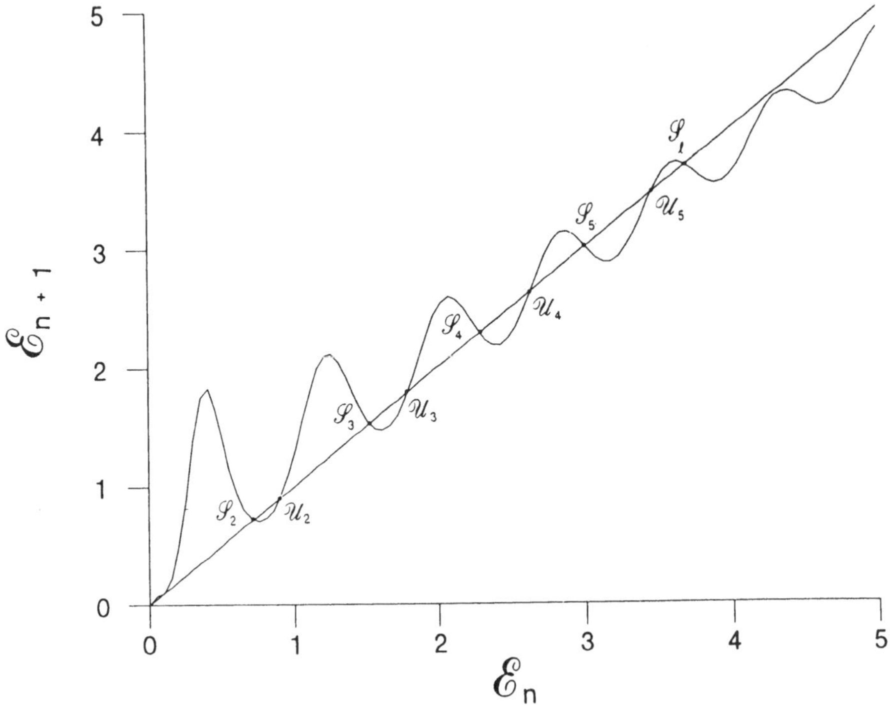

Figure 1: Return map $\mathcal{E}_{n+1} = \mathcal{D}(\mathcal{E}_n)$ for initially inverted atoms, $\tau_{int} = 9$, and $\alpha = 0.9$. $\mathcal{U}_i$ label the fixed points that are unconditionally unstable and $\mathcal{S}_i$ the conditionally stable fixed points of the map.

The map $\mathscr{D}(\mathscr{E})$ is not invertible, and its various fixed points have tightly interwoven basins of attraction. These can be found by seeking the successive preiterates of some domain of the map whose subsequent (forward) images are well understood. Grebogi *et al.* have used such techniques[7] to study multi-dimensioned interwined basin boundaries in a number of maps. They found that these boundaries can have different properties in different regions, and that these regions can be intertwined on an arbitrarily fine scale.

Consider for instance the map illustrated in Fig. 2, which resembles the map of the semiclassical micromaser. All initial conditions in the interval $[A,B[$ belong to the basin of attraction of S_1, and all points in $]C,D[$ certainly escape past $\mathscr{U}_2$. Constructing the successive preimages of these domains allows to determine their respective basins of attraction. This immediately leads to the realization that the basins of attraction of the micromaser are finely intertwined, and that they are Cantor sets at least in some domains. We have not yet determined the dimensions on the basins of attraction, nor have we tested the conjecture[7] that "basin boundaries have at most a finite number of possible dimension values."

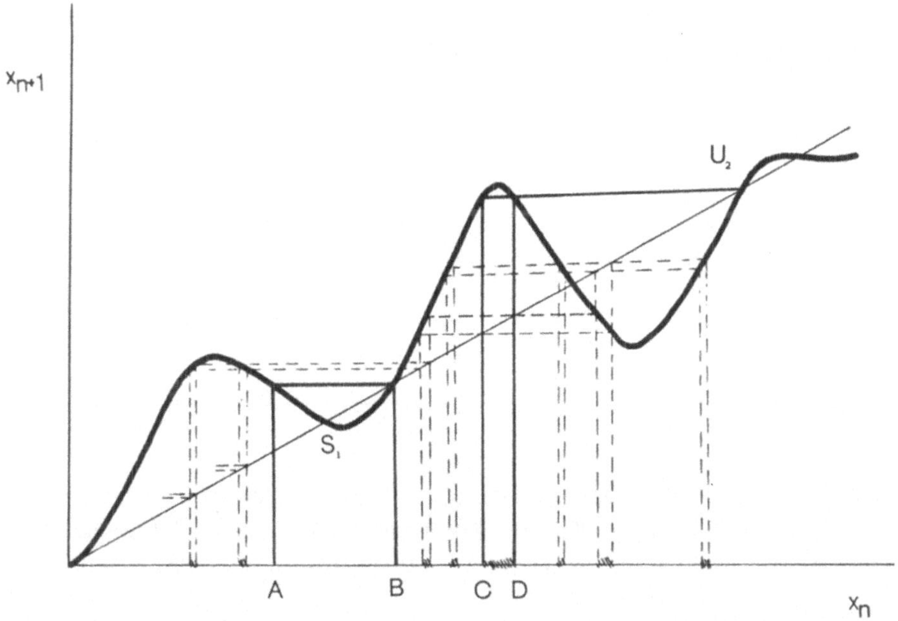

Figure 2: Map illustrating the determination of the basins of attraction of various fixed points. The shaded regions represent the initial conditions leading to an escape past U_2.

4. SEMICLASSICAL VERSUS QUANTUM MICROMASER

The semiclassical micromaser is generally chaotic and exhibits a number of coexisting and tightly intertwined basins of attraction. The quantum mechanical micromaser, on the other hand, always evolves towards a unique steady state. One would like to believe that the latter includes the former one as a limit, and yet, it appears that the semiclassical description allows for a much richer dynamical structure than allowed by quantum mechanics. Little is known on the quantum/classical correspondence in chaotic *dissipative* systems. We already mentioned the work of Graham and coworkers, who studied quantized versions of dissipative maps whose classical versions exhibit chaos[15], and found that quantum mechanical fluctuations and tunneling between different attractors lead, at least in the semi-classical limit $\hbar \to 0$, to the reduction of the quantum mechanical maps to classical maps with additional noise. Unfortunately the limit $\hbar \to 0$ has no meaning in the micromaser, which couples spin-1/2 systems to a boson field.

Although it is not clear if and how Graham's results can be formally related to the problem at end, the structure of the classical basins of attraction of the micromaser suggests that quantum fluctuations might be influential in forcing the system to jump from one of them to the other. Hence the addition of noise to the semiclassical problem might make it "look more quantum mechanical". To test this idea, we have performed preliminary simulations where instead of adding noise to the classical map, a precedure that is very computer intensive, we average the classical results over a random set of initial conditions.

The results of such a simulation are summarized in Fig. 3, along with the results of the quantum mechanical results reproduced from Ref.2. We find a striking qualitative agreement between the averaged semiclassical results and the quantum ones. The averaged semiclassical system now always reaches a steady state with $\langle \mathcal{E}_\infty \rangle = 0$ and an average intensity $\langle \mathcal{E}_\infty^2/4 \rangle(\Theta)$ (which corresponds to the

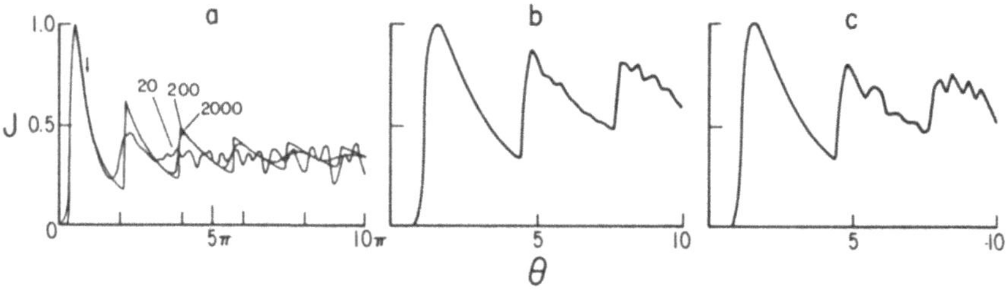

Figure 3: (*a*) Quantum mechanical normalized average photon number $I = \langle n \rangle / N_{ex}$ as a function of Θ for $N_{ex} = 20,\ 200,\ 2000$ (from Ref.2); (*b*) Averaged normalized semiclassical intensity $I = \langle \mathcal{E}_\infty^2 \rangle / 4N_{ex}$ for $N_{ex} = 50$; (*c*) Same for $N_{ex} = 100$.

quantum mechanical $\langle n \rangle$) essentially independent of N_{ex} for N_{ex} sufficiently large (see Figs. 3b and 3c).

This result is very unexpected: The scale parameters of the quantum mechanical problem are

$$N_{ex} = R/\gamma \tag{10}$$

and

$$\Theta = \sqrt{N_{ex}}\, \kappa t_{int}/2 \ , \tag{11}$$

while those of the semiclassical map are

$$\tau_{int} = \kappa t_{int}/4 = \Theta/\sqrt{N_{ex}} \tag{12}$$

and

$$\alpha = \exp(-\gamma t_0/2) = \exp(-1/2N_{ex}) \ . \tag{13}$$

It is not at all intuitively obvious that when averaged over initial conditions, the (α, τ_{int})-scaling of the classical map should turn essentially into a Θ-dependence only. In particular, the damping coefficient α, which is an *explicit* order parameter of the classical problem, and whose variation can produce[2] full Feigenbaum sequences, ceases to be important!

Just as striking as this result is the excellent qualitative agreement between the semiclassical dependence of $\mathit{l} = \langle \mathcal{E}_\infty^2 \rangle / 4N_{ex}$ on Θ and the corresponding quantum result for $\mathit{l} = \langle n \rangle / N_{ex}$. In particular, we recover the subsequent thresholds past the conventional maser threshold.[2] However, their position is slightly shifted to lower values of Θ. We do not know yet if this is an artifact resulting from our rather simplistic choice of a random set of initial conditions. Note that the rather "noisy" chatacter of $\mathit{l}(\Theta)$ is a numerical artifact due to the limited number of trajectories averaged over. This is evidenced in a comparison of Figs. 3b and 3c, which are the results of an average over 200 and 100 trajectories, respectively. In its present form, our program is not very efficient, a typical run taking several minutes on a supercomputer. Also, the convergence of the iterations of the classical map decreases drastically with increasing N_{ex}. Improving the program will certainly allow for better statistical averages and more complete numerical experiments, in particular on the steady-state intensity distribution $P(\mathcal{E}_\infty^2, \Theta)$.

5. CONCLUSIONS

The comparison between semiclassical and quantum descriptions of the micromaser is obviously far from complete. At this point, we do not have an explanation for the reported results. When we started this work, we merely wanted to convince ourselves once more that spin-1/2 systems are not good candidates to study the quantum-classical correspondence in situations exhibiting classical dynamic instabilities. Clearly, our results force us to revise at least temporarily this view, and open up more questions than they answer. Are they accidental or generic ? Can we recover more than just the average quantum mechanical average energy? What is the physical origin of the agreement, and in particular, what is the role of dissipation? What are its implications?

The qualitative agreement that we found might be just a coincidence, but then, it would have to be a rather remarkable one, specially since the quantum mechanical function $\mathit{l}(\Theta)$ is far from trivial. Whether the interpretation of our results will wind up being rather obvious or having any fundamental relevance remains to be seen.

ACKNOWLEDGMENTS

We have enjoyed numerous helpful discussions with Craig Savage. The comments by Robert Graham and Fritz Haake during a recent meeting in Como have also played an instrumental role in this work. This research is supported in part by the National Science Foundation Grant PHY-8603368 and by the Joint Services Optics Program.

REFERENCES

1. D. Meschede, H. Walther and G. Müller, Phys. Rev. Lett. **54**, 551 (1985); G. Rempe, H. Walther and N. Klein, Phys. Rev. Lett. **58**, 353 (1987).

2. P. Filipowicz, J. Javanainen, and P. Meystre, Optics Commun. **58**, 327 (1986); P. Filipowicz, J. Javanainen, and P. Meystre, Phys. Rev. **A34**, 3077 (1986).

3. P. Meystre and E. M. Wright, "Chaos in the Micromaser", to be published in the Proceedings of the ONR Workshop *Fractals and Chaos*, Como Italy 1986, Ed. by E. R. Pike (Adam Hilger, Bristol 1987).

4. H. Haken, *Synergetics: An Introduction* (Springer Verlag, Berlin 1978).

5. R. Graham, Z. Phys. **B59)**, 75 (1985); R. Graham and T. Tel, Z. Phys. **B60**, 127 (1985); T. Dittrich and R. Graham, Z. Phys. **B62**, 515 (1986).

6. E. T. Jaynes and F. W. Cummings, Proc. IEEE **51**, 89 (1963).

7. C. Grebogi, E. Kostelich, E. Ott and J. A. Yorke, "Multi-dimensioned interwined basin boundaries: Basin structure of the kicked double rotor", to be published in Physica **D**.

8. P. Filipowicz, J. Javanainen, and P. Meystre, J. Opt. Soc. Am. **B3**, 906 (1986).

9. See e.g. L. Allen L. and J. H. Eberly *Optical Resonance and Two-Level Atoms* (Wiley, New York 1975).

RYDBERG ATOMS TWO-PHOTON MICROMASER

J.M. Raimond, L. Davidovich (*), M. Brune and S. Haroche
Laboratoire de Spectroscopie Hertzienne de l'ENS
(unité associée au CNRS U.A. 18)
24, rue Lhomond, F-75231 Paris Cedex 05, France

(*) on sabbatical leave from Pontifica Universidade Católica
do Rio de Janeiro

Abstract

We show that the continuous-wave oscillation of a two-photon maser can be achieved with Rydberg atoms in a superconducting cavity. The maser should operate with only a few photons and a few atoms at a time in the cavity. Theoretical aspects of this new quantum device are presented. We describe briefly an experimental apparatus presently under construction in our Laboratory.

Lasers -or masers- operating on two-photon transitions between atomic levels of same parity have been the subject of a great number of theoretical papers in the last twenty years [1-8]. These lasers present interesting features, making them very different from "ordinary" one photon lasers. For instance, it has been pointed out that the field emitted by these new quantum devices might present interesting statistical properties (generation of "squeezed" states of light [5-6]).

Up to now, in spite of numerous attempts, there has been, to our knowledge, no realization of a continuous-wave two-photon oscillator. Only one report of two-photon amplification in a pulsed regime has been published so far [9]. This is due to the vanishingly small gain on a two-photon transition, at least for "ordinary" transitions between low-lying levels : two-photon amplification is masked by competing non-linear processes, such as multiple wave mixing or stimulated Raman effect.

Rydberg atoms are a very good tool for matter-field interaction experiments, because of their unusual properties. Among them, let us quote their very large coupling to radiation in the millimeter-wave range, where very high Q low order cavities are available. In the last few years, they made possible the realization of new quantum devices with fascinating properties, such as masers with a threshold down to one atom at a time in the cavity, operating on one photon transitions, either in a pulsed [10] or continuous-wave [11] regime. This "micro-maser" regime has been extensively studied theoretically [12-13].

We show in this paper that Rydberg atoms could also make possible the continuous-wave operation of a two-photon maser, in a regime where a few atoms and a few photons only are present, at a time, in the high Q superconducting cavity.

The outline of this paper is as follows : in the first section, we give an estimate of the two-photon maser threshold, and a simple semi-classical model of its dynamics. The second section will be devoted to a more realistic quantum description of the maser. The field density matrix master equation will allow us to investigate field statistics and phase diffusion. We describe briefly, in the last section, the experimental apparatus now under construction in our laboratory.

I. Semi-classical model

We evaluate first the order of magnitude of the threshold for a simple model of Rydberg atom two-photon maser.

Rydberg atoms are prepared by laser irradiation of an atomic beam in the upper level $|e\rangle$ of the two-photon transition. They then cross a resonant microwave cavity. The average atom-field interaction time is t_{int}. In the cavity, the atoms may undergo a transition to the lower level $|f\rangle$. We will estimate the minimum atomic flux $1/t_{at}$ required to sustain a non-vanishing field in the cavity (damping time $t_{cav} = Q/\omega$; Q : quality factor).

Typical levels configuration is depicted on fig. 1 : two-photon transition at frequency $\omega/2\pi$ occurs between $|e\rangle$ and $|f\rangle$ ($nS_{1/2}$ and $(n-1)S_{1/2}$ here). The transition amplitude is strongly enhanced by the occurence of a relay level $|i\rangle$ [$(n-1)P_{3/2}$ in fig. 1] close to the middle of the two-photon transition.

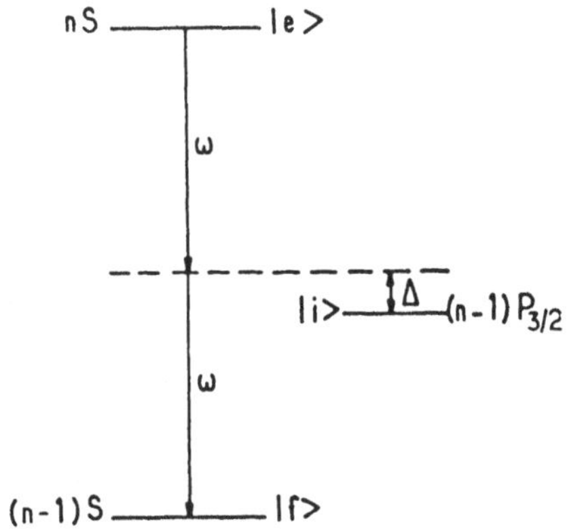

Fig. 1

An important quantity is thus the detuning Δ, given by :

$$\hbar\Delta = E_i - \frac{(E_e + E_f)}{2} \qquad (1)$$

In a cavity containing $\bar{N}$ photons, the atoms undergo a two-photon Rabi nutation at angular frequency $\Omega_2(\bar{N})$: the probability of finding the atom in state $|e\rangle$ a time t after its preparation in the same state is :

$$P_e(t) = \frac{1}{2}\left[1 + \cos \Omega_2(\bar{N})\, t\right] \qquad (2)$$

$\Omega_2(\bar{N})$ is approximately given by [14] :

$$\Omega_2(\bar{N}) = \frac{2\Omega_{ei}(\bar{N})\,\Omega_{if}(\bar{N})}{\Delta} \qquad (3)$$

where $\Omega_{ei}(\bar{N})$ and $\Omega_{if}(\bar{N})$ are the Rabi nutation angular frequencies for a field resonant on the $|e\rangle \rightarrow |i\rangle$ and $|i\rangle \rightarrow |f\rangle$ transitions respectively, with an energy corresponding to $\bar{N}$ photons :

$$\Omega_{ei} = \frac{D_{ei}\,\mathcal{E}_0}{\hbar}\sqrt{\bar{N}}\,; \quad \Omega_{if} = \frac{D_{if}\,\mathcal{E}_0}{\hbar}\sqrt{\bar{N}}$$

D_{ei} and D_{if} are the electric dipole matrix elements on these transitions, $\mathcal{E}_0 = \sqrt{\hbar\omega/2\epsilon_0 v}$ is the field per photon in the cavity (effective volume v). We assume here that atom-field coupling is constant. This is generally not the case when the atoms move in an actual cavity mode. However, this variation can be taken into account by a mere redefinition of $\mathcal{E}_0$.

Rydberg atoms present features which make $\Omega_2(N)$ unusually large : D_{ei} and D_{if} are huge and one can use, in the millimeter-wave range, a low order cavity having a large $\mathcal{E}_0$.

Moreover, the quantum defects of S and P levels differ by about 0.5 for all alkalis : the detuning Δ is thus rather small compared to ω. It is even exceedingly small for some selected transitions : the slow variation of quantum defects with n, measured in high-resolution spectroscopic experiments [15-18], entails that Δ crosses O around some n value. In Rubidium [17] and Cesium [16] spectra, this coïncidence occurs for easily excited levels with n ~ 40. As an example, $\Delta/2\pi$=-39MHz for the $40S_{1/2} \to 39S_{1/2}$ Rubidium transition at 2 x 68.416GHz ($39P_{3/2}$ relay level).

For this transition, we have D_{ei} = 1443 qa_0, D_{if} = 1479 qa_0. Assuming v=70mm^3 (cylindrical cavity in the TE_{121} mode, diameter D=7.8mm, length L=7.55mm), we get $\Omega_{ei} \sim \Omega_{if} \sim 7\ 10^7 \sqrt{\bar{N}}/s$ and finally

$$\Omega_2(\bar{N}) = B\,\bar{N} \quad \text{with } B = 4000\ s^{-1} \qquad (4)$$

The atoms experience thus a two-photon π pulse during t_{int} (30μs) in a field containing only 25 photons !

The maser threshold is easily obtained from these orders of magnitude. Each atom must leave a large part of its energy in the cavity : we must have

$$\Omega(\bar{N})\, t_{int} \sim \pi \qquad (5)$$

The $\bar{N}$ photon field is sustained by this energy income if

$$t_{at} \leqslant 2t_{cav} / \bar{N} \qquad (6)$$

The threshold condition is thus :

$$\frac{1}{t_{at}} \geqslant \frac{\pi}{2B\, t_{int}\, t_{cav}} \qquad (7)$$

If the cavity quality factor Q is close to the theoretical limit for a Niobium superconducting cavity at T = 2K (Q = 2 10^8), the threshold condition reads :

$$\frac{1}{t_{at}} \geqslant 3\ 10^4 \text{ at/s} \qquad (8)$$

and the cavity contains, at threshold, $\bar{N}$ = 30 photons only. This value can easily be obtained with a c.w. laser excitation. Moreover, the exceedingly low values of atomic flux (less than one atom at a time in the cavity) and of field intensity ensures that no competing process can overwhelm two-photon maser oscillation.

This model can be made more quantitative. Let us write the rate equation for the mean photon number $\bar{N}$:

$$\frac{d\bar{N}}{dt} = -\frac{\bar{N}}{t_{cav}} + \frac{2}{t_{at}}\left[\frac{1-\cos\Omega_2(\bar{N})\, t_{int}}{2}\right] \qquad (9)$$

The first term in r.h.s. of (9) describes the cavity losses at rate $1/t_{at}$, the second one the energy deposited by the atoms in the cavity.

The maser operating points are the intersections of the curves representing these loss and gain terms as a function of $\bar{N}$. Loss (straight line) and gain (sinusoïde) curves have been plotted on fig. 2 for four different values of t_{cav} and t_{at}.

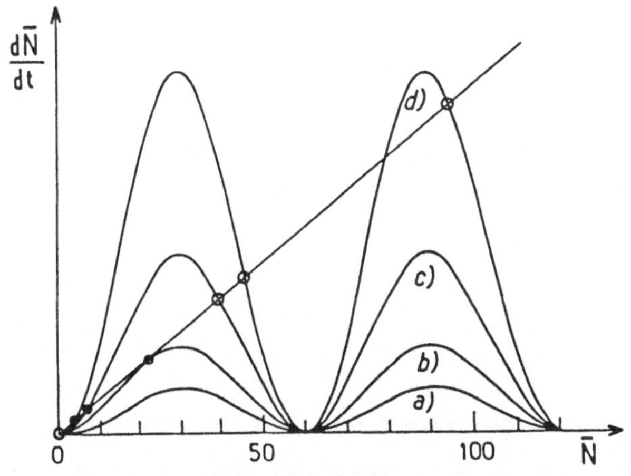

Fig. 2 : loss and gain terms for a function of $\bar{N}$ for a) $t_{cav} = 6t_{at}$, b) $t_{cav} = 12t_{at}$, c) $t_{cav} = 24t_{at}$, d) $t_{cav} = 48t_{at}$.

Stable operating points are open-circled. Even above threshold, there is a stable solution at $\bar{N}$=0 : in this semi-classical model, the two-photon maser does not start alone in an empty cavity. A large enough triggering field must be present initially to reach a large $\bar{N}$ operating point. This behaviour is very different from the one-photon maser one, which starts alone above threshold. At last, let us quote that eq. (9) admits, well above threshold, many stable solutions (see curve 2d).

This model is, of course, oversimplified. Many phenomena which might modify our conclusions are not taken into account, like spontaneous emission, blackbody field induced transitions, Maxwellian velocity spread, etc.

II. A simple quantum theory of the two-photon maser

We give here a simple quantum model of this maser, taking properly into account two-photon spontaneous emission. It will lead us to conclusions quite different from the semi-classical ones. The other effects mentioned above could be straightforwardly introduced in this model, but they make the algebra more complicated and do not modify drastically the system's behaviour [19].

We first derive a master equation for the field density matrix ρ, in a way reminiscent of the one used by Filipowicz et al [12-13] in the one-photon micromaser case.

We consider that, at most, one atom interacts with the cavity field at a time. Let t_i be the time when the ith atom enters the cavity. The field density matrix ρ at time t_{i+1} is expressed as a function of the one at time t_i by :

$$\rho\,(t_{i+1}) = e^{L\,t_{at}}\,F\,(t_{int})\,\rho\,(t_i) \tag{10}$$

$F\,(t_{int})$ is the operator describing the field change due to the interaction with a single atom. $\exp\,\{L\,t_{at}\}$ describes field relaxation during $t_{at} = t_{i+1} - t_i$. L is the well-known relaxation Liouville operator defined, in a T = 0K cavity, by :

$$L\rho = \frac{\omega}{2Q}\left[2a\rho a^{+} - a^{+}a\rho - \rho a^{+}a\right] \tag{11}$$

(a and a^+ are the photon annihilation and creation operators respectively).

In (10), we have taken into account separately the atom-field interaction and the field relaxation. This approximation is justified if both $e^{L\,t_{at}}$ and $F\,(t_{int})$ are close to unity, that is to say if $t_{cav} \gg t_{at}$, and if the field change due to the interaction with a single atom is small.

Let us now assume that t_{at} is a random variable, with a Poisson probability distribution :

$$P(t_{at}) = R\,e^{-R\,t_{at}} \qquad (12)$$

We take here into account the unavoidable pumping fluctuations. If the field change due to the interaction with a single atom is small enough, we can average (10) over a great number of interactions and write :

$$\rho(t_{i+1}) = \int_0^\infty d\,t_{at}\; Re^{-R\,t_{at}}\; e^{L\,t_{at}}\; F(t_{int})\,\rho(t_i) \qquad (13)$$

(this approximation is reminiscent of the coarse-grain averaging in standard laser theories [20]).

The integration in (13) can be performed, and we get :

$$\rho(t_{i+1}) = \frac{1}{1-L/R}\,F\,(t_{int})\,\rho\,(t_i) \qquad (14)$$

or else :

$$R\,[\rho(t_{i+1})-\rho(t_i)] = L\rho(t_{i+1}) + R\,[F(t_{int})\,\rho(t_i)-\rho(t_i)] \qquad (15)$$

As $t_{i+1} - t_i$ approaches zero (when compared to t_{cav}), one can replace the r.h.s of (15) by $\dot{\rho}$, and $L\rho(t_{i+1})$ by $L\rho(t_i)$. One thus gets a rate equation which turns out to be formally identical to the one of the Scully and Lamb model [10].

$F(t_{int})$ can be obtained from the study of two-photon Rabi oscillations. The calculation is straightforward in the dressed atom picture [21,22].

The master equation finally reads :

$$\dot{\rho}_{NM}(t) = -R\, \rho_{NM}(t)\, [1 - A(N,t_{int})\, A^*(M,t_{int})]$$

$$+ R\, \rho_{N-2,M-2}(t)\, B(N,t_{int})\, B^*(M,t_{int})$$

$$- \frac{\omega}{Q}(N+M)\rho_{NM}(t) + \frac{\omega}{Q}\sqrt{(N+1)(M+1)}\, \rho_{N+1,M+1}(t) \qquad (16)$$

with

$$A(N,t) = 1 + \frac{N+1}{2N+3}\left[e^{i\Omega(N)t} - 1\right] \qquad (17)$$

$$B(N,t) = \frac{\sqrt{N(N-1)}}{2N-1}\left[e^{i\Omega(N-2)t} - 1\right] \qquad (18)$$

The Rabi nutation angular frequency $\Omega(N)$ is given by :

$$\Omega(N) = \frac{\Omega_{ei}^2}{\Delta}\,(2N+3)$$

We assumed that $\Omega_{ei} = \Omega_{if}$. This is almost the case for the actual maser transition. Let us stress that $\Omega(0)$ does not vanish, because of two-photon spontaneous emission, and that $\Omega(N)$ and the semi-classical frequency $\Omega_2(N)$ coïncide for large N's.

The structure of (16) is very simple : the matrix elements ρ_{NM} are coupled only to $\rho_{N'M'}$ with $M-N = M'-N'$. We therefore get separate closed differential equations systems for the Fock states populations ρ_{NN} and for the coherences $\rho_{N,N-1}$ (only these elements will be used in the following discussion).

There are, at least, three ways to solve the equations set for the populations ρ_{NN}. First, it can be truncated at some N value much larger than the expected average photon number, and integrated numerically. On the other hand, when looking only for the steady state, one makes $\dot{\rho}_{NN}=0$. $\rho_{N+1,N+1}$ can therefore be expressed as a function of $\rho_{N,N}$ and $\rho_{N-2,N-2}$. One gets then recurrence equations allowing to calculate all the populations at equilibrium.

We show now that this master equation set can also be transformed into a Fokker-Planck equation valid for large N's.

First we introduce new notations : the reduced photon number $n = N/2N_{ex}$, where $N_{ex} = \omega/QR$ is the number of atoms crossing the cavity during its relaxation time, and $\delta = 1/2N_{ex}$. We will use 1/R as the time unit. The equation for ρ_{nn} now reads :

$$\dot{\rho}_{nn} = -\rho_{nn}\left[1-|\tilde{A}(n,\tau_{int})|^2\right] +$$

$$+ \rho_{n-2\delta,n-2\delta}\left[1-|\tilde{A}(n-2\delta,\tau_{int})|^2\right]$$

$$-2n\,\rho_{nn} + 2(n+\delta)\,\rho_{n+\delta,n+\delta} \qquad (19)$$

The reduced interaction time τ_{int} is given by :

$$\tau_{int} = 2N_{ex}\frac{\Omega_{ei}^2}{\Delta}t_{int} \qquad (20)$$

and

$$\tilde{A}(n,\tau_{int}) = A\left[2N_{ex}n,\ \frac{2N+3}{2N_{ex}}\ \frac{\tau_{int}}{\Omega(2N_{ex}n)}\right] \qquad (21)$$

We then consider ρ_{nn} as a function of a continuous variable n and expand (19) in powers of δ, a small parameter if $N_{ex} \gg 1$. We get :

$$\dot{\rho}_{nn} = -\delta\frac{\partial}{\partial n}\left[a_1(n)\rho_{nn}\right] + \frac{\delta^2}{2}\frac{\partial^2}{\partial n^2}\left[a_2(n)\rho_{nn}\right] + O(\delta^3) \qquad (22)$$

where

$$a_1(n) = 2\left[\sin^2 n\tau_{int} + \frac{3}{2}\delta\tau_{int}\sin 2n\tau_{int} - n\right] \qquad (23)$$

and

$$a_2(n) = 4\sin^2 n\tau_{int} + 2n \qquad (24)$$

Equation (22) is of the Fokker-Planck type. It is, of course, only an approximation to the original master equation, valid for $\delta \ll 1$ and $n \gg \delta$ (i.e. valid for high pumping rates N_{ex} and large fields).

The steady state solution of (22) is readily obtained as :

$$\rho^{S}_{nn} = \frac{C}{a_2(n)} \exp [- V(n)] \tag{25}$$

The effective potential V(n) being :

$$V(n) = - \frac{2}{\delta} \int_0^n \frac{a_1(n')}{a_2(n')} dn' \tag{26}$$

and C a normalization constant.

The behaviour of ρ^{s}_{nn} is thus mainly determined by V(n): ρ^{s}_{nn} is peaked around the absolute minimum of V(n). Let us stress that the extrema of V(n), corresponding to the zeroes of $a_1(n)$, coïncide with the steady state operating points in the classical model (unstable ones correspond to a maximum of V(n), stable ones to a minimum).

V(n) has been plotted, as a function of n, on fig. 3 for different τ_{int} values.

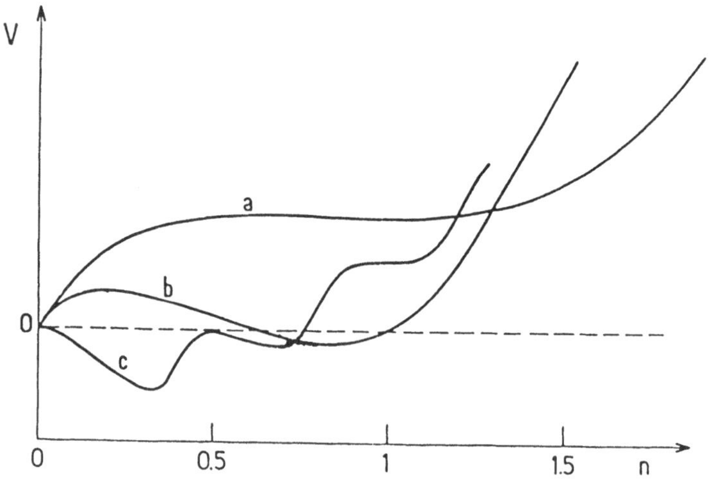

Fig. 3 : V(n) for $\tau_{int} = 0.9\ \pi/2$ (a), $\tau_{int} = 1.5\ \pi/2$ (b), $\tau_{int} = 5\ \pi/2$ (c)

The semi-classical model threshold value corresponds to $\tau_{int} \sim 0.9\ \pi/2$. For $\tau_{int} = 0.9\ \pi/2$, there is a minimum in $V(n)$ around $n = 1$, but its value is greater than $V(0) = 0$. ρ^s_{nn} is thus maximal for $n=0$, and the steady state mean photon number is close to zero (in fact, (25) and (26) cannot be used for $n \sim 0$: the approximations used to get the Fokker-Planck equation (22) are no longer valid in this case. However, direct integration of the master equation shows [19] that the conclusions obtained here are qualitatively correct).

The threshold condition is thus not the same as in the semi-classical model : a minimum of $V(n)$ for $n \neq 0$ must exist ($\tau_{int} \gtrsim 0.9\ \pi/2$, semi-classical threshold), but its value must be lower than zero ($\tau_{int} \gtrsim 1.4\ \pi/2$). Of course, this slight threshold modification does not change our conclusions on the maser's feasability. Let us stress at last that, for $2\ \pi/2 \lesssim \tau_{int} \lesssim 6.5\ \pi/2$, ρ^s_{nn} correspond to sub-Poissonian statistics for the field.

In the semi-classical model, the steady state solution depends on initial conditions (for instance on the presence of a triggering field). This is not the case here. To clarify this point, we consider qualitatively the time evolution of the system when ρ_{nn} is initially peaked around an arbitrary n_o value. Because of the drag term $a_1(n)$ in (22), the mean photon number $\bar{n}$ will reach in a short time a value close to the nearest local minimum of V (the time scale of this process turns out to be t_{cav}). Generally, this minimum is not the lowest one : the state reached is only metastable. Because of the fluctuations (term $a_2(n)$), ρ_{nn} will escape from this potential well and reach a more stable one. Standard techniques [23] allow us to determine the time scale of this process. Passage times are typically found around $10^2 t_{cav}$ (0.1s in an actual experiment).

In the case of a maser starting in an empty cavity ($n_o = 0$), for $t_{int} \gtrsim 1.4\ \pi/2$, a field will build up in the cavity with a typical time constant of this order of magnitude : maser triggering is no longer needed when spontaneous emission processes are included in the model. For $\tau_{int} \gtrsim 5\ \pi/2$, $n=0$ is even no longer a minimum of $V(n)$: for such high pumping rates, the maser starts alone with a time constant of the order ot t_{cav}.

The structure of the master equation for field coherences $\rho_{N,N-1}$ is very similar to the one for ρ_{NN}. It may also be expanded in powers of δ, and its solution reads, letting $g(n) = \rho_{N,N-1}$,

$$g(n,t) = g_S(n)\, e^{-\mu t} \qquad (27)$$

where g_S is the steady-state solution of a Fokker-Planck equation. μ reads :

$$\mu=\delta^2\left[\tau_{int}^2+\frac{1}{4n_S}\right]+i\delta\,\tau_{int}\left[1-\frac{\delta}{2n_S}\left[1-\frac{\sin 2n_S\tau_{int}}{2n_S\tau_{int}}\right]\right] \qquad (28)$$

where n_S is the mean steady state value of n, determined by (25).

The real part of μ describes phase diffusion, with a typical time constant $4N_{ex}T_{cav}$ (for $\tau_{int} \sim n \sim 1$). The term $\delta^2/4n_S$ in (28) corresponds to the usual phase diffusion rate due to dissipation in one photon lasers models [20].

The imaginary part of μ corresponds to a very weak frequency shift (of the order of $1/t_{cav}$). It can be interpreted as the effect of the refractive index of the atomic medium.

The same techniques can be applied to the rate equations for elements like $\rho_{n,n-2}$. It is thus possible to calculate the squeezing properties of the cavity field in this model. This discussion is out of the scope of this paper, and will be given elsewhere [19].

III. Outline of the experiment

We are presently building an experiment in order to realize a two-photon maser. We plan to use the $40S_{1/2} \rightarrow 39S_{1/2}$ transition in Rubidium. We take advantage of the weak detuning ($\Delta/2\pi$ = −39MHz), and of a convenient excitation scheme [24]. 40S levels are prepared by a stepwise continuous-wave laser excitation. Transitions $5S\text{-}5P_{3/2}$ (7802Å) and $5P_{3/2} \rightarrow 5D_{5/2}$ (7759Å) are induced by temperature tuned AlGaAs diode lasers. The transition from $5d_{5/2}$ to $40P_{3/2}$(1.26μ) is excited by a liquid nitrogen cooled InGaAsP diode laser. $40S_{1/2}$ is finally reached by a microwave transition (62GHz), easily saturated by a Yig 10GHz source, frequency multiplied in a non-linear high frequency diode [25]. In spite of the numerous steps involved, this excitation scheme is easily realized, since it involves only unexpensive and easy-to-use solid state sources. The atomic flux (10^7at/s) actually obtained is well above the theoretical maser threshold.

The most critical part in this experiment is the high Q microwave superconducting cavity at 68.416GHz [26]. It is precisely machined from high purity Niobium, chemically polished to remove the damage layer, electron beam welded and finally baked at high temperatures in ultra-high vacuum environment. These operations are performed at CERN (E.F. division), and the obtained Q is close to the theoretical B.C.S. limit.

The very precise (~ 1kHz) tuning to the atomic frequency will be performed by elastic deformation of the cavity's walls.

Maser operation will be detected indirectly by monitoring the populations of 40S and 39S levels at the exit from the cavity, using the well-known field ionization technique. This detection will also provide us with a clear-cut test of a true two-photon maser operation : the population of the relay level $39P_{3/2}$ should always remain close to zero.

References

[1] P.P. Sorokin, N. Braslau, IBM J. Research Develop. 8, 177 (1964)

[2] A.M. Prokhorov, Science, 149 (1965)

[3] V.S. Letokhov, JETP Lett. 7, 221 (1968)

[4] L.M. Narducci, W.W. Eidson, P. Furcinitti, P.C. Eteson, Phys. Rev. A 16, 1665 (1977)

[5] H.P. Yuen, Phys. Lett. 51A, 1 (1975)

[6] H.P. Yuen, Phys. Rev. A 13, 226 (1976)

[7] T. Hoshimiya, A. Yamagishi, N. Tanno, H. Inaba, Japanese J. Appl. Phys. 17, 2177 (1978)

[8] N. Nayak, B.K. Mohanty, Phys. Rev. A19, 1204 (1979)

[9] B. Nikolaus, D.Z. Zhang, P.E. Toschek, Phys. Rev. Lett. 47, 171 (1981)

[10] P. Goy, J.M. Raimond, M. Gross, S. Haroche, Phys. Rev. Lett. 50, 1903 (1983)

[11] D. Meschede, H. Walther, G. Müller, Phys. Rev. Lett. 54, 551 (1985)

[12] P. Filipowicz, J. Javanainen and P. Meystre, Phys. Rev. A34, 3077 (1986)

[13] P. Filipowicz, J. Javanainen, P. Meystre, Opt. Comm. 58, 327 (1986)

[14] D. Grischkowsky, M.M.T. Loy, P.F. Liao, Phys. Rev. A12, 2514 (1975)

[15] C. Fabre, S. Haroche, P. Goy, Phys. Rev. A22, 778 (1980)

[16] P. Goy, J.M. Raimond, G. Vitrant, S. Haroche, Phys. Rev. A26, 2733 (1982)

[17] D. Meschede, JOSA B, to be published

[18] J. Liang, M. Gross, P. Goy, S. Haroche, Phys. Rev. A 34, 2889 (1986)

[19] L. Davidovich, M. Brune, J.M. Raimond, to be published

[20] M. Sargent III, M.O. Scully, W.E. Lamb, "Laser Physics", Addison, Reading (1974)

[21] M. Brune, J.M. Raimond, S. Haroche, Phys. Rev. A 35, 154 (1987)

[22] H.I. Yoo, J.H. Eberly, Phys. Reports, 116, 239 (1985)

[23] N.G. Van Kampen, Stochastic Processes in Physics and Chemistry, North-Holland (1981)

[24] P. Filipowicz, P. Meystre, G. Rempe, H. Walther, Optica Acta 32, 1105 (1985)

[25] P. Goy, Int. J. of Infrared and mm-waves, 3, 221 (1962)

[26] P. Goy, M. Brune, J.M. Raimond, E. Chiaveri, J. Tuckmantel, to be published

PART V: Cooling and Trapping of Particles

PREPARATION OF COLD ATOMS FOR PRECISION MEASUREMENTS

K.Cloppenburg, G.Hennig, J.Nellessen and W.Ertmer
Institut für Angewandte Physik, Universität Bonn
Wegelerstr. 8, D-5300 Bonn 1

Introduction

Many high resolution spectroscopic experiments are ultimately limited by the random motion of atoms or velocity related effects. This motion causes e.g. transit time effects or second order Doppler shifts and broadening of spectral lines, which are not canceled by most of the so-called Doppler free spectroscopic techniques. Atomic collision experiments often suffer from the broad velocity distribution and the angle spread of the crossing beams too.

A thorough solution of these general problems would be a direct velocity manipulation of free atoms aiming at a reduced average velocity and a strong reduction of the temperature respectively the width of the resulting velocity distribution.
This velocity reduction can be achieved by the light pressure force from a resonant laser beam. In the basic scheme atoms absorb photons from a laser beam changing the atomic momentum by $\hbar\vec{k}$ ($\vec{k}$=wavevector). After a short time - typically 10 ns - the atoms spontaneously reemit a fluorescence photon. Because of the random direction of emission the momentum recoil from the emitted photons averages to zero whereas the atomic momentum change by the absorptions of the photon momenta adds up constructively to a velocity change of $n\cdot\hbar\vec{k}/M$ (M= mass of the atom, n = number of absorptions) on average [1].
The resulting averaged "spontaneous" force $\vec{f}$ (atomic momentum change per time) reads:

$$\vec{f} = \hbar\vec{k}\,\frac{1}{\tau}\,\frac{s}{1+2s} \qquad (1)$$

τ : natural lifetime of the "upper" cooling level
s: saturation parameter.

The expression $\tau\cdot(2+1/s)$ is the cycle time, which depends on the laser intensity and the detuning between the laser frequency and the Doppler shifted absorption frequency of the atoms. This force saturates with increasing saturation to $\vec{f}_{sat}$:

$$\vec{f}_{sat} = \hbar\vec{k}\,\frac{1}{2\tau} \tag{2}$$

The corresponding accelerations $\vec{a}$ and $\vec{a}_{sat}$ read

$$\begin{aligned} \vec{a} &= \frac{\hbar\vec{k}}{M}\,\frac{1}{\tau}\,\frac{s}{1+2s} \\ \vec{a}_{sat} &= \frac{\hbar\vec{k}}{M}\,\frac{1}{2\tau} \end{aligned} \tag{3}$$

For simplification this interaction of photons with atoms should take place within an atomic beam and not within a gas. In an atomic beam almost one degree of freedom only - the longitudinal velocity distribution - has to be cooled and the atoms move in a high vacuum environment collision free without heating by walls or other molecules.
The basic scheme of most cooling experiments consits of an atomic beam and a counterpropagating laser beam. The different experimental techniques differ mainly in the experimental solution of two fundamental problems arising from the neccesary large number of absorptions and the accompanying large change of Doppler shift.

Atomic beam cooling

When the atoms are decelerated by the successive absorptions of photon momentum from the counterpropagating laser beam, their Doppler shift changes very fast and thus the atoms run out of resonance after the accumulated shift of a few homogeneous line widths. On the other hand the stopping process needs very many photons (e.g.~20.000, if sodium atoms shall be stopped from an initial velocity of about 600 m/s, using the Na-D-line) and thus optical pumping has to be totally avoided or counteracted.
In the first successful cooling experiments, which produced really a slow atomic sodium beam [2], a longitudinal magnetic field with a decreasing field strength along the beam axis solved both problems. Using circularly polarized sodium-D_2 light the atoms are relatively fast optically pumped into the magnetic sub-level $^2S_{1/2}$ (F=2, m_F=2) of the ground state, when they enter the magnetic field. This state is only excitable to the $^2P_{3/2}$ (F=3, m_F=3) level by the circularly polarized cooling laser beam, forming an nearly ideal two-level system.

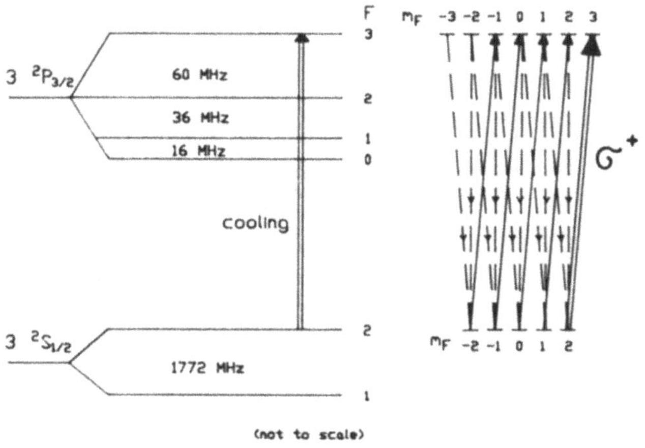

Fig.1: Level scheme of ^{23}Na with cooling transition

When the magnetic field strength is strong enough (e.g. by a bias field), it can compensate for imperfect polarization of the light field and will avoid resulting optical pumping. When the laser frequency is tuned into resonance with fast, Doppler- and Zeeman-shifted atoms at the entrance of the solenoid, they will stay in resonance within the magnet during the slowing process if the magnetic field strength decreases in such way that the changing Zeeman shift compensates for the changing Doppler shift.
Assuming a constant deceleration $\vec{a}$ within the magnet the velocity v(z) changes like (starting with v_0):

$$v(z) = (v_0^2 - 2az)^{1/2} \tag{4}$$

The corresponding change of the magnetic field strength B(z) to compensate the z-dependent Doppler shift by the changing Zeeman shift of the cooling transition reads then (for details see, e.g. [2,3]):

$$\vec{B}(z) = \vec{B}_0 \ (1-2az/v_0^2)^{1/2} + \vec{B}_b \tag{5}$$

$\vec{B}_b$: bias field

The difference between the magnetic field strengths at both ends of the solenoid defines the magnitude of the slowed down velocity interval, and the laser detuning defines the final velocity of the decelerated atoms. Atoms outside this velocity interval - faster atoms and atoms slower than the final velocity - will not be affected (almost).

As each location z within the magnet corresponds, as given in eq. (4), to a resonant velocity, every atom within the affected velocity interval will find its resonant point z and will stay in resonance till the exit of the magnet. All atoms of the decelerated interval will thus have nearly the same velocity at the exit of the magnet. Therefore the resulting velocity distribution is highly compressed to some m/s (s.below). The final velocity may be choosen over a broad range including zero. In order to get the atoms at rest outside the magnet, the laser frequency is tuned into resonance with a sufficiently slow velocity group at the exit of the magnet, allowing the atoms to slow down further, when they walk out of resonance [3].
In first experiments using this technique, the production of a steady flow of cooled sodium atoms with a temperature below 100 mK and a density of about $10^5 cm^{-3}$ was demonstrated (for details see e.g.[3]).

The second general cooling scheme uses fast frequency modulation techniques to compensate optical pumping and to keep the resonance condition (without magnetic fields) [4].
In alkali spectra (e.g. Na) the transition $^2S_{1/2}(F=2) \Rightarrow {}^2P_{3/2}(F=3)$ does not provide a completely ideal two-level system in zero magnetic field; because of the relatively small hyperfine splitting of the upper level and the limited perfection of the circular polarization of the laser light, the atoms can also make the transition $^2S_{1/2}(F=2) \Rightarrow {}^2P_{3/2}(F=2)$ and the upper level can decay to the level $^2S_{1/2}(F=1)$ which is out of resonance. To compensate for this optical pumping a second frequency in the cooling laser beam inducing the transition $^2S_{1/2}(F=1) \Rightarrow {}^2P_{3/2}(F=2)$ can repump the atoms into the level $^2S_{1/2}(F=2)$ via the transition $^2P_{3/2}(F=2) \Rightarrow {}^2S_{1/2}(F=2)$.
This second frequency can be provided as one of two sidebands of a frequency-modulated laser beam; in the case of the sodium D_2 line the difference frequency would be 1712 MHz. For this purpose the laser beam is, for example, sent through an electro-optic phase modulator [4] that is driven at half the desired difference frequency. In case of a sufficient modulation index about 35% of the incoming intensity can be transfered to the first-order sidebands.
The problem of maintaining the resonance condition for the decelerating atoms is solved in this scheme by a fast tuning of the laser frequency synchronously with the rapidly changing Doppler shift. This again can be achieved by electro-optic modulation techniques, if the laser beam with the two frequencies is sent through a second electro-optic modulator, the driving frequency of which is chirped in the right way - in the sodium experiment, for example, from 5 MHz to 1000

MHz within about 1.5 ms. This will produce a pair of sidebands which stays in resonance with the decelerating atoms in both hyperfine sublevels of the ground state, if the carrier frequency is chosen correctly [4].

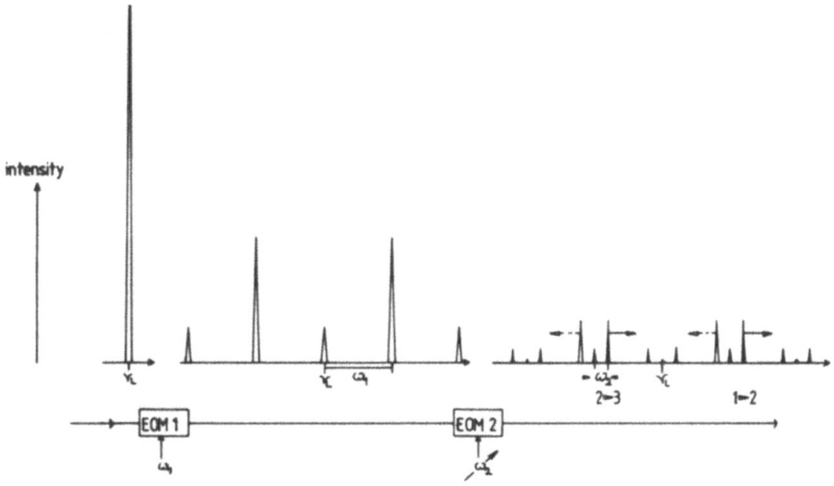

Fig. 2: Schematic frequency spectrum of the cooling laser beam behind the two electro-optic modulators (EOM). The cooling and repumping sidebands are marked 2⇒3 and 1⇒2.

Fig. 3 shows the experimental scheme for this experiment, the sideband spectrum is schematically shown in Fig. 2.
In the scan method the deceleration $\vec{a}$ has to match the scan speed $\dot{\nu}_L$ of the laser frequency; assuming constant deceleration the scan speed is almost constant,

$$\dot{\nu}_L = \frac{|\vec{a}|}{\lambda} \qquad (6)$$

and the frequency varies linearly in time, starting periodically red shifted at the laser frequency ν_s.

$$\nu_L(t) = \nu_s(1+\alpha t) \qquad (7)$$

This frequency ν_s is in resonance with fast atoms v_s at the beginning of each cooling cycle. During the sideband is swept over a frequency interval $\Delta\nu_L$ these fast atoms stay in resonance and slower atoms will get into resonance. Thus at the end of each cooling cycle the corresponding velocity interval $\Delta v = \Delta \nu_L / \lambda$ is compressed into a narrow velocity distribution (s.below) at the final velocity $v_s - \Delta v$.

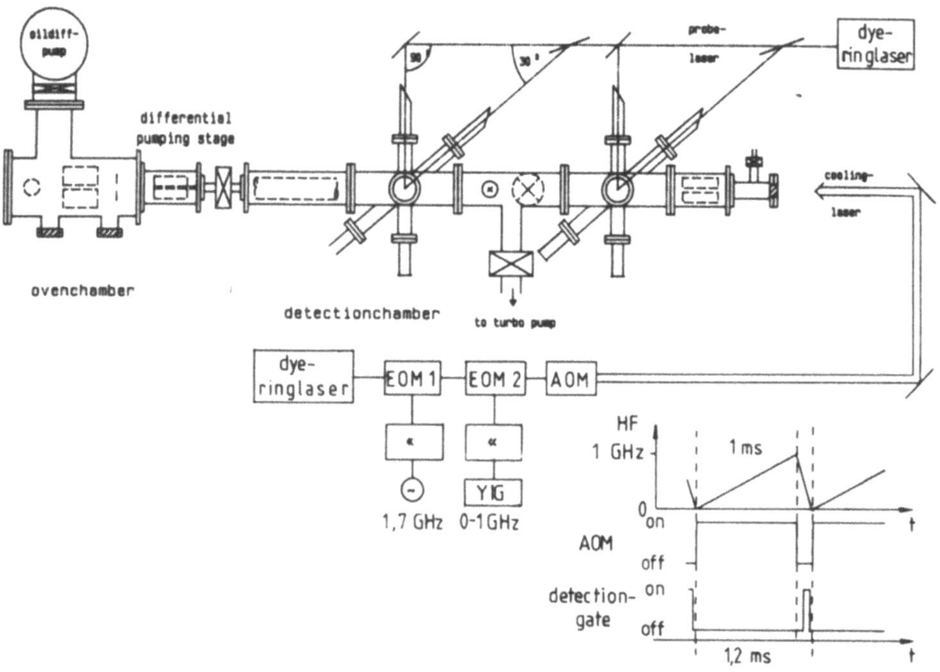

Fig.3: Schematic of atomic beam cooling experiment. The cooling laser output is fed through an acousto-optic shutter, a $LiTaO_3$ travelling-wave electro-optic modulator (EOM) which provides the frequency-swept sidebands, and an additional EOM to provide the F=1 atom recovery sideband. The σ-polarized cooling laser beam is carefully "mode matched" to the weakly diverging atomic beam (~3 mrad full angle). The detection system for the fluorescence light is not shown.

In our experiments the scan interval was limited by our microwave equipment to ~ 1GHz corresponding to a velocity interval of ~600 m/s. After each cooling scan, which takes about 1 ms, the swept sideband moves very fast within ~200 µs back into the starting position beginning the next cooling cycle. During this time fresh incoming, uncooled atoms fly only ~1 m before they get into resonance in the next cooling cycle. Thus all atoms slower than the starting velocity v_s and within Δv will be cooled.

In the first experiments with this second scheme [4,5] the resulting temperature within the cooled atomic beam was below 50mK with a density of 10^6 atoms per cm^3. As the slow atoms move only a short distance during the cooling cycles, the resulting pile-up of slow atoms forms a nearly constant flow of cold atoms as in the previous scheme.

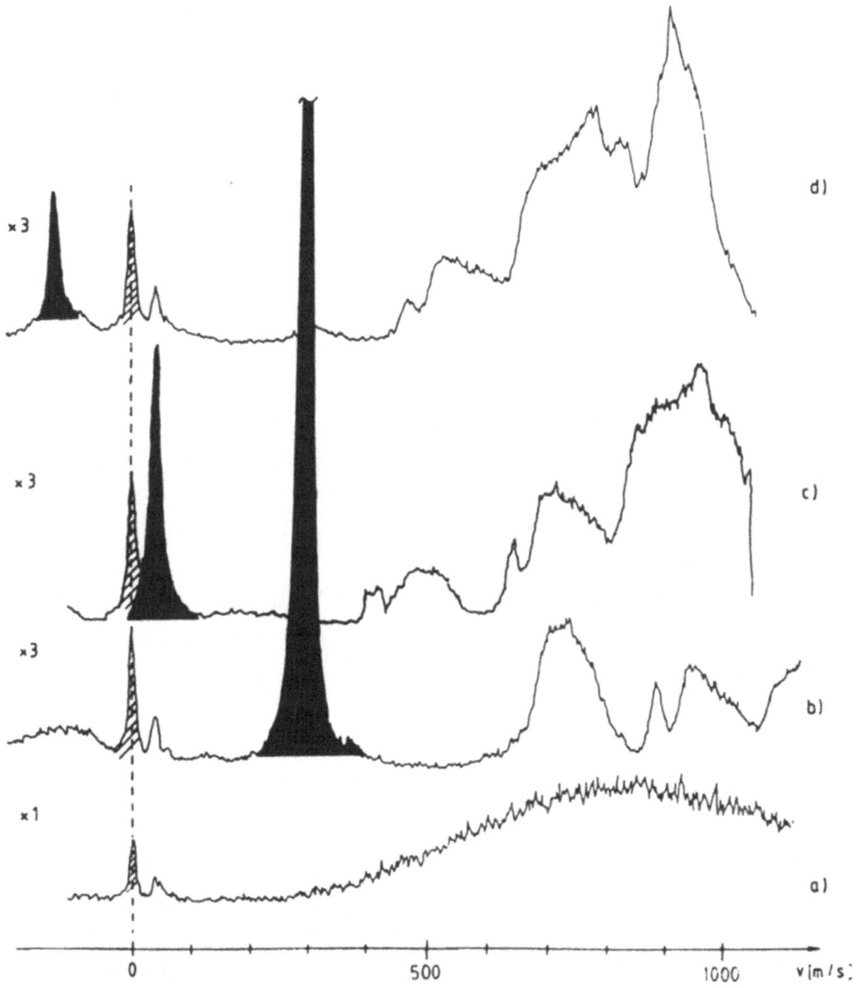

Fig.4: Sodium-atomic-beam cooling using a frequency-chirped laser. Trace a), cooling laser off. The D_2 transition shows the velocity distribution of the F=2 ground state as well as the frequency markers from the perpendicular probe beam. The vertical dashed line marks the position of the F=3 resonance with zero velocity atoms. Trace b) - d), cooling laser sideband is swept ~1GHz (Δv=590 m/s), carrying atoms to lower velocities where they are left (blackened peaks) when the cooling laser is cut off and the velocity distribution is measured. The final velocity in trace b) is ~290 m/s, in trace c) ~40 m/s, and in trace d) ~ -130 m/s (!). The apparent weakening of the slow-atom peak is partly because fewer atoms are available, when the sweep starts below the velocity distribution maximum and for geometrical reasons (s.below). The figures also show "actions" by the other (unused) sidebands.

For probing the velocity distribution we used a second dye laser beam, which was split into a Doppler free beam - perpendicular to the atomic beam - and a beam crossing the atomic beam with a small angle (≈30°,

see Fig.3). The frequency of this probing laser was slowly tuned over the Doppler profil of the $^2S_{1/2}$ (F=2) ⇒ $^2P_{3/2}$ (F=3) transition. The laser induced fluorescence signal was only periodically counted within a ~50 μs time interval during the fly-back time of the cooling side-band. During this time the cooling laser beam was shut off by the acusto-optic modulator. Thus optical pumping and transient effects were avoided.
Because of the 30° crossing angle our velocity readout scale is compressed by a factor cos30° =0.866. Fig. 4 shows the result. Because of the saturation in the probe beam being 2...3 and the Doppler free beam being <<1 the whole width of the probe beam can be explained - in comparison to the width of the Doppler free width - by saturation effects (within the momentary velocity resolution). As 10 MHz natural width corresponds to a Doppler broadening by 6 m/s, we can estimate a residual velocity width of less than 5 m/s (for the more complicated details see [4,6]).
This result is in good agreement with our Monte-Carlo simulation of the cooling process [7,8] and the numerical integration of the Fokker-Planck equation [8].

Relay cooling

A scan width of 1 GHz corresponds to a cooled velocity internal of about 600 m/s. If the fixed sideband for the transition F=2 ⇒ F=3 is placed near the average velocity (~600 m/s) it is possible to cool twice the single sideband cooling interval by the so called "relay cooling" scheme.
For this purpose the frequency sweeps of the fast tuned sidebands (cooling and flyback) have to be almost symmetric and should start at zero or not more than 1/2 of the natural linewidth from zero. Then both fast tuned sidebands have - alternating - the right tuning sign and tuning speed.
Fig. 5 explains schematically the situation.

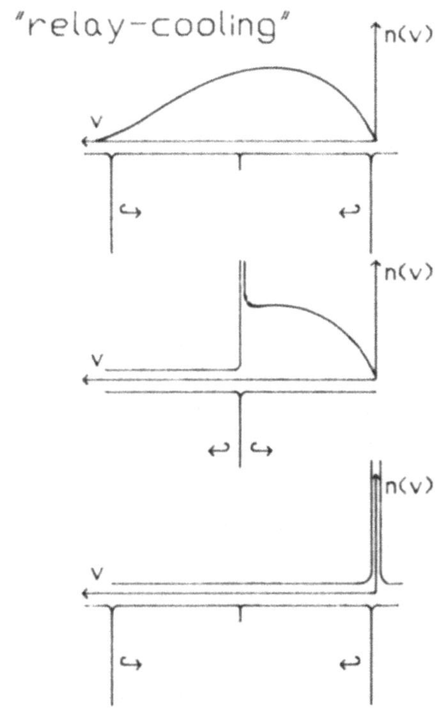

Fig. 5:Scheme for the combined action of both sidebands and the carrier, to double the cooling intervall. It is one cooling period show starting at the top The fresh incoming atoms are omited in this representation.

The "left" sideband decelerates the fast atoms - e.g. 1300m/s ⇒ 700m/s - down to a medium velocity, resonant with the carrier, and during the "fly-back" of the "left" sideband the "right sideband" takes them over and cools them further down together with the slower atoms - e.g. 700m/s ⇒ 100m/s. In case the atomic beam apparatus is long enough (for ^{23}Na ~ 2m) relay cooling stops or decelerates so essentially the whole atomic beam (for ^{23}Na $\Delta v \approx 1200$m/s) with 1 GHz scan width. Experiments showed that this scheme works satisfactorily, if the atomic beam is long enough and when the laser power at this longer beam line is still sufficient. [6].

Higher Order Sideband Cooling

In case of enough laser and microwave power it is also possible to use the second order sideband for cooling.

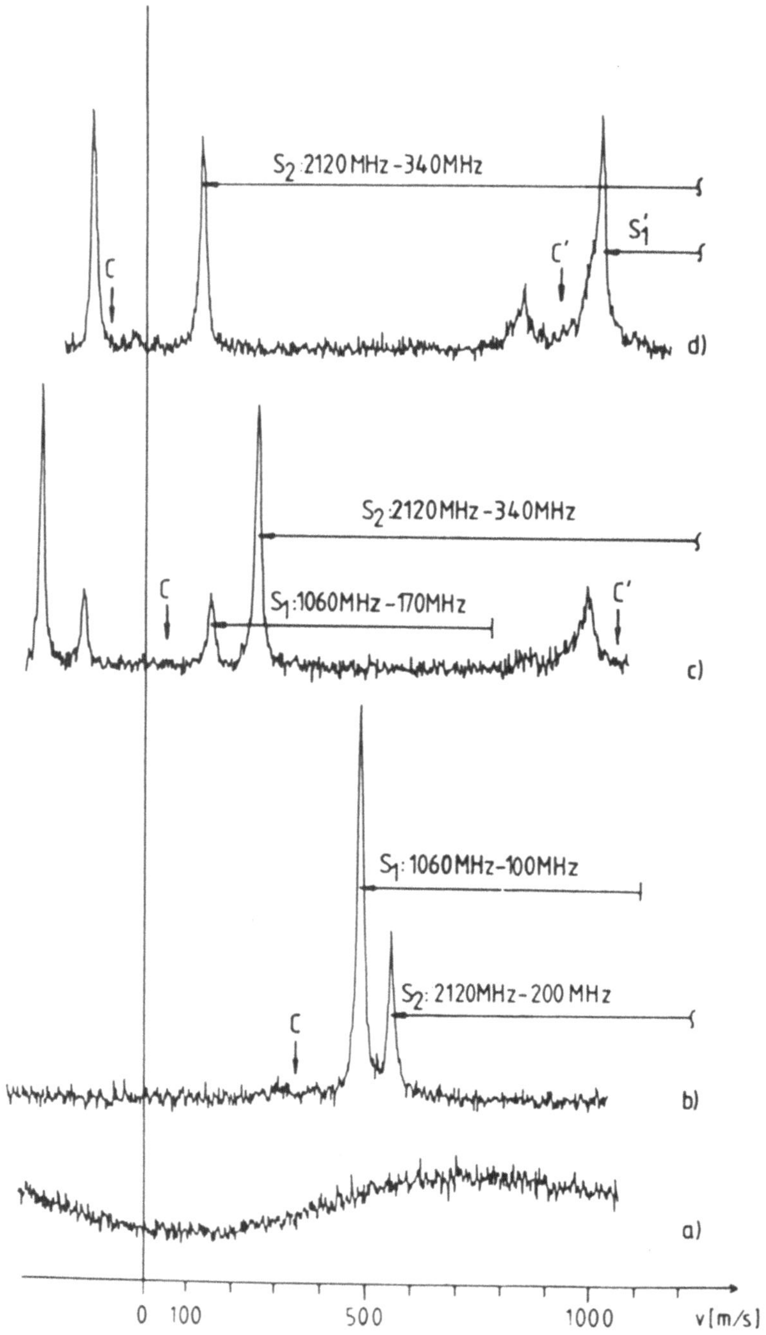

Fig.6: Velocity distribution of a laser cooled sodium atomic beam Trace a,) cooling laser off; only the F=2 part of the velocity distribution is shown. The vertical line indicates zero velocity. Trace b) cooling laser on. C marks the position of the carrier frequency and S_1 and S_2 mark the chirping interval and the final position of the 1st

oder (S_1) and 2nd order (S_2) sideband. The distance between S_1 and S_2 final positions corresponds to the 100 MHz final offset (100 MHz corr. 60m/s). In trace c) this offset is 170 MHz for demonstration and the carrier position is drastically shifted to ~80 m/s. C marks the position of the left 1,7GHz sideband of the carrier acting on the ~ 1000 m/s faster F=2 atoms (C is in resonance with the some velocity group in F=1 and slow F=2 atoms!) Therefore the same situation as shown in trace b) and c) (left part) repeats at higher velocities as shown in trace d).

Applying a 1 GHz chirping interval for the microwave driving field of the EOM, the 1st order sideband is swept 1 GHz and the 2nd order sideband is swept 2 GHz. Of cause the scan speed of the 2nd order sideband is doubled too. So one has to take care that the laserpower in this sideband is sufficient to match the conditions in eq. (3) and (6).
In order to demonstrate that this scheme doubles the decelerated velocity interval too, we performed the following experiment: the laser frequency was split into a triplet by the first EOM: the "carrier" (~80%) and two 1st order sidebands 1.712 GHz apart (~10% each). The carrier was located at some slow velocity (velocities are given for the F=2 population). But the chirping interval for the second EOM was choosen to be 100-1060 MHz (170-1060 MHz). Therefore the 1st order sideband stops 100 MHz (170 MHz) apart from the carrier and the 2nd order sideband 200 MHz (340 MHZ) apart from the carrier producing so two final velocity groups.
The apparatus used for this experiment was similar to the one shown in Fig.3; but instead of using the Doppler free probe beam as zero marker, we retroreflected the ordinary probe beam. This produces "mirror images" in the readout of the F=2 velocity distribution (on top of the small F=1 population). The center between two "mirror images" defines zero velocity. Fig. 6 shows the experimental results. From this Figure we see that both sidebands - the 1st and the 2nd order sideband - decelerate the atoms. That means that the 2nd order sideband actually decelerates essentially the whole atomic beam.
The density of the two final velocity groups depends on the probing location in the atomic beam and the position of the carrier frequency. In trace b) of Fig.6 the carrier is in resonance with 350m/s atoms and the 1st order sideband catches most of the atoms and the velocity distribution gaining so the higher density.
A Monte-Carlo simulation [7,8] can calculate the various dependencies. Fig.7 shows our numerical result for a similar situation as given in Fig.6. In the experiment the probing position was located at about s_2=150cm. The sideband C'and its "subsidebands" were omitted during the calculation. The calculation shows the main features in satisfac-

tory agreement.

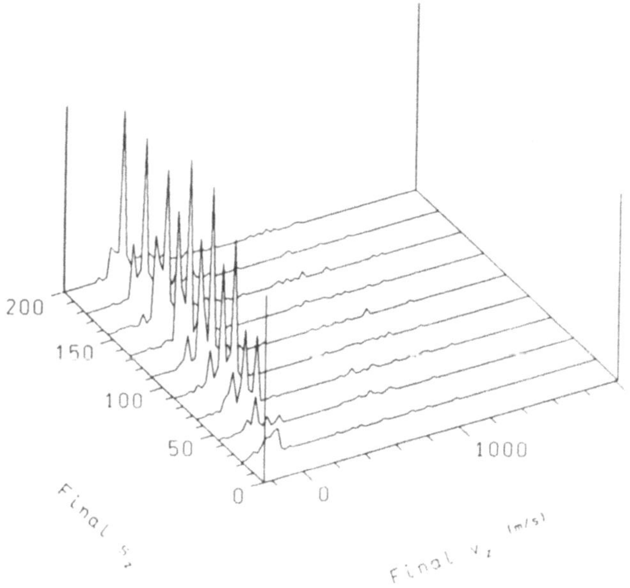

Fig.7: Plot of the longitudinal velocity distribution (in arb. units) along the atomic beam as a function of the longitudinal position s_z (in cm). The plot shows the steady state result for an experimental situation as given in Fig.6.

Properties and Handling of Cooled Atomic Beams

As result of the numerous spontaneous emmissions, some longitudinal momentum is transfered into transverse momentum. The resulting velocity width v_T(rms) of this "transverse heating effect" by n spontaneous emmissions reads

$$v_T(\text{rms}) = v_r\ (n/3)^{1/2} \tag{10}$$

$$v_r = \hbar k/M \tag{11}$$

with

$$n = \frac{1}{v_r}\ (2k_B T/M)^{1/2}$$

K_B : Boltzmann-constant

T: evaporation temperatur

this transverse velocity spread becomes

$$v_T(\text{rms}) = \left(\frac{2k_B T\hbar^2 k^2}{9M^3}\right)^{1/4} \tag{12}$$

The transverse velocity distribution of this beam may be further reduced by transverse laser beams, the frequency of which is tuned about one halfwidth to the red of the center frequency. Thus the saturation of, and correspondingly the spontaneous force on, the slow atoms by the transverse laser beams increases for the atoms which are moving towards a transverse laser beam [9,10]. The effect will be at maximum when the red detuning $\delta\omega$ is [8]

$$\delta\omega = \frac{\gamma}{2} \left(\frac{1+2s_0}{3} \right)^{1/2} \tag{13}$$

γ: natural linewidth (FWHM)

s_0: saturation on center frequency.

Another possibility of transverse cooling using dipole-forces is given in [11].

The velocity-dependent force will also reduce the final longitudinal velocity distribution [4]: a counter-propagating laser beam, which is tuned slightly red of the absorption frequency of the slow atoms, will decelerate atoms, which remained faster than the average, more than slower ones, when they move out of resonance. This laser beam could be, for example, the carrier frequency of the cooling laser, if the chirping method is used for cooling. This frequency may be permanent in the cooling beam. This velocity-controlling effect is also responsible for the longitudinal velocity width just after cooling [8].

For precision experiments it will be of special interest to separate the cold atoms from the hot (fast) atoms.

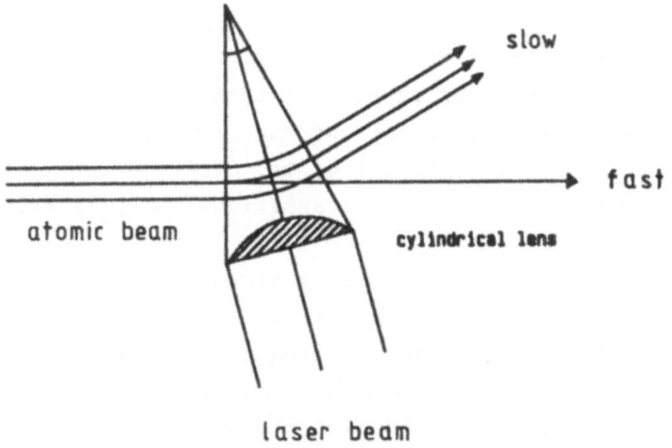

Fig.8: Geometry for atomic beam deflection by a transverse laser beam

One way of separation is simply to apply a second "two-frequency" transverse laser beam as shown in Fig. 8. If the intensity profile of this laser beam has the correct shape, only the slow atoms will be deflected and these atoms will show no further increase in velocity spread. The transverse laser beam providing the necessary spontaneous force to match the radial acceleration $a_r = \nu_o^2/r$ (ν_o: average velocity of the slow atomic beam) has to have a saturation profil like

$$s(r) = \frac{1}{\frac{r v_r}{\tau v_0^2} - 2} \tag{14}$$

or for s <<1

$$s(r) = \frac{\tau\ v_0^2}{v_r} \cdot \frac{1}{r} \tag{15}$$

This means that a cylindrical lens gives the proper intensity profile for the purpose of bending an atomic beam of slow atoms. For a beam with 50 m/s average velocity and 5 mm diameter a lens with a focal length of 50 mm and a transverse laser beam of 20 mm waist only a few milliwatts are sufficient to bend the slow atomic beam ~40° from the axis [6,12]. Additionally, the transverse velocity spread will be reduced in the bending plane as a by-product of the bending mechanism. The result of this bending scheme is thus a slow atomic beam, the direction of which is unconstrainedly selectable and which is unperturbed by fast atoms.

One application of cold atomic beams is clearly its use for optical or microwave frequency standards.
The great interest in cold atoms for optical frequency standards is twofold: reduction of Doppler effects and prolonged interaction times. For cold atoms the second-order Doppler effect for optical transitions may be less than 10 mHz. The interaction time of the light field with cold atoms may well exceed the interval of 1s, offering spectral widths in the sub-Hz regime for selected transitions, Doppler-free schemes presupposed.
There are two main concepts for optical frequency standards using cold atomic beams. The first one uses the cold atomic beam mainly for filling atom-traps, which form the essential frequency discriminator of an optical frequency standard. The second one makes full use of the monochromatic velocity distribution of cold atomic beams.
The monochromatic atomic beam may be deflected through 90° with a

dipole magnet or a transverse laser beam (s.above). The resulting "Zacharias fountain" atomic beam would give an ideal opportunity for two-zone Ramsey excitation with very slow atoms going a second time through the same interaction region, when they are free falling down again [4,12]. But also the horizontal, separated, and cooled beam is of great interest for precision measurements in the $\Delta\nu/\nu<10^{-15}$ regime. Some candidate atoms are given in [10].

Other applications for cold atomic beams are clearly collision physics, surface physics, photon statistics, quantum effects (Bose condensation), polarized targets or isotope separation.

This work was supported by the Deutsche Forschungsgemeinschaft.
We gratefully acknowledge the experimental assistance by H. El`Kashef.

References

[1] See, e.g., V.S.Letokhov and V.G.Minogin, Phys.Rep.73, 1 (1981)

[2] J.V.Prodan, W.D.Phillips, H.Metcalf, Phys.Rev.Lett.49,1149 (1982)

[3] J.V.Prodan, A.Migdall, W.D.Phillips,I.So.,H.Metcalf, J.Dalibard, Phys.Rev.Lett. 54,992 (1985)

[4] W.Ertmer, R.Blatt, J.L.Hall, M.Zhu, Phys.Rev.Lett.54,996 (1985)

[5] S.Chu, L.Holberg, J.E.Bjorkholm, A.Cable, A.Ashkin, Phys.Rev.Lett. 55, 48 (1985)

[6] W.Ertmer, to be published

[7] R.Blatt, W.Ertmer, P.Zoller, and J.L.Hall, Phys.Rev.A34,3022 (1986)

[8] M.Bordach, W.Ertmer, H.Wallis, to be published

[9] V.I.Balykin, V.S.Letokhov, A.I.Sidorov, Pis'ma Zh. Eksp. Teor. Fiz.,40, 251 (1984)

[10] W.Ertmer, R.Blatt, J.L.Hall, Prog.Quantum Electron 8, 249 (1984)

[11] A.Aspect,J.Dalibard,A.Heidmann,C.Salomon, and Cohen-Tannoudji, Phys.Rev.Lett. 57, 1688 (1986)

[12] W.Ertmer, S.Penselin, Metrologia 22, 195 (1986)

ATOMIC MOTION IN A RESONANT LASER STANDING WAVE

J. Dalibard, A. Heidmann, C. Salomon, A. Aspect,
H. Metcalf* and C. Cohen-Tannoudji
Laboratoire de Spectroscopie Hertzienne de
l'Ecole Normale Supérieure et Collège de France
24 rue Lhomond, F-75231 Paris Cedex 05, FRANCE

These last few years have seen major advances in the possible ways to control atomic motion using laser light. It is now possible to slow down and even to stop an atomic beam [1-3], and the first observation of optically trapped atoms has been reported [4]. It is therefore an important task to try to describe in the most accurate way the basic phenomena which are at the origin of exchange of momentum between atoms and resonant light, and to interpret them in terms of elementary processes as absorption and spontaneous or stimulated emission of photons by the atom.

In a plane running wave, this description is simple : stimulated emission plays no role, so that the atom only undergoes fluorescence cycles involving the absorption of a laser photon (gain of momentum $\hbar\vec{k}_L$ for the atom), and the emission of a fluorescence photon (zero average change of atomic momentum). The average force acting on the atom is then :

$$\vec{f} = n\,\hbar\vec{k}_L$$

where n is the average number of fluorescence cycles per unit time, which for a two level atom saturates to the value $\Gamma/2$ at high laser intensity (Γ^{-1} is the lifetime of the atomic excited level in seconds). This force is the so called "radiation pressure force" or "scattering force" (see for ex. [5]).

When several plane running waves are simultaneously present, the situation becomes more complicated. Stimulated emission can play an important role, since redistribution of photons between the waves may occur, via cycles involving absorption of a photon from one wave (momentum $\hbar\vec{k}_1$), and stimulated emission of a photon into a second wave (momentum $\hbar\vec{k}_2$). This redistribution occurs at a rate given by the Rabi frequency ω_1 of the atom driven by the laser light, and then leads to forces whose order of magnitude is given by

$$f' \sim \omega_1 \; \hbar|\vec{k}_1 - \vec{k}_2|$$

Since the Rabi frequency does not saturate with light intensity, this force f' may be considerably larger than the force f found in a single plane running wave.

Our goal in this paper is to present various possible ways to study atomic motion in presence of both stimulated and spontaneous processes. We will consider here the simple problem of a one-dimensional motion along z axis, in two plane counterpropagating running waves. We will discuss the possibility of using such a standing wave for damping the atomic motion, and we will study the limits of this cooling mechanism.

At this date, three experimental groups have published results exhibiting evidence for a cooling of free atoms in a standing wave [6-9]. In the two first experiments (which were actually 2-D [6] and 3-D [7] experiments), the situation was chosen such as stimulated processes play a weak role : the Rabi frequency ω_1 was smaller than the natural width Γ. The cooling has then been observed for a negative detuning $\delta = \omega_L - \omega_A$ (ω_L : laser frequency, ω_A atomic frequency, all in rad/sec), $|\delta|$ being of the order of the natural width. In the second experiment [8] which has been realized in our laboratory, the situation was reversed : the Rabi frequency was two orders of magnitude larger than the natural width, and the cooling then occured for positive detuning, much larger than the natural width. We will see in this paper how to interpret these results, and in particular, the change of sign of the detuning leading to cooling when the laser power is increased.

This paper is organized as follows. The first part A is devoted to a study of the problem using Optical Bloch Equation (O.B.E.). We show how it is possible to derive from this equation the expression of the radiative force to first order in the atomic velocity. We also give some examples of results obtained by a numerical calculation of this force, valid for any atomic velocity, and based on a continued fraction expansion of the O.B.E. solution [12-13]. The second part deals with the dressed atom approach to this problem, which appears to be very efficient to describe the high intensity limit $\omega_1 \gg \Gamma$. We present a Monte-Carlo simulation of the atomic motion of this dressed atom, which has the advantage of handling easily fluctuations caused by spontaneous emission processes. This simulation then gives us an estimation of the final energy of an atom cooled by a strong standing wave. It also suggests the possibility of observing a channelization of the atoms in the nodes of the intense standing wave.

1. THE OPTICAL BLOCH EQUATIONS APPROACH

We shall treat here classically the atomic motion, assuming that the atom is localized around a point $\vec{r}$, with a spatial spread $\Delta\vec{r}$ much smaller than the light wavelength λ. The average force acting on the atom at point $\vec{r}$, with a velocity $\vec{v}$, is then related to the gradient of the atom-laser coupling $V_{AL}(\vec{r})$ [5,14] :

$$\vec{f} = \langle - \vec{\nabla}(V_{AL}(\vec{r})) \rangle_{\vec{r}\vec{v}} \tag{1}$$

1.1 The Optical Bloch Equations

The atom-laser coupling can be written, for a plane standing wave parallel to the z axis, and using the RWA and electric dipole approximations :

$$V_{AL}(\vec{r}) = - \vec{d}.\vec{\mathcal{E}}(z) \left(|e\rangle\langle g| \, e^{-i\omega_L t} + |g\rangle\langle e| \, e^{i\omega_L t} \right) \tag{2}$$

This coupling is characterized by the Rabi frequency $\omega_1(z)$:

$$\frac{\hbar\omega_1(z)}{2} = \frac{\hbar\tilde{\omega}_1}{2} \cos kz = - \vec{d}.\vec{\mathcal{E}}(z) \tag{3}$$

g is the atomic ground level and e the excited one. We have noted $\vec{d}$ the atomic dipole moment, and we have assumed that the laser is in a coherent state so that we can describe it classically. The coupling (3) is then zero at the nodes of the standing wave, and maximal at the antinodes. Introducing now the three quantities :

$$\begin{cases} S_1 = \left(\rho_{eg}\, e^{i\omega_L t} + \rho_{ge}\, e^{-i\omega_L t}\right)/2 \\ S_2 = \left(\rho_{eg}\, e^{i\omega_L t} - \rho_{ge}\, e^{-i\omega_L t}\right)/2i \\ S_3 = (\rho_{ee} - \rho_{gg})/2 \end{cases} \tag{4}$$

where ρ is the reduced atomic density matrix, the average force $\bar{f}(z,v)$ can be written :

$$\bar{f}(z,v) = \hbar k\tilde{\omega}_1 \,.\, \sin kz.\ \bar{S}_1(z,v) \tag{5}$$

$\bar{S}_1(z,v)$ is the stationary value of S_1 for an atom at point z, with a velocity v along 0z.

The evolution of the three quantities S_1, S_2, S_3 is given by the following Optical Bloch Equations (O.B.E.) [15] :

$$\begin{cases} \dot{S}_1 = -\,S_1\,\Gamma/2 + \delta\, S_2 \\ \dot{S}_2 = -\,\delta\, S_1 - S_2\,\Gamma/2 - \omega_1(z)\, S_3 \\ \dot{S}_3 = \omega_1(z)\, S_2 - \Gamma\, S_3 - \Gamma/2 \end{cases} \tag{6}$$

From these equations, one can in principle determine $\bar{S}_1(z,v)$ for any z and v and therefore calculate the value of $\bar{f}(z,v)$.

1.2 The force on an atom at rest : the dipole force

Let us consider first the case of an atom at rest. The solution of (6) is then easily obtained :

$$\bar{S}_1(z,0) = \frac{\delta\,\omega_1(z)/2}{\delta^2 + \Gamma^2/4 + \omega_1^2(z)/2} \tag{5}$$

so that

$$\bar{f}(z,0) = -\frac{d}{dz}\left(\frac{\hbar\delta}{2}\,\mathrm{Log}\left(1+\frac{\omega_1^2(z)/2}{\delta^2+\Gamma^2/4}\right)\right) \tag{6}$$

This is the so called "dipole force", which derives from a periodic potential, so that the average over a wavelength of $\bar{f}(z,0)$ is zero. It has been suggested to use such a force to trap atoms near the nodes or the antinodes of the standing wave, depending on the sign of δ [16-18].

1.3 The force on a slow atom : radiative cooling

We consider now low velocity atoms. The force f(z,v) can then be obtained by an adiabatic expansion of the forced solution of O.B.E. The parameter ε of the expansion is the relaxation time Γ^{-1} of the atom, divided by the time λ/v needed for the atom to travel over a wavelength :

$$\varepsilon = \frac{\Gamma^{-1}}{\lambda/v} = \frac{1}{2\pi}\,\frac{kv}{\Gamma} \tag{7}$$

If ε is small compared to 1, one can assume that the internal state of the atom follows quasi-adiabatically the external motion. The force can then be written to first order in v :

$$\bar{f}(z,v) = \bar{f}(z,0) - m\gamma\, v \tag{8}$$

where γ is given by [9] :

$$m\gamma = \int_0^\infty d\tau\, \frac{i\tau}{\hbar}\, \langle [F(z,\tau)\ ,\ F(z,0)]\rangle \tag{9}$$

$F(z,\tau)$ denotes the force operator $-dV_{AL}/dz$ (in Heisenberg point of view), and the correlation function of F contributing to eq. 9 can be evaluated by means of quantum regression theorem. The result of the calculation is [5] :

$$m\gamma = -\hbar k^2 \frac{s}{(1+s)^3}\left(\frac{\delta\Gamma}{\delta^2+\Gamma^2/4}(1-s) - 2\frac{\delta}{\Gamma}s^2\right) tg^2\, kz \qquad (10)$$

where the saturation parameter s is :

$$s = \frac{\omega_1^2(z)/2}{\delta^2+\Gamma^2/4} = 4\, s_0 \cos^2 kz \qquad (11)$$

s_0 denotes the saturation parameter obtained if only one of the two travelling waves is present. This damping coefficient $m\gamma$ can now be averaged on a wavelength, and we obtain [13] :

$$\overline{m\gamma} = -\hbar k^2 \frac{1}{(1+4s_0)^{3/2}}\left\{\frac{2\delta\Gamma}{\delta^2+\Gamma^2/4} s_0(1+2s_0) - 2\frac{\delta}{\Gamma}\left(1+6s_0+6s_0^2-(1+4s_0)^{3/2}\right)\right\} \qquad (12)$$

This average coefficient describes the efficiency of a standing wave for damping the atomic motion.

In the low intensity case ($s_0 \ll 1$), one gets :

$$\overline{m\gamma} \simeq \hbar k^2 s_0 \frac{2\delta\Gamma}{\delta^2+\Gamma^2/4} \leqslant \hbar k^2 \qquad (13)$$

This is a well known result which can be obtained by adding simply the two radiation pressures of the two counterpropagating waves [20]. It leads to a cooling when the detuning δ is negative (optimal value $-\Gamma/2$) which can be interpreted by noting that due to the Doppler effect, the atom in such a configuration is more sensitive to the counterpropagating running wave than to the copropagating one : it is therefore decelerated.

At high intensities ($s_0 \gg 1$) and high detunings ($|\delta| \gg \Gamma$), the force changes its sign : it becomes a damping force for positive detunings, and an accelerating one for negative detunings. Furthermore one notes that the order of magnitude of $\overline{m\gamma}$ changes. The damping coefficient can be much higher than the limit obtained in (13). One indeed gets :

$$\omega_1 \gg \delta \gg \Gamma \quad \Longrightarrow \quad m\gamma \sim \hbar k^2 \frac{|\omega_1|}{\Gamma} \gg \hbar k^2 \tag{14}$$

This change of sign of the force, and the increase of the damping coefficient is an evidence for the predominant role of stimulated emission in this regime.

1.4 Case of arbitrarily high velocities

When the atomic velocity is arbitrary, it is no longer possible to get an analytical expression for the force. But it is possible, using a continued fraction expansion method for (6), to get a numerical estimation of the force (5) [12-13]. Using this method, we have plotted some results in fig. 1. At low intensity (fig. 1.a : $\delta = +\Gamma/2$, $\tilde{\omega}_1 = \Gamma/3$), one can check that the force is nearly equal to the sum of the radiation pressures of the two counterpropagating waves. In particular, it is never a damping force for such a positive detuning. When the intensity increases, this conclusion is reversed, at least for low velocities, as it can be seen in fig. 1.b,... 1.e, plotted for increasing detunings δ and Rabi frequencies $\tilde{\omega}_1$. For $\tilde{\omega}_1$ and δ large compared to the natural linewidth Γ, the curves have the following characteristics :

(i) For low velocities ($kv \ll \Gamma/3$), as expected, the force varies linearly with the velocity, with a proportionnality coefficient given in eq. (12).

(ii) The force is maximum for velocities such as $kv \sim \Gamma/3$, and can be much higher than the radiation pressure $\hbar k\Gamma/2$ found if only one of the two counterpropagating waves is present. For example in fig. 1.e, for $\omega_1 = 10\ \delta = 500\ \Gamma$, the maximal force is $\sim 40\ (\hbar k\Gamma/2)$.

(iii) The force then decreases as $1/v$ until v reaches a critical value v_c, whose value increases with δ and ω_1.

(iv) When v is larger than v_c, resonances appear in $\bar{f}(v)$ (Dopplerons resonances [21]) and the sign of the force changes : it becomes a cooling force for $\delta < 0$ and an accelerating one for $\delta > 0$.

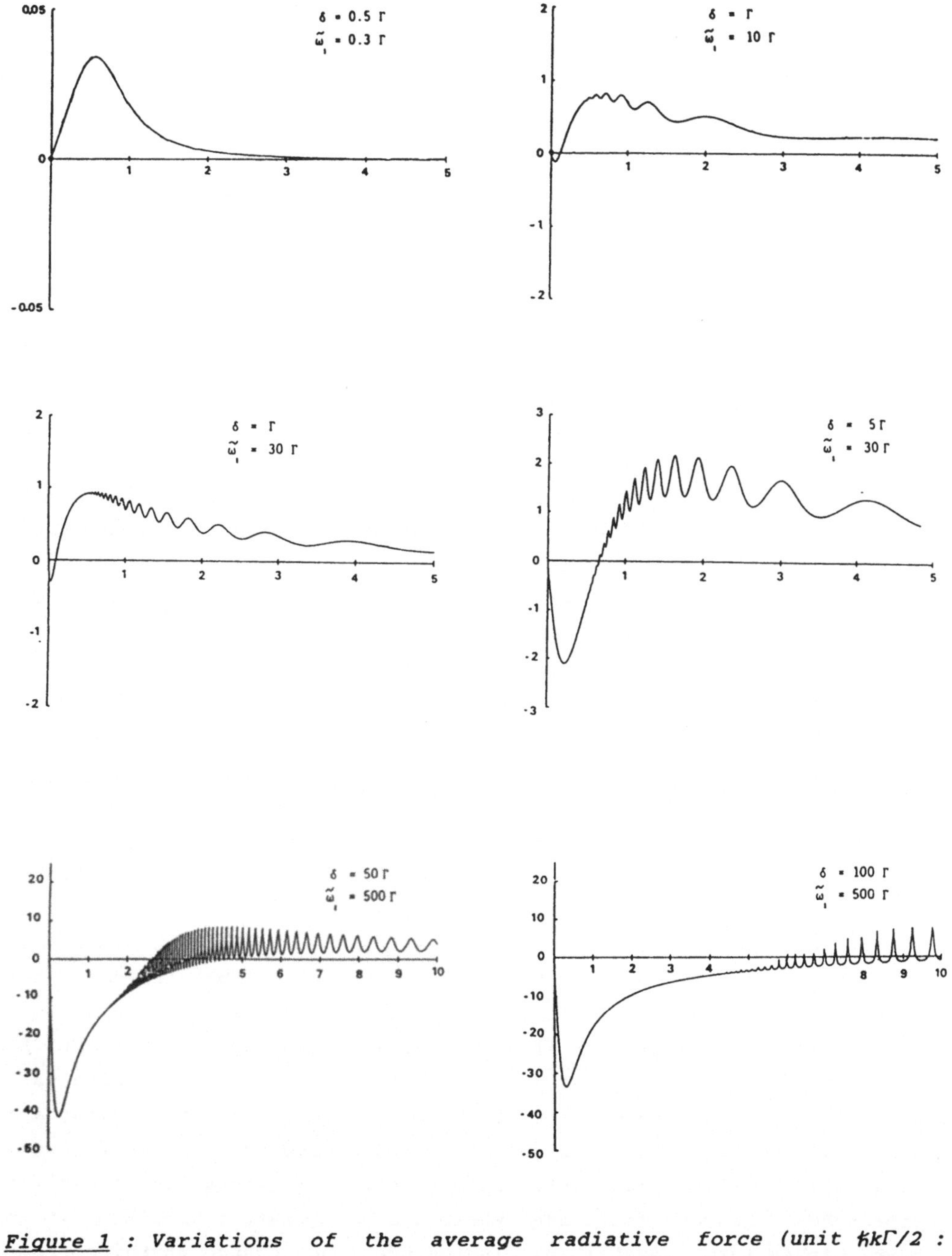

__Figure 1__ : Variations of the average radiative force (unit $\hbar k\Gamma/2$: resonant radiation pressure) acting on an atom in a standing wave as a function of the atomic velocity (unit Γ/k : Doppler effect equal to the natural linewidth). These curves have been realized for increasing detunings δ and Rabi frequencies $\tilde{\omega}_1$.

It is difficult to interpret all these characteristics in the framework of this theory based on a continued fraction expansion. On the contrary, we are going to see now that the dressed atom point of view gives in the strong intensity limit a clear interpretation for the various characteristics of the force.

2. THE DRESSED ATOM POINT OF VIEW [22,23]

2.1 The microscopic Sisyphus myth

In a strong standing wave, the energies of the dressed levels, i.e., the eigenstates of the atom plus laser-field system, oscillate periodically in space, as the Rabi frequency $\omega_1(z)$ in (eq. 3). Figure 2 represents these dressed states for a positive detuning ($\omega_L > \omega_0$). At a node [$\omega_1(z) = 0$], the dressed states $|1,n\rangle$ and $|2,n\rangle$ respectively coincide with the unperturbed states $|g,n+1\rangle$ and $|e,n\rangle$ (an atom in the ground state g or in the excited state e, in the presence of n+1 or n laser photons). Out of a node [$\omega_1(z) \neq 0$] the dressed states are linear combinations of $|g,n+1\rangle$ and $|e,n\rangle$ and their splitting $\hbar[\delta^2+\omega_1^2(z)]^{1/2}$ is maximum at the antinodes of the standing wave. Consider now the effect of spontaneous emission. An atom in level $|1,n\rangle$ or $|2,n\rangle$ -each containing some admixture of $|e,n\rangle$- can emit a spontaneous photon and decay to level $|1,n-1\rangle$ or $|2,n-1\rangle$- each containing some admixture of $|g,n\rangle$. The key point is that the various rates for such spontaneous processes vary in space. If the atom is in level $|1,n\rangle$, its decay rate is zero at a node where $|1,n\rangle = |g,n+1\rangle$ and maximum at an antinode where the contamination of $|1,n\rangle$ by $|e,n\rangle$ is maximum. In contrast, for an atom in level $|2,n\rangle$, the decay is maximum at the nodes, where $|2,n\rangle$ is equal to $|e,n\rangle$. We can now follow the "trajectory" of a moving atom starting, for example, at a node of the standing wave in level $|1,n+1\rangle$ (fig. 2). Starting from this valley, the atom climbs uphill until it approaches the top (antinode) where its decay rate is maximum. It may jump either into level $|1,n\rangle$ (which does not change anything from a mechanical point of view) or into level $|2,n\rangle$, in which case the atom is again in a valley. It has now to climb up again until it reaches a new top (node) where $|2,n\rangle$ is the most unstable, and so on. It is clear that the atomic velocity is decreased in such a process, which can be

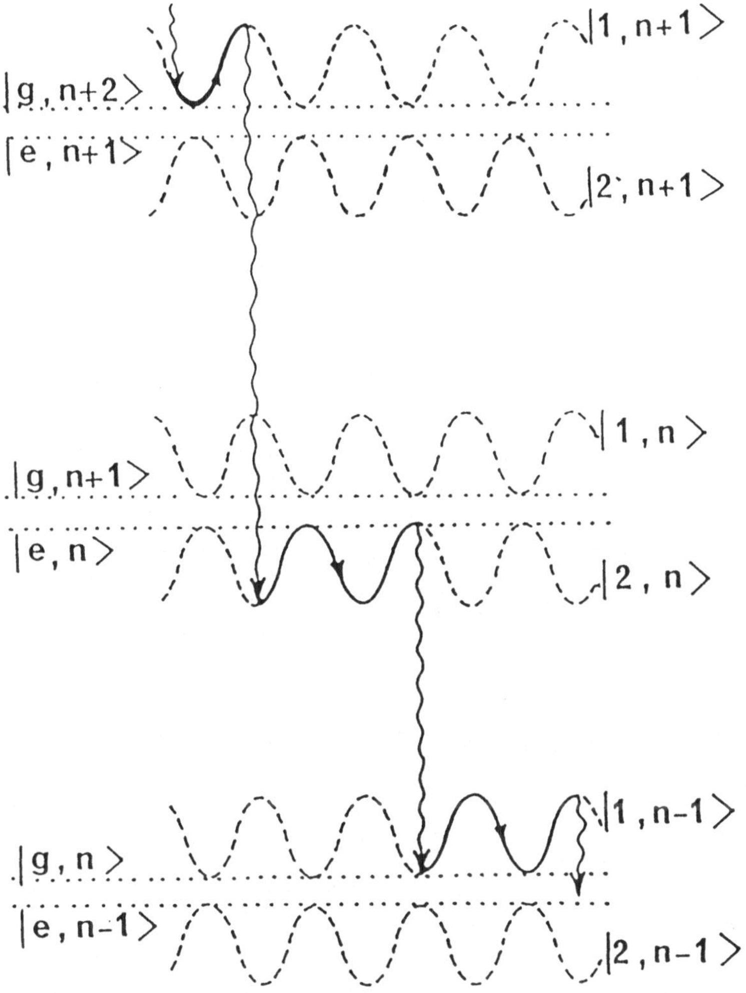

Figure 2 : Laser cooling in a strong standing wave. The dashed lines represent the spatial variations of the dressed-atom energy levels which coincide with the unperturbed levels (dotted lines) at the nodes. The solid lines represent the "trajectory" of a slowly moving atom. Because of the spatial variation of the dressed wave functions, spontaneous emission occurs preferentially at an antinode (node) for a dressed state of type 1 (2). Between two spontaneous emissions (wavy lines), the atom sees, on the average, more uphill parts than downhill ones and is therefore slowed down.

viewed as a microscopic realization of the "Sisyphus myth" : Every time the atom has climbed a hill, it may be put back at the bottom of another one by spontaneous emission.

Such a picture can be used to derive quantitative results for the average force acting on the atom [22]. These results appear to be in perfect agreement with the ones obtained by the continued fractions method, provided that the atomic velocity is smaller than v_c (results i, ii and iii). For velocities higher than v_c, one has to modify the previous picture and to take into account the Landau-Zener transitions, which can occur at each node of the standing wave, from one level $|i,n\rangle$ to the adjacent level $|j,n\rangle$. For certain velocities, the probability amplitudes for such transitions interfere constructively for two successive nodes, and the radiative force then exhibits a resonance. In ref. [24], using a method close to the dressed atom approach, the authors show how to calculate, with a relatively good precision, the position of these resonances.

2.2 Monte-Carlo simulation of atomic motion in a strong standing wave

In order to determine all the characteristics of the cooling of atoms in a strong standing wave, in particular the final temperature of cold atoms, we have developped a numerical simulation of the atomic motion in the standing wave. This simulation consists of doing an integration step by step of the motion presented on fig. 2. Each step consists in two phases : we calculate first the new position and the new velocity of the atom as a function of the old position and velocity and of the force seen on the dressed level occupied at this time. Then, we randomly decide an eventual change of dressed level, with a probability law given by the dressed atom theory. Note that here the atomic velocity is not constant but it varies when the atom goes uphill or downhill. This is in contrast to the continued fraction method in which one imposes a constant velocity to the atom in order to calculate the force. Actually the two methods give similar results as soon as the atomic kinetic energy $mv^2/2$ is much larger than the height of the hills $\sim \hbar\omega_1/2$. On the contrary, when the two parameters are on the same order, the numerical simulation of the dressed atom picture is a priori much closer to reality than the continued fraction method.

Examples of results of this numerical simulation are given in fig. 3 for Cesium atoms irradiated by a laser wave resonant with the resonance line (λ = 852 nm, Γ^{-1} = 30 ns). The parameters of the standing wave are $\tilde{\omega}_1 = 6\ \delta = 80\ \Gamma$. This value of $\tilde{\omega}_1$ is close to the one we had in the experiment of ref. 8, and the value of δ is then chosen in order to optimize the force for velocities around $kv = \Gamma$ (see fig. 1). Initial velocities are randomly chosen with a gaussian law, centered on $v = 0$, and with a full-width-half maximum (FWHM) of 5 m/s. One sees (fig. 3.a) that in a time of the order of 10 μs, these velocities concentrate in a peak of FWHM 1.2 m/s, which corresponds precisely to a kinetic energy of the order of the height of the hills. The typical time constant of this damping is 2 μs, which is in good agreement with the one deduced from eq. 12 (1.8 μs). Atoms are then more or less trapped in the valleys of fig. 2, and the cooling mechanism described above becomes much less efficient. The atomic kinetic energy $m\bar{v^2}/2$ then satisfies :

$$\frac{m\bar{v^2}}{2} \simeq \frac{\hbar\tilde{\omega}_1}{2} \tag{15}$$

This has to be compared to the well-known limit of radiative cooling in a weak standing wave [5] :

$$\frac{m\bar{v^2}}{2} \simeq \frac{\hbar\Gamma}{2} \tag{16}$$

2.3 Channelization of atoms

Equation (15) gives the atomic energy after a time of few $\bar{\gamma}^{-1}$ (eq. 12). Actually, we have observed numerically (fig. 3.b) a new decrease of the atomic energy, with a much larger time constant ($\simeq$ 100 μs). Simultaneously atoms accumulate at the nodes of the standing wave, as it can be seen on position profiles of fig. 3.c. The final curve, after an interaction time of 500 μs, gives $\Delta z \sim \lambda/40$ and $\Delta v \sim 0.2$ m/s (the product $\Delta x.\Delta p$ is then 10 $\hbar$ so that we are at the limit of a classical treatment of atomic velocity).

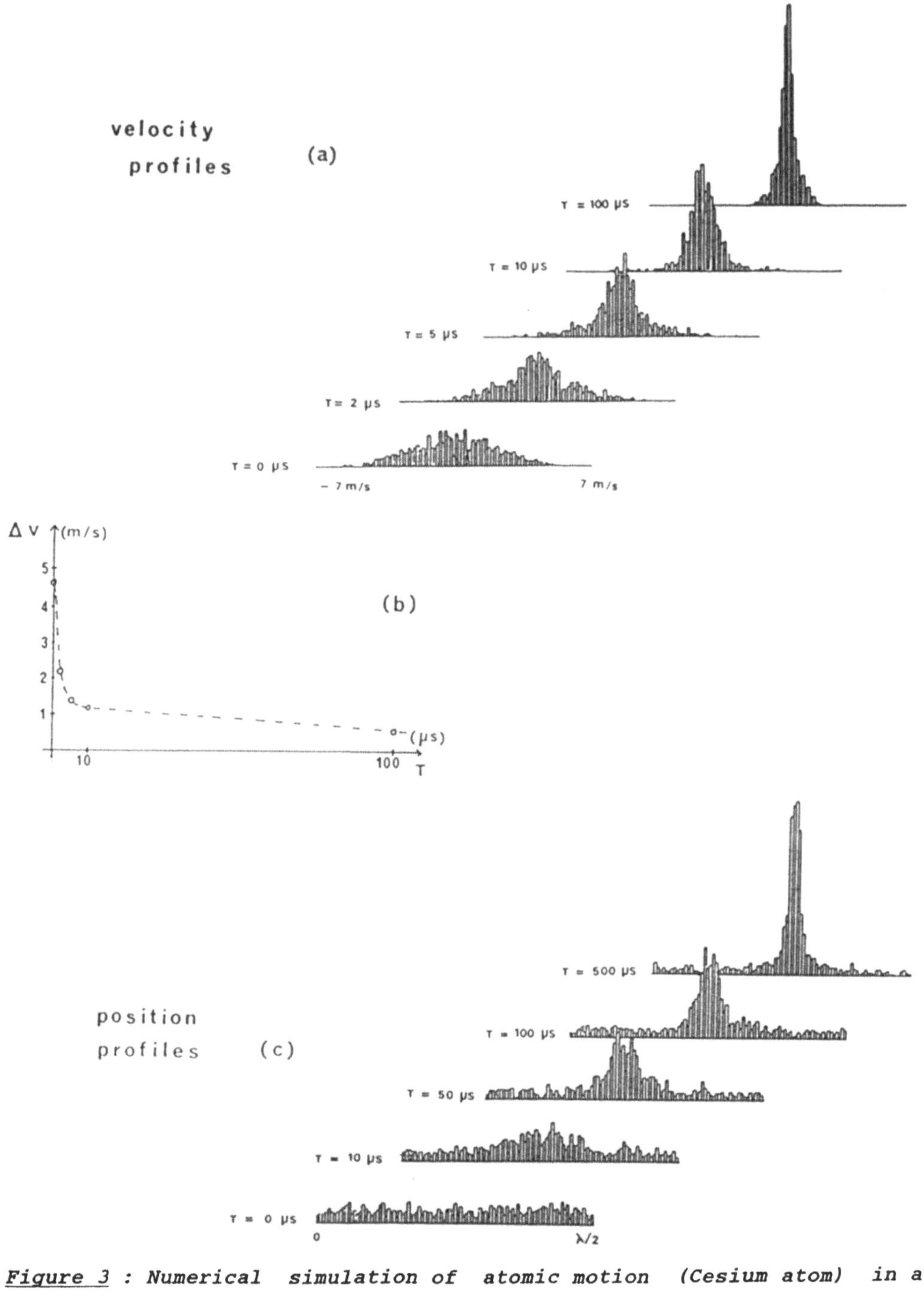

Figure 3 : Numerical simulation of atomic motion (Cesium atom) in a strong standing wave ($\tilde{\omega}_1 = 6\ \delta = 80\ \Gamma$).

3.a. Evolution of velocity profiles

3.b. Evolution of average quadratic velocities

3.c. Evolution of position profiles (modulo one wavelength).

This accumulation of atoms at the nodes of the standing wave, that one can call "channelization", is not strictly speaking a trapping, but rather a dynamical equilibrium. Atoms oscillate around a node on a level of type 1, then jump on a level of type 2 on which they "travel" over a fraction of wavelength to fall back on another level of type 1, near another node.

The first part of this cooling process has been observed in our laboratory, with experimental conditions very close to the ones chosen here [8]. The experimental results confirm the theoretical predictions exposed here. On the contrary, the channelization of atoms in the standing wave has not yet been observed. Several techniques could be considered for this. First an observation of the atomic spectrum of fluorescence or absorption should give a hint for the spatial distribution of the atoms. Second, one could think of a Bragg diffraction technique for probing such a spatial ordering of atoms.

To conclude, let us emphasize the potentialities of these "stimulated" forces for many applications, for instance for slowing down an atomic beam : as we have seen, the forces that can be realized are much greater than the usual radiation pressure which has been used up to now for this purpose. The stopping distance could then be reduced by at least one order of magnitude, which would be of special interest for the realization of compact atomic clocks using slow atoms.

* Permanent address : Department of Physics, State University of New York, Stony Brook, New York 11790, U.S.A.

[1] J.V. Prodan, A. Migdall, W.D. Phillips, I. So., H. Metcalf and J. Dalibard, Phys. Rev. Lett. 54, 992 (1985).

[2] W. Ertmer, R. Blatt, J. Hall and M. Zhu, Phys. Rev. Lett. 54, 996 (1985).

[3] V.I. Balykin, V.S. Letokhov and A.I. Sidorov, Opt. Commun. 49, 248 (1984).

[4] S. Chu, J. Bjorkholm, A. Ashkin and A. Cable, Phys. Rev. Lett. 57, 314 (1986).

[5] J.P. Gordon and A. Ashkin, Phys. Rev. A 21, 1606 (1980).

[6] V.I. Balykin, V.S. Letokhov and A.I. Sidorov, J.E.T.P. Letters 40, 1027 (1984).

[7] S. Chu, L. Hollberg, J.E. Bjorkholm, A. Cable and A. Ashkin, Phys. Rev. Lett. 55, 48 (1985).

[8] A. Aspect, J. Dalibard, A. Heidmann, C. Salomon and C. Cohen-Tannoudji, Phys. Rev. Lett. 57, 1688 (1986).

[9] Similar techniques of cooling have also been used for trapped ions ; see for example refs. 10 and 11.

[10] D.J. Wineland, R.E. Drullinger and F.L. Walls, Phys. Rev. Lett. 40, 1639 (1978).

[11] W. Neuhauser, M. Hohenstatt, P. Toschek and H.G. Dehmelt, Phys. Rev. Lett. 41, 233 (1978).

[12] V.G. Minogin, Opt. Commun. 37, 442 (1981).

[13] V.G. Minogin and O.T. Serimaa, Opt. Commun. 30, 373 (1979).

[14] R.J. Cook, Phys. Rev. A 20, 224 (1979) and Phys. Rev. Lett. 44, 976 (1980).

[15] L. Allen and J.H. Eberly, Optical Resonance and Two-Level Atoms, New York, Wiley (1975).

[16] V.S. Letokhov and V.G. Minogin, Phys. Lett. 61 A, 370 (1977) and Appl. Phys. 17, 99 (1977).

[17] V.S. Letokhov, V.G. Minogin and B.D. Pavlik, Opt. Commun. 19, 72 (1976) and Sov. Phys. J.E.T.P. 45, 698 (1978).

[18] V.S. Letokhov and B.D. Pavlik, App. Phys. 9, 229 (1976).

[19] J. Dalibard and C. Cohen-Tannoudji, J. Phys. B 18, 1661 (1985).

[20] T.W. Hänsch and A. Schawlow, Opt. Commun. 13, 68 (1975).

[21] E. Kyrölä and S. Stenholm, Opt. Commun. 22, 123 (1977).

[22] J. Dalibard and C. Cohen-Tannoudji, J.O.S.A. B2, 1707 (1985).

[23] A very similar appraoch has been developped by A.P. Kazantsev, Sov. Phys. J.E.T.P. 36, 861 (1973), 39, 784 (1974) et Sov. Phys. Usp. 21, 58 (1978). See also A.P. Kazantsev, V.S. Smirnov, G.I. Surdutovich, D.O. Chudesnikov and V.P. Yakovlev, J.O.S.A. B2, 1731 (1985).

[24] A.P. Kazantsev, D.O. Chudesnikov and V.P. Yakovlev, to be published.

LOW-TEMPERATURE PHYSICS WITH LASER COOLING

Juha Javanainen
Department of Physics and Astronomy
University of Rochester
Rochester, NY 14627
USA

1. Introduction

Experimental breakthroughs such as cooling of one electromagnetically trapped ion [1] and realization of magnetic [2] and all-optical [3] traps for neutral atoms have demonstrated tangible mechanical effects of laser light on atomic particles. The recent "quantum jump" observations [4-6] relying on laser cooling are examples of the ultimate sensitivity of spectroscopy, a single atomic particle. It may be, though, that most applications of light pressure will instead strive for high resolution, utilizing the low temperature to overcome the second-order Doppler shift [7]. To increase the signal-to-noise ratio, such experiments would normally be carried out on many particles. But, aside from applications, the very combination of many particles and low temperature may in itself give rise to novel phenomena. These are the theme of our paper.

Numerical simulations of a one-component plasma [8,9] have quite a while ago indicated that, at low enough temperatures, the Coulomb interactions force it to crystallize. Analogous crystallization of ions in electromagnetic traps [10,11] and in heavy-ion storage rings [12] have been discussed. The ensuing "Wigner crystal" would represent a new form of condensed matter. Its properties may be unusual, moreover the interaction between the particles of the frozen plasma are known and can conceivably be handled theoretically. Rigid few-ion clusters would also facilitate experiments on collective interactions with light of a small number of non-moving atoms. Until now such experiments have simply not been possible in the optical domain, an unfortunate circumstance as individual field quanta and associated quantum effects can be detected easily only at optical and higher frequencies.

At this writing clusters in ion traps have, however, not been observed. Whether unideal conditions of the experiments or fundamental physics are responsible, is not clear. Only very recently have we initiated [13] the theoretical study of laser cooling of interacting many-particle systems using the hypothetical periodic crystal as the first example. We have shown that, owing to the interference of light scattered from different ions, the laser only couples to a small fraction of the phonon modes. In exchange, the coupling is so strong that the cooling rate per particle in the crystal is qualitatively the same as for a single trapped ion. In this paper we generalize the model of Ref. 13 to an arbitrary cluster with small-vibration excitations around the equilibrium. In particular, the statement that the cooling is "qualitatively the same" as for a single ion, is made precise. The analysis emphasizes the importance of the interactions between the small-vibration modes, but these remain unexplored.

The prospect of a nearly ideal Bose condensate to quantitatively test and sharpen the theoretical understanding of many-body physics has driven intensive investigations into spin-polarized hydrogen [14], so far without decisive success. Magnetic and all-optical traps for neutral atoms, which offer sub-mK temperatures and density not curbed by the Coulomb repulsion, may in the long run deliver a new boost to the pursuit toward Bose condensation [15-17].

In this paper we analyze Bose condensation of trapped atoms. We also reiterate our suggestion [18] that, when two traps containing Bose condensates are brought close enough to each other, an oscillatory exchange of atoms between the traps emerges which is governed by the phases created in the spontaneous symmetry breaking associated with the Bose condensation. Finally, we discuss the Bose condensate in a trap as a test bench of the very fundamentals of statistical mechanics.

In the rest of this paper we expand on these synopses. Crystallization of trapped ions is covered in section 2, and Bose condensation in a trap is the subject of section 3.

2. Crystallization of trapped ions

Assume a cloud of N ions resides in a harmonic trap in which the oscillation frequencies of a single ion along the directions of the three coordinate axes would be ν_i, i=1, 2, or 3. Denoting the ith orthogonal coordinate of the ion $\alpha = 1,\ldots,N$ with $x^{\alpha i}$ and the full coordinate vector with $\mathbf{x}^\alpha$, we write the total potential energy of the ion system

$$V(\{x^{\alpha i}\}) = \frac{e^2}{8\pi\varepsilon_0} \sum_{\alpha \neq \beta} \frac{1}{\left|\mathbf{x}^\alpha - \mathbf{x}^\beta\right|} + \frac{1}{2} M \sum_{\alpha,i} \nu_i^2 \,(x^{\alpha i})^2. \tag{1}$$

When the temperature is lowered toward zero, the ions settle to a configuration which gives the global minimum of the potential energy. As an example, the projection onto the 1-2 plane of the configuration for N = 18 ions that we believe represents the global minimum (but we cannot prove it) is shown in Fig. 1. Here we have set $e^2/4\pi\varepsilon_0 = M = 1$, and chosen $\nu_1 = 0.7$, $\nu_2 = 1.3$, and $\nu_3 = 2$. Remarkably enough, the configuration does not have the reflection and inversion symmetries of the trap itself.

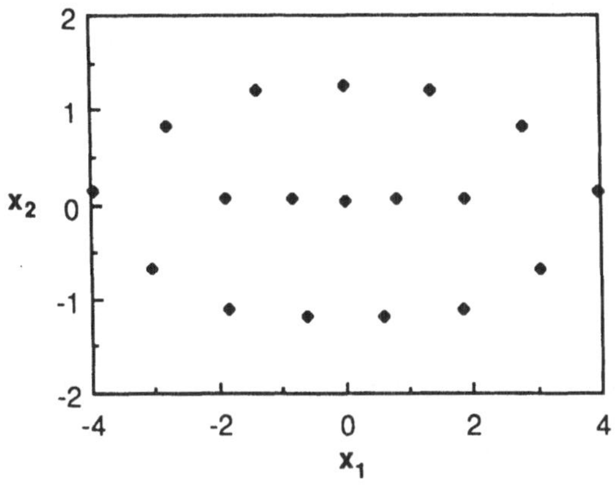

Figure 1. Projection onto the 1-2 plane of the equilibrium configuration of 18 ions, for $e^2/4\pi\varepsilon_0 = M = 1$, $\nu_1 = 0.7$, $\nu_2 = 1.3$, $\nu_3 = 2$. The configuration does not have all inversion and reflection symmetries of the trap.

Nonetheless, rigid clusters of ions have never been observed. It is an empirical fact that cooling of many-ion clouds stops at temperatures higher than cooling of a single ion, too high for crystallization. Spurious coupling of the kinetic energy between the micromotion and the secular motion of the ions, which is believed to take place as a result of the imperfections of the trap geometry [19], may be the reason. But it may also be that the collective motion owing to the Coulomb interactions of the particles at some point simply prevents the cooling, so that crystallization is impossible as a matter of principles.

The latter is the issue we shall address, albeit indirectly: We assume that the ions have crystallized already, and study the dynamics of light-pressure cooling [20]. Clearly it is at least a necessary condition for crystallization that laser cooling can maintain the crystalline state.

Let us assume that the remaining thermal excitations in the ion cluster can be treated as *small vibrations*. We thus expand the potential energy of the ion cluster up to second order in the displacements from the equilibrium $\{X^{\alpha i}\}$,

$$V(\{x^{\alpha i}\}) = V(\{X^{\alpha i}\}) + \frac{1}{2}\sum_{\beta j,\gamma k} D_{\beta j,\gamma k}(x^{\beta j}-X^{\beta j})(x^{\gamma k}-X^{\gamma k}) + \dots, \tag{2}$$

with

$$D_{\beta j,\gamma k} = \frac{\partial^2}{\partial X^{\beta j}\,\partial X^{\gamma k}} V(\{X^{\alpha i}\}). \tag{3}$$

As a symmetric matrix D has got real and, for a stable configuration, non-negative eigenvalues. We denote them by $M\nu^2(k)$, where M is the mass of the ion and $\nu(k)$ is the frequency of the eigenmode k = 1,..., 3N. The corresponding orthonormal eigenvectors $\xi^{\alpha i}(k)$, in particular, satisfy the completeness relation

$$\sum_k \xi^{\alpha i}(k)\,\xi^{\beta j}(k) = \delta_{\alpha,\beta}\,\delta_{i,j}\,. \tag{4}$$

We finally quantize the small vibrations of the ion cluster by writing the coordinate i of the ion α as

$$x^{\alpha i} = X^{\alpha i} + \sum_k \sqrt{\frac{\hbar}{2M\nu(k)}}\,\xi^{\alpha i}(k)\,(b_k + b_k^\dagger)\,, \tag{5}$$

where b_k is the annihilation operator of the excitation of the mode k. We are now in a position to write down the complete Hamiltonian for the system ions+light:

$$\frac{H}{\hbar} = \omega \sum_\alpha |2\rangle_{\alpha\,\alpha}\langle 2| + \sum_q \Omega_q\, a_q^\dagger a_q + \sum_k \nu(k)\, b_k^\dagger b_k$$

$$- \frac{\mathbf{d}}{\hbar}\cdot \sum_\alpha \left[\, |2\rangle_{\alpha\,\alpha}\langle 1|\, \mathbf{E}^{(+)}(\mathbf{x}^\alpha) + \mathrm{h.\,c.} \right]. \tag{6}$$

Here the first term gives the internal energy of the two-level ions with upper and lower levels 2 and 1 and optical transition frequency ω, the second term is the energy of the photon modes labeled by the index q which stands for both the wave vector $\mathbf{q}$ and the polarization state, the third term is the energy of the small vibrations, and the final term is the dipole interaction. $\mathbf{E}^{(+)}$ denotes the positive frequency part of the quantized electric field, and $\mathbf{d}$ is the dipole moment matrix element of the ion between the states 1 and 2. As the electric field in the Hamiltonian (6) is evaluated at the positions of the ions $\mathbf{x}^\alpha$ which in this theory are quantized dynamical variables, the possibility of affecting the center-of-mass motion of the ions by light, i.e., recoil effects, are fully included.

We take the light intensity to be low enough that a photon absorption is always followed by spontaneous emission. Corresponding to the two interactions, second-order perturbation theory is a suitable method for attacking (6). We are especially interested in the changes of the vibrational state. As the recoil effects proportional to the photon momenta are generically small, we shall only consider processes that are lowest nonzero order also in the photon momenta. These are cycles of absorption and spontaneous emission aided by absorption or emission of precisely one vibrational quantum in one of the modes k.

We do not work out the details of the perturbation theory [13,21], but merely cite the final result [22]: The rate of a transition where the ion system starts out with n_k vibrational quanta in the mode k and makes a transition to a state with $n_k + 1$ quanta is found to be

$$R(n_k \rightarrow n_k+1) = (n_k+1)\frac{3\hbar\kappa^2\gamma\,\omega^2}{8\pi M\,\nu(k)\,c^2} \times \left[\frac{A_{[nn]}}{\Delta^2+\gamma^2} + \frac{A_{[ee]}}{(\Delta+\nu(k))^2+\gamma^2} - 2\,\mathrm{Re}\frac{A_{[en]}}{(\Delta+i\gamma)(\Delta+\nu(k)-i\gamma)}\right]. \qquad (7)$$

Here κ is the Rabi frequency corresponding to the laser field strength, γ the natural linewidth of the optical two-level transition of the ion, $\Delta = \omega - \Omega$ is the detuning between the laser frequency Ω and the ions' transition frequency ω, $\mathbf{e}$ stands for the propagation direction of the laser beam, and the symbolic functional notation A specifies the treatment of the various directions in the integral over the directions of the scattered photons,

$$A_{[uv]} = \int d\mathbf{n}\,[1 - (\mathbf{n}\cdot\frac{\mathbf{d}}{|\mathbf{d}|})^2] \sum_{\alpha,\beta} e^{i(\mathbf{X}^\alpha-\mathbf{X}^\beta)\cdot(\mathbf{e}-\mathbf{n})\,\omega/c}\; \mathbf{u}\cdot\xi^\alpha(k)\;\; \mathbf{v}\cdot\xi^\beta(k) \quad . \qquad (8)$$

The rate for a process in which the vibrational quantum number decreases, $R(n_k \rightarrow n_k-1)$, is obtained from (7) by replacing n_k+1 with n_k and $\nu(k)$ with $-\nu(k)$. The rates $R(n_k \rightarrow n_k\pm1)$ represent within our assumptions all transitions where the center-of-mass motion of the ions may change, hence they completely determine both the time evolution of the cooling and the ensuing steady state.

It is clear from (7) that all modes k can be cooled efficiently at the same time only if $\nu(k) \leq \gamma$ for all k. Cooling of any mode k depends on transitions where the two-level resonance is aided by absorption or emission of vibrational quanta, and near-resonance is possible for all modes only if the range of the mode frequencies falls within the linewidth of the transition.

Except for the resonance factors, the second significant ingredient of (7) are the functionals A. The spatial phases $\propto \exp(-i\mathbf{X}^\alpha\cdot\mathbf{q})$ in (8) indicate that they convey the interference of light scattered from different ions in the cluster. The functionals A also contain the vibrational eigenmodes $\xi^{\alpha i}(k)$. Although more apparent in the omitted derivation of (7) than in the result itself, the eigenmodes determine how the photon recoil kicks on individual ions add coherently to produce a change of the state of the collective mode k. All told, the factors A depend in a complicated manner on the equilibrium configuration of the ions. In view of the shapes of the clusters, see Fig. 1, they can usually be analyzed only numerically, with great labor.

However, using the properties of the coefficients A one very general conclusion can be drawn. Let us first assume that all vibrational frequencies are below the transition linewidth, so it is meaningful to expand (7) into a power series of $\nu(k)$. Second, as a technical assumption we ignore the angular distribution of fluorescence from each individual ion. In practice this implies that we replace the factor in the square bracket inside the integral in (8) with its average over all angles, 2/3. Now, whatever is the distribution of the vibrational quantum numbers in the mode, the average energy is

$$E_k = \langle \hbar\nu(k) n_k \rangle = \hbar\nu(k) \langle n_k \rangle . \tag{9}$$

Because the rates R are linear in n_k, the rate of change of energy of the mode k may be written

$$\begin{aligned} \frac{dE_k}{dt} &= \langle \hbar\nu(k)[\, R(n_k \to n_k+1) - R(n_k \to n_k-1)\,] \rangle \\ &= \frac{\hbar^2 \kappa^2 \gamma \omega^2}{4\pi M c^2 (\Delta^2+\gamma^2)} \left\{ A^k_{nn} + A^k_{ee} - 2A^k_{en} - \frac{4\Delta E_k}{\hbar\,(\Delta^2+\gamma^2)} \left[A^k_{ee} - A^k_{en} \right] \right\} . \end{aligned} \tag{10}$$

The new coefficients A are closely related to the old ones,

$$A^k_{uv} = \int dn \sum_{\alpha,\beta} e^{i(\mathbf{x}^\alpha - \mathbf{x}^\beta)\cdot(\mathbf{e} - \mathbf{n})\,\omega/c}\, \mathbf{u}\cdot\xi^\alpha(k)\; \mathbf{v}\cdot\xi^\beta(k) . \tag{11}$$

The *crucial* assumption we are going to make is that the total excitation energy is at every instant of time divided equally between the modes. Hence, the energy in any mode k satisfies

$$E_k = E = \frac{1}{3N} \sum_k E_k . \tag{12}$$

The closure of the eigenmodes ξ^α, Eq. (4), gives

$$\sum_k A^k_{ee} = \sum_k A^k_{nn} = 4\pi N ; \quad \sum_k A^k_{en} = 0, \tag{13}$$

and summing both sides of Eq. (10) like in (12) we thus obtain

$$\frac{dE}{dt} = \frac{2\hbar^2\kappa^2\gamma\omega^2}{3Mc^2(\Delta^2+\gamma^2)}\left[1-\frac{2\Delta E}{\hbar(\Delta^2+\gamma^2)}\right] . \tag{14}$$

Equation (14) makes no reference to the mode structure, so it applies equally well to *collective modes* and to *vibrations of a single ion.* It basically states that if a single ion can be cooled, then so can an ion cluster. This was already the point of Ref. 13; here we have presented a more general and precise statement and proof.

Leaving aside a number of technicalities, the necessary condition for (14) is that the mode-mode interactions keep all vibrations at the same temperature, $E_k = E$. As a counterexample we pointed out in Ref. 13 that, in the absence of mode interactions and for a periodic crystal where the crystal momentum is conserved, a traveling laser wave would only couple to a fraction of the phonon modes which scales with the number of ions as $N^{-1/3}$. Besides, half of the modes would be cooled and half of them *heated.* We reiterate our conjecture that the mode-mode interactions, or the absence thereof, is a major factor affecting the cooling of interacting many-ion systems.

In summary, two features of the fundamental physics of ion clusters may be detrimental to cooling to temperatures low enough to achieve crystallization: First, the range of the collective excitation frequencies of the ions may be so broad ($\geq \gamma$) that all modes do not couple efficiently to light. Second, owing to the interference of light scattered from different ions, the cooling is distributed unevenly among the modes, and some modes may even be heated. It is relatively easy to check [22] that for the present-day experimental parameters the range of vibrational frequencies should not be a source of concern. In contrast, the mode-mode interactions cannot be discussed within the harmonic model of small vibrations where the modes are independent,

3. Bose condensates in light-pressure traps

Let us consider a noninteracting Bose gas in an isotropic three-dimensional harmonic oscillator potential which gives the atoms the oscillation frequency ν.

The Bose-Einstein statistics determines the number of atoms in the state $\mathbf{n} = (n_1, n_2, n_3)$ with energy $\varepsilon_{\mathbf{n}} = \hbar\nu\,(n_1+n_2+n_3)$, at temperature T, as

$$N_n = \frac{1}{e^{(\varepsilon_n - \mu)/kT} - 1} . \tag{15}$$

To ensure positive occupation numbers, the chemical potential μ has to remain negative . As $N_n(\mu)$ are monotonously increasing functions of μ, the total number of atoms outside the ground state $\mathbf{n} = (0,0,0)$ is limited from above by

$$N_c = \sum_{n_1, n_2, n_3 > 0} N_n(\mu = 0) , \tag{16}$$

a *finite* number. In the "high-temperature" limit $kT/\hbar\nu \gg 1$ the sum can be carried out by converting it to an integral and choosing ε_n as one of the integration variables. The result is

$$N_c = 1.202 \left(\frac{kT}{\hbar\nu} \right)^3 . \tag{17}$$

If the number of particles exceeds N_c , $N_0 = N - N_c$ atoms are packed in the ground state which in the limit $\mu \to 0$ can accommodate an arbitrary number of them. This is known as Bose condensation.

Our argument is a simple variation of the treatment of Bose condensation of a gas of free atoms as offered in elementary statistical-mechanics textbooks, except for the problem of the thermodynamic limit [23, 24]. Rigorous phase transitions are traditionally thought to take place in the limit $N \to \infty$, in apparent contradiction to (17). For instance, in a gas of free atoms one usually lets $N \to \infty$ keeping the density $n = N/V$ constant, and obtains a condition for the *density* rather than *particle number*.

Using the limit of the harmonic oscillator wave functions for large quantum numbers ($\psi_n(0) \propto (n_1 n_2 n_3)^{-1/2}$, [25]), the density of the gas at the trap center at the Bose condensation threshold is easily found to scale with the dimensional quantities as

$$n_c = \sum_n N_n(\mu = 0) \, |\psi_n(0)|^2 \propto \left(\frac{MkT}{\hbar^2} \right)^{\frac{3}{2}} , \tag{18}$$

precisely like the critical density of the free gas. Obviously one can express the condition for Bose condensation in terms of the intensive quantities n and T and simultaneously let $N \to \infty$ only if the thermodynamic limit is defined by demanding that $N \to \infty$, $\nu \to 0$ in such a way that

$N\nu^3$ remains constant. The thermodynamic limit involves tampering with the trap parameters themselves. Another unpleasant price to be paid is that the spatial dimension of the ground-state wave function scales with the frequency ν as $\nu^{-1/2}$, and hence the density of the Bose condensate $\propto N\nu^{3/2} = N\nu^3 \cdot \nu^{-3/2}$ diverges.

The temperature that can be reached using laser cooling is of the order of $T \sim \hbar\gamma/k$, and the order of magnitude of the oscillation frequency in the recent all-optical trap for sodium [3] was $\nu \sim 2\pi \times 100$ kHz, thus the critical atomic number becomes $N_c \sim 10^6$. For the reported number of atoms, 500, Bose condensation is not expected. A severe problem with sodium, and probably all other elements except hydrogen as well, is that it will not remain a gas in thermal equilibrium at µK temperatures. On the other hand, spin-polarized atomic hydrogen is metastable against formation of H_2 molecules, but the transition wavelengths do not permit laser cooling or trapping at today's state of the technology. It may take quite a while before Bose condensation in atom traps will be observed.

Nevertheless, we disregard the experimental odds and proceed to pursue our theme. In elementary statistical mechanics the macroscopic population of the ground state,

$$\langle b_0^\dagger b_0 \rangle = N_0 \,, \tag{19}$$

is adopted as the signature of Bose condensation. However, like in laser theories which predict the Poissonian photon statistics [26] but not the coherent field commonly employed to model the output of an ideal laser, we suggest that (19) is not the whole story. We assume that the annihilation and creation operators of the atoms in the ground state of the trap themselves acquire nonzero expectation values,

$$\langle b_0 \rangle = e^{-i\phi}\sqrt{N_0} \,, \langle b_0^\dagger \rangle \cong e^{i\phi}\sqrt{N_0} \,. \tag{20}$$

The random phase ϕ is attributed to spontaneous symmetry breaking, a concept familiar from the theory of (presumably) another Bose condensate, ^{4}He below the λ transition [27].

The issue is, does the phase ϕ have observable consequences. In the laser analog we would ask, how can we tell the field from a laser has a phase. There appears to be no simple answer. It is well known that the usual square-law detectors do not make any distinction between different fields having a Poissonian photon statistics, so photon counters are not sensitive to the phase.

But *two* lasers can beat against each other, revealing their *phase difference.* Guided by the analog, we imagine that two traps containing Bose condensates are first far apart, and are then brought close enough to each other to exchange atoms. When the traps are separated, their ground-state wave functions ψ_l and ψ_r are degenerate. The interaction of the traps at the close distance d is taken to be strong enough to lift the degeneracy, but too weak to mix oscillator states that were not degenerate initially. The relevant Hamiltonian for the ground state thus reads

$$H = \hbar\kappa(b_l^\dagger b_r + b_r^\dagger b_l) . \tag{21}$$

Here b_l, b_r are boson operators that annihilate atoms with the wave functions ψ_l, ψ_r, and κ is a parameter characterizing the interaction strength. A qualitative estimate for κ is obtained by multiplying the overlap of the wave functions ψ_l and ψ_r by the oscillator frequency ν, giving

$$\kappa \sim \nu \exp\left[-\frac{M\nu d^2}{4\hbar} \right] . \tag{22}$$

The frequency κ decreases very rapidly with increasing distance between the traps; e.g., for sodium and with $\nu = 2\pi\times 100$ kHz, the time scale $1/\kappa$ is one hour for $d = 3$ μm.

The Heisenberg equations of motion of the boson operators under the Hamiltonian (21) can be solved trivially. We adopt for each trap separately an initial condition of the form (20), and obtain for the number of atoms in the left trap the expression

$$< b_l^\dagger(t)\, b_l(t) > \; = \; N_l \cos^2\kappa t + N_r \sin^2\kappa t + \sqrt{N_l N_r}\, \sin(\phi_l - \phi_r) \sin 2\kappa t . \tag{23}$$

The result shows that if the initial number of atoms is different ($N_l \neq N_r$), the atoms start oscillating between the traps, as one might expect. Much more surprising is that the atoms start oscillating even though their number is the same. The phenomenon is governed by the phases ϕ; the amplitude of the oscillations of the number of atoms measure the phase difference ϕ_l-ϕ_r. We are describing a macroscopic quantum phenomenon associated with spontaneous symmetry breaking, very much analogous to the (DC) Josephson effect [28].

Experimental demonstration of Bose condensation may be very difficult, and construction of the double trap is probably even more demanding. Also, the trap scheme operates on neutral particles which do not directly couple to electromagnetic fields. Analogs of the practical applications of the Josephson effect are thus not obvious. The trapped atoms couple to gravity and acceleration, and the oscillation frequency κ is extremely sensitive to the separation of the traps; but we do not envisage new-generation laser gyros or meter sticks.

Instead, we forward both a single trap and the double trap as minilabs for theoretical, and later maybe experimental, studies of the very fundamentals of statistical mechanics. The interactions of spin-polarized hydrogen atoms are weak. However, in a trap the density of the Bose condensate tends to infinity at least in the thermodynamic limit we have described, so even weak interactions may profoundly affect the Bose condensate. In fact, the thermodynamic limit itself is a most interesting problem. The number of particles in the trap may be varied, and may be small in the many-particle standards. What happens to Bose condensation and to spontaneous symmetry breaking under such conditions is at present to a large extent unknown.

References

1. W. Neuhauser, M. Hohenstatt, P. E. Toschek, and H. G. Dehmelt, Phys. Rev. A **22**, 1137 (1980).
2 A. Migdall, J. V. Prodan, W. D. Phillips, T. H. Bergeman, and H. J. Metcalf, Phys. Rev. Lett. **54**, 2596 (1985).
3. S. Chu, J. E. Bjorkholm, A. Ashkin, and A. Cable, Phys. Rev. Lett. **57**, 314 (1986).
4. W. Nagourney, J. Sandberg, and H. Dehmelt, Phys. Rev. Lett. **56**, 2727 (1986).
5. Th. Sauter, W. Neuhauser, R. Blatt, and P. E. Toschek, Phys. Rev. Lett. **57**, 1696 (1986).
6. J. C. Bergquist, R. G. Hulet, W. M. Itano, and D. J. Wineland, Phys. Rev. Lett. **57**, 1699 (1986).
7. J. J. Bollinger, J. D. Prestage, W. M. Itano, and D. J. Wineland, Phys. Rev. Lett. **54**, 1000 (1985).
8 J. P. Hansen, Phys. Rev. A **8**, 3096 (1973).
9 W. L. Slattery, G. D. Doolen, and H. E. DeWitt, Phys Rev. A **21**, 2087 (1980).
10 J. J. Bollinger and D. J. Wineland, Phys. Rev. Lett. **53**, 348 (1984).
11. J. Mostowski and M. Gajda, Acta Phys. Pol. A **67**, 783 (1985).
12. J. P. Schiffer and O. Poulsen, Europhysics Lett. **1**, 55 (1986).
13. J. Javanainen, Phys. Rev. Lett. **56**, 1798 (1986).
14. T. J. Greytak and D. Kleppner, in *New trends in Atomic Physics*, edited by G. Grynberg and R. Stora (North-Holland, Amsterdam, 1984).
15. R. V. E. Lovelace, C. Mehanian, T. J. Tommila, and D. M. Lee, Nature **318**, 30 (1985).
16. T. J. Tommila and R. V. E. Lovelace (to be published).
17. V. Bagnato and D. E. Pritchard (to be published).
18 J. Javanainen, Phys. Rev. Lett. **57**, 3164 (1986).
19. H. G. Dehmelt, Adv. At. Mol. Phys. **3**, 53 (1967).
20. An authoritative review of theories of cooling of *a single particle*, trapped or free, is given by S. Stenholm, Rev. Mod. Phys. **58**, 699 (1986).
21. The perturbative method is a derivative of that in D. J. Wineland and W. M. Itano, Phys. Rev. A **20**, 1521 (1979).
22. J. Javanainen (to be published).
23. S. R. de Groot, G. J. Hooyman, and C. A. ten Seldam, Proc. Royal Soc. London, Ser. A **203**, 266 (1950).
24. R. Masut and W. J. Mullin, Am. J. Phys. **47**, 493 (1979).
25. J. T. M. Walraven and I. F. Silvera, Phys. Rev. Lett **44**, 168 (1980).
26. M. Sargent III, M. O. Scully, and W. E. Lamb, Jr., *Laser Physics* (Addison-Wesley, Reading, MA, 1974).
27. D. Forster, *Hydrodynamic Fluctuations, Broken Symmetry and Correlation Functions*, Frontiers is Physics Vol. 47 (Benjamin, Reading, MA, 1975).
28. B. D. Josephson, Phys. Lett. **1**, 251 (1962).

PART VI: Other Fundamentals

EXPECTATION VALUES, Q-FUNCTIONS AND EIGENVALUES FOR DISPERSIVE OPTICAL BISTATBILITY

H. Risken, K. Vogel
Abteilung für Theoretische Physik, Universität Ulm
D-7900 Ulm, Federal Republic of Germany

1 Introduction

Optical bistability has become an important field in quantum optics, see for instance [1-4] for reviews. A fully quantum mechanical treatment of optical bistability requires the solution of the master equation, i. e. the equation of motion for the density operator. For the model of Drummond and Walls (DW) [5] describing dispersive optical bistability we discuss two methods for solving this master equation. The DW model has the advantage that only the operators of the cavity light mode enter in the equation of motion of the density operator. Besides this simplicity it is a nonlinear and nontrivial model.

The most important solution is the stationary solution of the master equation. This solution was already obtained by DW using a complex P-representation of the density operator. The complex P-function as well as the positive P-function have been introduced and further investigated by Gardiner [6]. As was shown by DW expectation values of the light operators can be expressed in terms of generalized Gauss hypergeometric series.

The next important quantity of the master equation for the density operator is its lowest nonzero eigenvalue. Without fluctuations we have two stable states for appropriate driving fields and system parameters. With the inclusion of fluctuations transitions between these two stable states are possible. The lowest nonzero eigenvalue determines the transition rates between the two states. For the model of DW no analytic results are known for the eigenvalues. Therefore a numerical procedure is needed for determining the lowest nonzero eigenvalue. In this manuscript two methods are presented by which some of the lowest real eigenvalues are calculated. (An extension to the calculation of complex eigenvalues seems also to be possible). In the first method [7] we expand the density operator into eigenstates of the system Hamilton operator. Then one obtains a system of coupled differential equations for the matrix elements of the density operator. If we restrict ourselves to the slow motion of the density operator only the diagonal elements of the density matrix need to be taken into account in the small cavity damping limit. The lowest nonzero eigenvalues as well as some other low real ones follow from the equation of motion of the diagonal matrix elements.

In the second method [8] we transform the master equation into a Fokker-Planck equation (FPE) for the Q-function of the density operator, see (4.7). Because the diffusion matrix is not positive definite or semidefinite it is not an ordinary FPE which can be interpreted as describing the Brownian motion of a particle in a suitable potential. We have termed such a FPE a quantum-Fokker-Planck equation (QFPE). For a non positive definite diffusion matrix we may still have a stable stationary solution. An illustrative example is the equation

$$\frac{\partial W}{\partial t} = \left[\frac{\partial}{\partial x_1}(x_1-\alpha x_2) + \frac{\partial}{\partial x_2}(x_2+\alpha x_1) + \frac{\partial^2}{\partial x_1^2} - q\frac{\partial^2}{\partial x_2^2} \right] W , \tag{1.1}$$

which has a stable stationary solution if the conditions $q < 1$ and $(1+q)^2/(1-q)^2 < 1+\alpha^2$ are fulfilled. By applying the matrix continued fraction (MCF) method for solving two variable FPEs [9] we obtain the stationary solutions as well as the lowest nonzero real eigenvalues and some other low real eigenvalues of the QFPE for the Q-function. The eigenfunction corresponding to the lowest nonzero eigenvalue is also calculated. The method is not only applicable for pure quantum fluctuations where the number of thermal quanta n_{th} is zero. It is also applicable to the case $n_{th} > 0$ where the detailed balance condition for the complex P-function is no longer valid and therefore a stationary solution is hard to obtain.

The present paper is organized as follows. In Chap. 2 we present the DW model as well as the classical equation of motion without fluctuations. Next in Chap. 3 we shortly review the procedure for the small cavity damping limit. In Chap. 4 we present the QFPE for the Q-function and outline in Chap. 5 the MCF method for solving it. Finanlly in Chap. 6 stationary Q-functions, some real eigenvalues, the eigenfunction for the lowest nonzero eigenvalue as well as some expectation values are given. It is further shown that expectation values as well as the eigenvalues can also be obtained by applying the MCF-method to the Glauber-Sudarshan P-function.

2 Model and Basic Equations

By expanding the polarization up to third order, by including a coherent classical driving field, by adding losses due to cavity damping and by making the rotating-wave approximation DW obtained a master equation for the density operator of the light field inside the cavity. In a slightly different notation this master equation takes the form

$$\dot{\rho} = -i[H,\rho] + \kappa L_{ir}[\rho] , \tag{2.1}$$

where H and L_{ir} are given by

$$H = -\Omega a^{\dagger}a + \chi a^{\dagger 2}a^{2} - F(a+a^{\dagger}) \tag{2.2}$$

$$L_{ir}[\rho] = 2a\rho a^{\dagger} - \rho a^{\dagger}a - a^{\dagger}a\rho + 2n_{th}[[a,\rho],a^{\dagger}] \ . \tag{2.3}$$

Here $\Omega = \omega_1 - \omega_c$ is the difference between the frequency of the classical driving field F and the cavity frequency, χ is the imaginary part of the third order susceptibility, κ is the cavity damping constant, n_{th} is the number of thermal quanta and $a^{\dagger}$ and a are the creation and annihilation operators for the light field inside the cavity.

From (2.1) we obtain for the complex amplitude

$$\alpha = \mathrm{Tr}(a\rho) \ , \qquad \alpha^{*} = \mathrm{Tr}(\alpha^{\dagger}\rho) \tag{2.4}$$

$$\dot{\alpha} = i\Omega\alpha - \kappa\alpha - 2i\chi\,\mathrm{Tr}(a^{\dagger}a^{2}\rho) + iF \ . \tag{2.5}$$

By replacing $\mathrm{Tr}(a^{\dagger}a^{2}\rho)$ in terms of the expectation value (2.4), i. e. by $\alpha^{*}\alpha^{2}$ we arrive at the classical equation whithout fluctuations

$$\dot{\alpha} = \left[i\Omega - \kappa - 2i\chi\alpha^{*}\alpha\right]\alpha + iF \ . \tag{2.6}$$

By using the normalized time $\tilde{t}$, amplitude $\tilde{\alpha}$, intensity $\tilde{I} = \tilde{\alpha}^{*}\tilde{\alpha}$, damping constant $\tilde{\kappa}$ and driving field $\tilde{F}$ defined by $(\Omega > 0)$

$$\tilde{t} = \Omega t, \qquad \tilde{\alpha} = \sqrt{\chi/\Omega}\,\alpha, \qquad \tilde{I} = (\chi/\Omega)\,I, \qquad \tilde{\kappa} = \kappa/\Omega, \qquad \tilde{F} = \frac{1}{\Omega}\sqrt{\chi/\Omega}\,F \tag{2.7}$$

(2.6) is transformed to the normalized form

$$d\tilde{\alpha}/d\tilde{t} = [i(1 - 2\tilde{\alpha}^{*}\tilde{\alpha}) - \tilde{\kappa}]\,\tilde{\alpha} + i\tilde{F}. \tag{2.8}$$

It follows from (2.8) that the following connection between $|\tilde{F}|$ and $\tilde{I}$ is valid for the stationary state ($d\tilde{\alpha}/d\tilde{t} = 0$, see also [5])

$$|\tilde{F}| = \sqrt{\tilde{I}[\tilde{\kappa}^{2} + (1-2\tilde{I})^{2}]} \ . \tag{2.9}$$

According to DW the stationary solution is unstable between the turning points, see Fig. 1. Bistabilty occurs if we have turnig points, i. e. for

$$\tilde{\kappa} < 1/\sqrt{3}, \qquad |\tilde{F}| < \sqrt{1 + 9\tilde{\kappa}^{2} + (1-3\tilde{\kappa}^{2})^{3/2}}\Big/(3\sqrt{3}) \ . \tag{2.10}$$

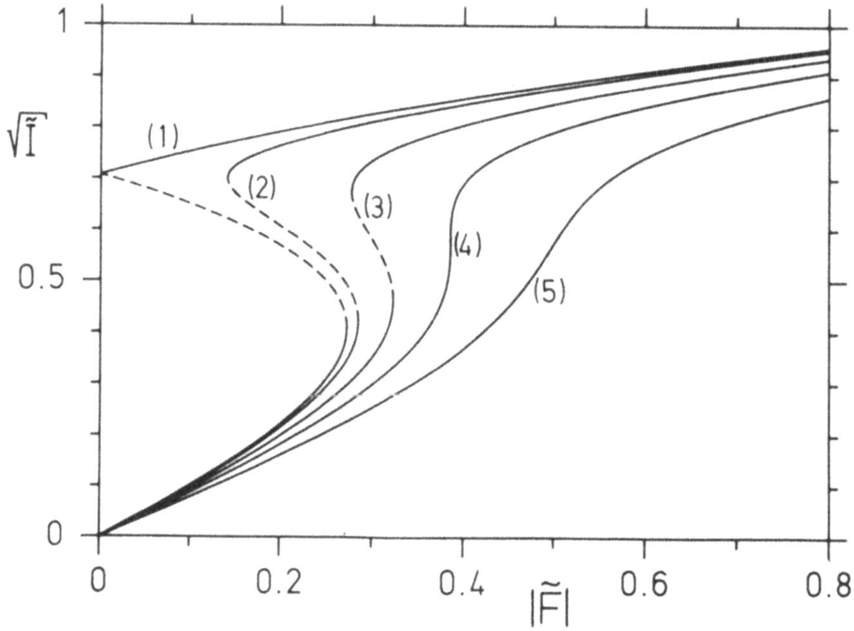

Fig. 1: Stable (solid lines) and unstable solutions $(\tilde{I})^{1/2} = |\tilde{\alpha}|$ (broken lines) of (2.8) as a function of the normalized driving field $\tilde{F}$ for $\tilde{\kappa} = 0$ (1), $\tilde{\kappa} = 0.2$ (2), $\tilde{\kappa} = 0.4$ (3), $\tilde{\kappa} = 1/\sqrt{3}$ (4) and $\tilde{\kappa} = 0.8$ (5)

3 Small Cavity Damping Limit

For vanishing κ the system will remain in an eigenstate of H if it was initially in such a state. Therefore the eigenstates of H denoted by $|m)$, i. e.

$$H|m) = E_m|m) \tag{3.1}$$

will play an important role. By expanding $|m)$ into Fock-states $|n\rangle$ these eigenstates can be calculated [7]. As was further shown in this reference the slow motion of the density operator ρ is given by the diagonal elements of the density operator

$$p_m = (m|\rho|m) \tag{3.2}$$

in the small cavity damping limit for $\kappa \ll |E_m - E_n|$. From (2.1) one can derive the following Pauli master equation for p_m [7]

$$\dot{p}_m = 2\kappa\left[\sum_{\ell} w(\ell\to m)p_{\ell} - \sum_{\ell} w(m\to\ell)p_m\right] = 2\kappa \sum_{\ell} W_{m\ell}\, p_{\ell} \ , \tag{3.3}$$

where the transition probabilities $w(\ell\to m)$ are given by

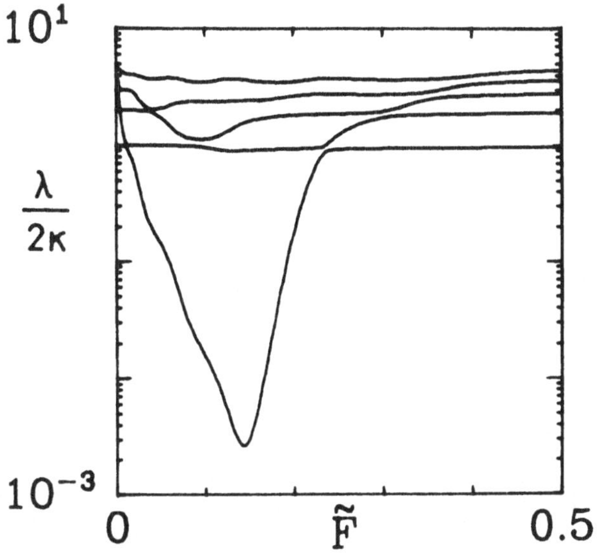

Fig. 2: Plot of the real eigenvalues of the matrix $W_{m\ell}$ in a logarithmic scale as a function of the normalized driving field $\tilde{F}$ for $\chi/\Omega = 0.11$ and $n_{th} = 0$ (taken from [7])

$$w(\ell \to m) = |(m|a|\ell)|^2(1+n_{th}) + |(m|a^\dagger|\ell)|^2 n_{th} \ . \tag{3.4}$$

These transition rates are easily calculated by using the expansion of the eigenstates $|m)$ in Fock states [7]. The real eigenvalues, which scale with 2κ, are the eigenvalues of the matrix

$$W_{m\ell} = w(\ell \to m) - \sum_n w(m \to n)\delta_{m\ell} \tag{3.5}$$

occuring in (3.3). In Fig. 2 some of these lowest real eigenvalues are plotted as a function of the driving field. As was shown in [7] stationary expectation values as well as the Q-function can also be obtained from (3.3) in the small cavity damping limit.

4 Quantum-Fokker-Planck Equation

Any normally or antinormally expectation value of the light field operators a and $a^\dagger$ may be obtained from the characteristic functions

$$\tilde{P}(\beta) = \tilde{P}^+(\beta) = \mathrm{Tr}\{e^{i\beta^* a^\dagger} e^{i\beta a} \rho\} \ , \qquad \tilde{Q}(\beta) = \tilde{P}^-(\beta) = \mathrm{Tr}\{e^{i\beta a} e^{i\beta^* a^\dagger} \rho\} \tag{4.1}$$

by appropriate differentiation with respect to β and β^*. The Fourier-transforms of these characteristic functions

$$P^{\pm}(\alpha) = \pi^{-2}\int e^{-i\alpha\beta - i\alpha^*\beta^*}\,\tilde{P}^{\pm}(\beta)\,d^2\beta \tag{4.2}$$

are the Glauber-Sudarshan P-function and the Q-function, see for instance [10]. Because the equations of motion for the P- and the Q-function derived later on differ only by ± signs, we have used in (4.1,4.2) and will use later on the notation

$$P(\alpha) = P^+(\alpha)\ ; \qquad Q(\alpha) = P^-(\alpha)\ . \tag{4.3}$$

The Q-function can be expressed by the density operator according to

$$Q(\alpha) = P^-(\alpha) = \langle\alpha|\rho|\alpha\rangle/\pi\ , \qquad |\alpha\rangle = \text{coherent state} \tag{4.4}$$

whereas the density operator ρ itself can be expressed by the Glauber-Sudarshan P-function [11,12] by the relation

$$\rho = \int |\alpha\rangle\langle\alpha|\, P(\alpha)\, d^2\alpha\ . \tag{4.5}$$

Normally and antinormally ordered expectation values are obtained from the P- and Q-function by integration

$$\langle a^{\dagger n}a^m\rangle = \int \alpha^{*n}\alpha^m P(\alpha)\,d^2\alpha\ ; \qquad \langle a^m a^{\dagger n}\rangle = \int \alpha^{*n}\alpha^m Q(\alpha)\,d^2\alpha\ . \tag{4.6}$$

Because of squeezing [13] the P-function does not exist in general. As will be seen later on the expansion coefficients of the P-function into a complete set, however, do exist. (Also its Fourier transform $\tilde{P}(\beta)$ does exist.) In order to derive an equation for these expansion coefficients we may nevertheless use the equation of motion for the P-function.

From the master equation (2.1) the following equation for the $P=P^+$- and $Q=P^-$-function was derived [5]

$$\begin{aligned}\frac{\partial P^{\pm}}{\partial t} = &- \frac{\partial}{\partial\alpha}\left(-\kappa\alpha + i\Omega\alpha + 2i(1\mp 1)\chi\alpha - 2i\chi\alpha^2\alpha^* + iF\right)P^{\pm} \\ &- \frac{\partial}{\partial\alpha^*}\left(-\kappa\alpha^* - i\Omega\alpha^* - 2i(1\mp 1)\chi\alpha^* + 2i\chi\alpha^{*2}\alpha - iF\right)P^{\pm} \\ &\mp i\chi\frac{\partial^2}{\partial\alpha^2}\alpha^2 P^{\pm} + 2\kappa\left(n_{th} + \frac{1}{2} \mp \frac{1}{2}\right)\frac{\partial^2 P^{\pm}}{\partial\alpha^*\partial\alpha} \pm i\chi\frac{\partial^2}{\partial\alpha^{*2}}\alpha^{*2}P^{\pm}\ .\end{aligned} \tag{4.7}$$

The upper signs are valid for the P-function, the lower ones for the Q-function. In real notation with

$$\alpha = \alpha_1 + i\alpha_2 \tag{4.8}$$

(4.7) takes the form (sumation convention for the 1 and 2 index)

$$\frac{\partial P^{\pm}}{\partial t} = -\frac{\partial}{\partial \alpha_i} D_i^{\pm} P^{\pm} + \frac{\partial^2}{\partial \alpha_i \partial \alpha_j} D_{ij}^{\pm} P^{\pm} , \tag{4.9}$$

where the drift and diffusion coefficients are given by

$$D_1^{\pm} = -\kappa\alpha_1 - \Omega\alpha_2 - 2(1\mp 1)\chi\alpha_2 + 2\chi\alpha_2(\alpha_1^2 + \alpha_2^2) ,$$

$$D_2^{\pm} = -\kappa\alpha_2 + \Omega\alpha_1 + 2(1\mp 1)\chi\alpha_1 - 2\chi\alpha_1(\alpha_1^2 + \alpha_2^2) + F ,$$

$$D_{11}^{\pm} = \pm\chi\alpha_1\alpha_2 + \frac{\kappa}{2}(n_{th} + \frac{1}{2} \mp \frac{1}{2}), \quad D_{12}^{\pm} = \mp\frac{\chi}{2}(\alpha_1^2 - \alpha_2^2), \quad D_{22}^{\pm} = \mp\chi\alpha_1\alpha_2 + \frac{\kappa}{2}(n_{th} + \frac{1}{2} \mp \frac{1}{2}). \tag{4.10}$$

It is easily derived from (4.10) that the diffusion matrix is not positive definite if the intensity is large enough, i. e.

$$D_{11}^{\pm} D_{22}^{\pm} - (D_{12}^{\pm})^2 < 0 \qquad \text{for } \alpha^*\alpha = \alpha_1^2 + \alpha_2^2 > \frac{\kappa}{\chi}(n_{th} + \frac{1}{2} \mp \frac{1}{2}) . \tag{4.11}$$

Because the diffusion matrix is not positive definte or positive semidefinite everywhere, (4.9) cannot be interpreted as describing the Brownian motion of a particle in a suitable potential and therefore no simple simulation of (4.9) is possible. For this reason (4.9) was termed quantum-Fokker-Planck equation (QFPE). By doubling the phase space [6] it may be possible to derive a FPE with a positive definite diffusion matrix and a simulation is then possible [14]. (By addding Langevin noise forces to (2.6) Graham and Schenzle [15] and Haug et al. [16] obtained a FPE for dispersive optical bistability with a positive definite diffusion matrix. It is similar to (4.7) for $n_{th} >> 1$.) To apply the matrix continued fraction method (MCF method) it is more appropriate to use the intensity I and the phase ϕ defined by

$$I = \alpha^*\alpha = \alpha_1^2 + \alpha_2^2, \qquad \alpha = \alpha_1 + i\alpha_2 = \sqrt{I}\, e^{i\phi} \tag{4.12}$$

as variables. Eqs. (4.7,9) are then transformed to

$$\frac{\partial P^{\pm}}{\partial t} = \Big\{ -\frac{\partial}{\partial I}\big(-2\kappa I + \kappa(2n_{th} + 1 \mp 1) + 2F\sqrt{I}\sin\phi\big) - \frac{\partial}{\partial \phi}\big(\Omega - 2\chi(I-1) \mp \chi + \frac{F}{\sqrt{I}}\cos\phi\big)$$

$$+ \kappa(2n_{th} + 1 \mp 1)\frac{\partial^2}{\partial I^2} I \mp 2\chi \frac{\partial^2}{\partial\phi\partial I} I + \frac{\kappa}{2I}(n_{th} + \frac{1}{2} \mp \frac{1}{2})\frac{\partial^2}{\partial \phi^2} \Big\} P^{\pm} . \tag{4.13}$$

5 Solution in Terms of Matrix Continued Fractions

For the $P=P^+$- and $Q=P^-$-function we use the real expansion

$$P^{\pm}(I,\phi) = \sum_{m=0}^{\infty} \Big\{ a_m^0 \exp(-I/I_s)\, L_m^0(I/I_s)$$

$$+ \sum_{n=1}^{\infty} \frac{2}{\sqrt{n!}} \left(a_m^n \cos n\phi - b_m^n \sin n\phi\right) \exp(-I/I_s)\,(I/I_s)^{n/2} L_m^n(I/I_s) \Big\} , \quad (5.1)$$

where $L_m^n(I/I_s)$ are the generalized Laguerre polynomials. (An expansion of this type was already used in [16,17].) The scaling intensity I_s is arbitrary but will be chosen such to achieve good numerical convergence of the expansion (5.1). The factor $1/\sqrt{n!}$ was added to reduce the numerical errors for the inversion of the matrices. In the next step we insert (5.1) into (4.13) and obtain the following recurrence relations for the expansion coefficients

$$\dot{a}_m^0 = \left(2F/\sqrt{I_s}\right) m\, b_{m-1}^1 - \frac{2\kappa}{I_s}\left(n_{th} + \frac{1}{2} \mp \frac{1}{2} - I_s\right) m\, a_{m-1}^0 - 2\kappa m\, a_m^0$$

$$\dot{a}_m^n = \left(F/\sqrt{I_s(n+1)}\right) m\, b_{m-1}^{n+1} + F\sqrt{n/I_s}\; b_m^{n-1} - \frac{2\kappa}{I_s}\left(n_{th} + \frac{1}{2} \mp \frac{1}{2} - I_s\right) m\, a_{m-1}^n - 2\kappa(m+n/2)\, a_m^n$$

$$- n\left\{-\Omega + \chi(2I_s \pm 1)(2m+n) + \chi(2I_s - 2 \pm 1)\right\} b_m^n + 2nm\chi(I_s \pm 1)\, b_{m-1}^n + \frac{2\chi}{I_s} n(m+n+1)\, b_{m+1}^n$$

$$\dot{b}_m^n = -\left(F/\sqrt{I_s(n+1)}\right) m\, a_{m-1}^{n+1} - F\sqrt{n/I_s}\; a_m^{n-1} - \frac{2\kappa}{I_s}\left(n_{th} + \frac{1}{2} \mp \frac{1}{2} - I_s\right) m\, b_{m-1}^n - 2\kappa(m+n/2)\, b_m^n$$

$$+ n\left\{-\Omega + \chi(2I_s \pm 1)(2m+n) + \chi(2I_s - 2 \pm 1)\right\} a_m^n - 2nm\chi(I_s \pm 1)\, a_{m-1}^n - \frac{2\chi}{I_s} n(m+n+1)\, a_{m+1}^n . \quad (5.2)$$

The normally or antinormally ordered moments (4.6) can be expressed by the coefficients of the expansion (5.1) for the P- and Q function and vice versa. We have for instance for the P-function

$$\langle a \rangle = \pi I_s^{3/2}\left(a_0^1 - i b_0^1\right) , \qquad \langle a^\dagger a \rangle = \pi I_s^2 \left(a_0^0 - a_1^0\right) \quad (5.3)$$

and for the Q-function

$$\langle a \rangle = \pi I_s^{3/2}\left(a_0^1 - i b_0^1\right) , \qquad \langle a^\dagger a \rangle = \langle a a^\dagger \rangle - 1 = \pi I_s^2 \left(a_0^0 - a_1^0\right) - 1 . \quad (5.4)$$

It should be noted that the normalization

$$\int P(\alpha)d^2\alpha = \int Q(\alpha)d^2\alpha = \int P^{\pm}(\alpha)d^2\alpha = 1 \tag{5.5}$$

requries $a_0^0 = 1/(I_s\pi)$ in both cases. If squeezing occurs the expansion (5.1) for the P-function does not exist, though the expansion coefficients of the P-function do exist. The equation of motion for the expansion coefficients of the P-function (5.2) determine these coefficients. In principle, the P-function and the corresponding QFPE can be avoided by only using the the master equation (2.1) and the connection between moments and expansion coefficients for the derivation of (5.2).

In (5.2) only coefficients with adjacent indices are coupled. By introducing column vectors of the form

$$\mathbf{c}_m = (a_m^0,\ a_m^1,\ b_m^1,\ a_m^2,\ b_m^2, \ldots) \tag{5.6}$$

we can cast the recurrence relations (5.2) into the tridiagonal vector recurrence relation

$$\dot{\mathbf{c}}_m = \mathbf{Q}_m^-\mathbf{c}_{m-1} + \mathbf{Q}_m\mathbf{c}_m + \mathbf{Q}_m^+\mathbf{c}_{m+1}\ , \tag{5.7}$$

where $\mathbf{Q}_m^{\pm}$ and $\mathbf{Q}_m$ are matrices following from (5.2). By investigating the Brownian motion problem in tilted periodic potentials Vollmer and one of us had derived tridiagonal vector recurrence relations of the form (5.7) [18]. In these references the stationary solution, eigenvalues, eigenfunctions as well as some other instationary solutions of an equation of the form (5.7) have been obtained by calculating appropriate matrix continued fractions (MCF), see also [9] Chaps. 9 and 11 for a review. The same MCF method can be applied to (5.7). In order to calculate the matrix continued fractions, the expansion (5.1) has to be truncated at a large but finite index $n = L$, so that the matrices $\mathbf{Q}$, $\mathbf{Q}^{\pm}$ in (5.7) have the dimension $(2L+1)*(2L+1)$. Furthermore, the infinite continued fractions have to be replaced by their Mth approximants. This means that the expansion (5.1) is also truncated at the index $m = M$. The truncation indices L and M have to be chosen such that the final results do not change within a given accuracy if L and M is increased. The explicit structure of the matrices $\mathbf{Q}_m$, $\mathbf{Q}_m^{\pm}$ and the details of the numerical procedure will be presented in a future publication.

6 Results

First we discuss some stationary expectation values. The average amplitude $|\langle a\rangle|$ and the classcial amplitude $\sqrt{I} = |\alpha|$ as a function of the driving field are plotted in Fig. 3. The average value $\langle a\rangle$ calculated with the P-function (5.3) agrees with the

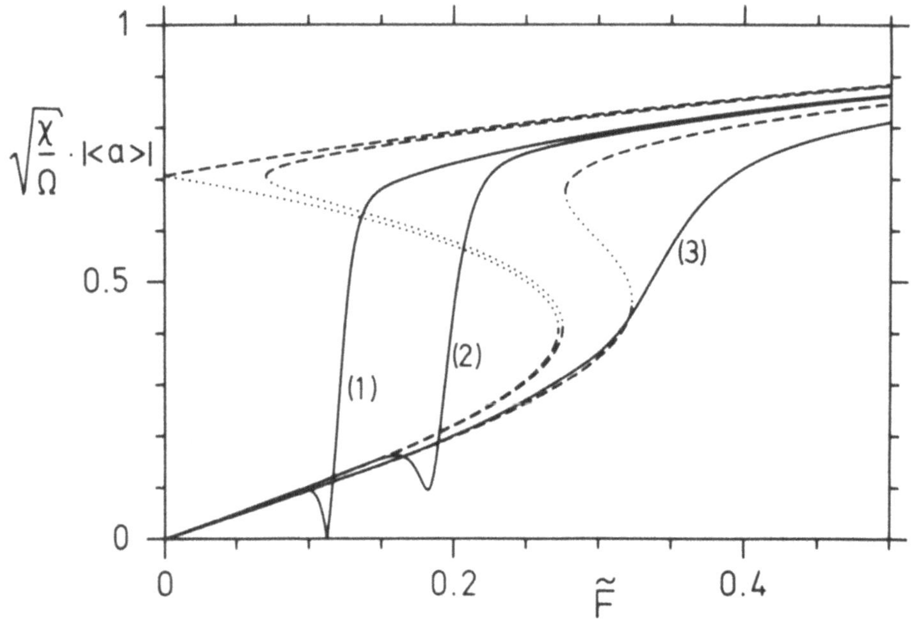

Fig. 3: The absolute amount of the average amplitude (solid lines) and of the classical amplitude, i. e. $|\tilde{\alpha}|$ of Fig. 1 (broken lines: stable, dotted lines: unstable) as a function of $\tilde{F}$ for $\chi/\Omega = 0.1$, $n_{th} = 0$, $\tilde{\kappa} = 0.001$ (1), $\tilde{\kappa} = 0.1$ (2) and $\tilde{\kappa} = 0.4$ (3) (all quantities are normalized according to (2.7))

same value (5.4) calculated with the Q-function and with the analytic expression derived by DW.

In order to decide whether squeezing occurs we introduce the quadrature phases

$$x_1 = \frac{1}{2}(a + a^\dagger), \quad x_2 = \frac{1}{2i}(a - a^\dagger), \quad [x_1, x_2] = \frac{i}{2} \tag{6.1}$$

and calculate their variances

$$\sigma_{ik} = \frac{1}{2}\langle x_i x_k + x_k x_i \rangle - \langle x_i \rangle \langle x_k \rangle . \tag{6.2}$$

The matrix elements of σ can be expressed in terms of $\langle a \rangle = \langle a^\dagger \rangle^*$, $\langle a^2 \rangle = \langle a^{\dagger 2} \rangle^*$ and $\langle a^\dagger a \rangle = \langle a a^\dagger \rangle - 1$, see (5.3). Instead of (6.1) we can also choose new operators y_1 and y_2 connected to x_1 and x_2 via a 'rotation'

$$y_1 = x_1 \cos\phi + x_2 \sin\phi, \qquad y_2 = -x_1 \sin\phi + x_2 \cos\phi \tag{6.3}$$

in such a way that σ is diagonal. This choise is equivalent to the diagonalization of the 2*2 matrix σ. Let us call the lower eigenvalue of σ Δy_1^2 and the higher one Δy_2^2.

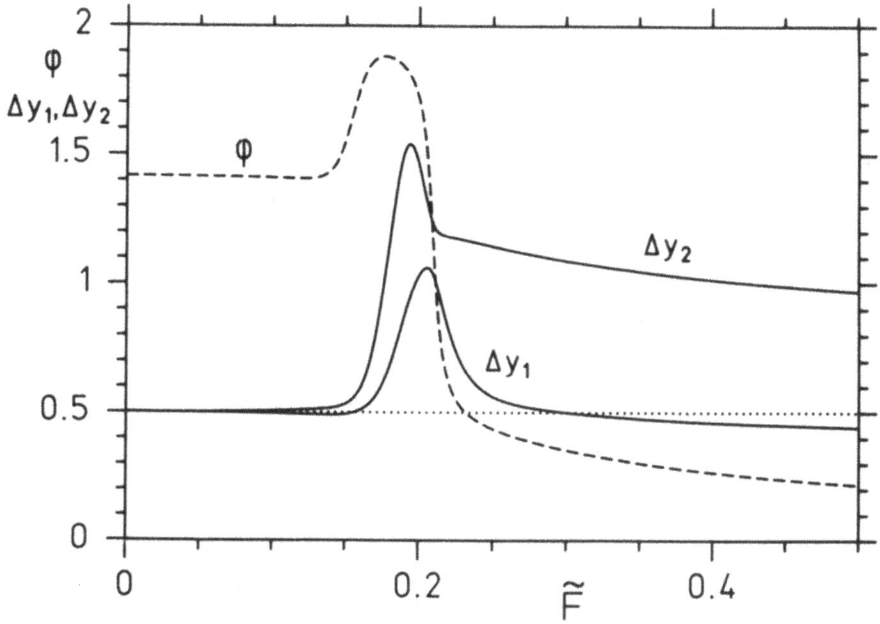

Fig. 4: Variances Δy_1, Δy_2 and the rotation angle ϕ as a function of the driving field $\tilde{F}$ for $\chi/\Omega = 0.1$, $\tilde{\kappa} = 0.1$ and $n_{th} = 0$

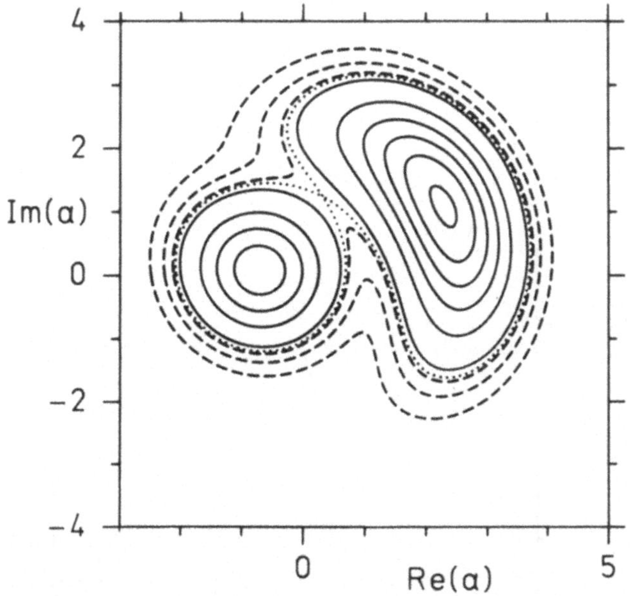

Fig. 5: Stationary Q-function for $\tilde{F} = 0.2$, $\chi/\Omega = 0.1$, $\tilde{\kappa} = 0.1$ and $n_{th} = 0$. The contour lines are $Q = 0.02$, $0.04, \ldots, 0.12$ (solid lines); $Q = 0.005$, 0.010, 0.015 (broken lines); $Q = 0.01665$ (line through saddle point, dotted line)

If for a state we have $\Delta y_1^2 < 1/4$ this state is called a squeezed state. In this case Δy_2^2 must be greater than 1/4 because the uncertainty relation $\Delta y_1^2 \, \Delta y_2^2 \geq 1/16$ must be valid for any state. The rotation angle ϕ as well as the variances Δy_1 and Δy_2 as a function of the driving field F are shown in Fig. 4. As seen we get squeezed states for $\tilde{F} \geq 0.3$. The eigenvalues of the variance matrix have a maximum approximately for those F where the Q-function has two peaks of equal heights, compare Fig. 5

Fig. 5 shows the stationary distribution $Q(\alpha)$ for a field strength F where two peaks are clearly visible. The results are similar to those obtained by the method discussed in [7] for the small cavity damping limit. In contrast to that case an asymmetry with respect to the real axis occurs for finite damping. Because of the squeezing the P-function does not exist. If one tries to sum up the expansion (5.1) for the P-function, the result oscillates and depends on the scaling intensity I_s as well as on the truncation indices L and M.

In Fig. 6 we have plotted some of the lowest nonzero eigenvalues calculated from (5.2) with the MCF method. For the value $\kappa/\Omega = 0.001$ and $\chi/\Omega = 0.11$ we got within the accuracy of the plot the same result as in Fig. 2 which was calculated by diagonalizing the matrix (3.5). Even the small ripples on the left part of Fig. 2 have been reproduced. We also get the same eigenvalues for both P-and Q-function. The lowest nonzero eigenvalue is connected with the escape rate [7]. Thus also the decay

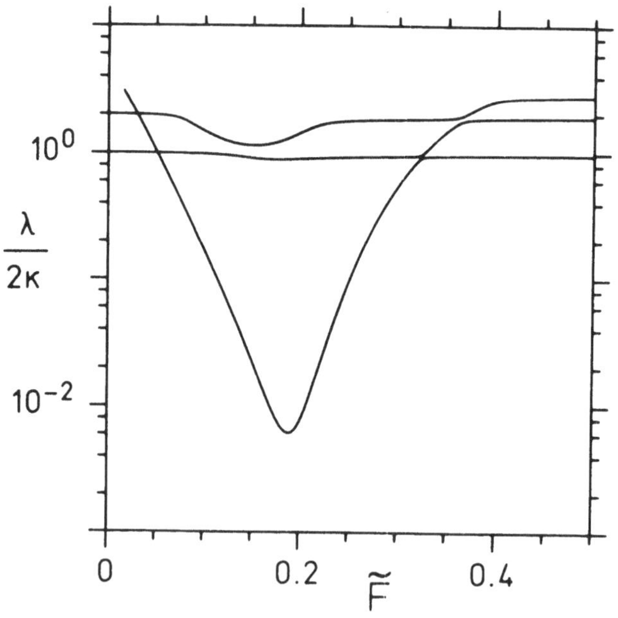

Fig. 6: Plot of the lowest real nonzero eigenvalues divided by 2κ in a logarithmic scale as a function of $\tilde{F}$ for $\chi/\Omega = 0.1$, $\tilde{\kappa} = 0.1$ and $n_{th} = 0$

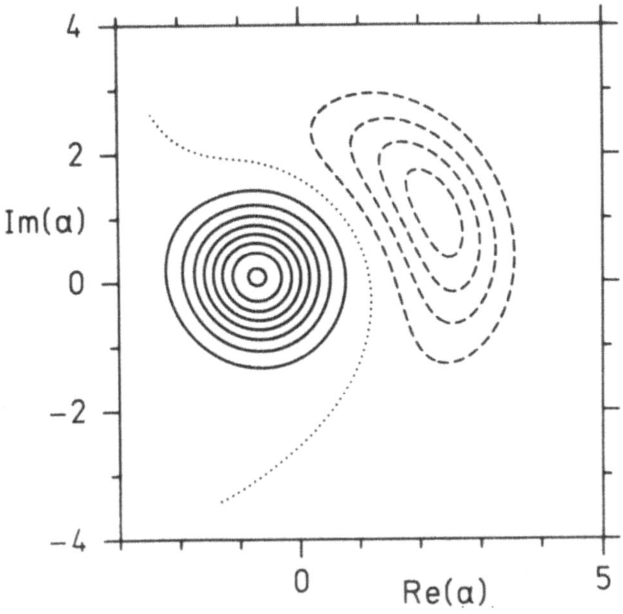

Fig. 7: Unnormalized eigenfunction for the Q-function belonging to the lowest nonzero eigenvalue for the parameters of Fig. 5. The contour lines are equidistantly spaced. The $Q = \text{const} > 0$ lines are the solid ones, the $Q = \text{const} < 0$ lines are the broken ones and the nodal curve $Q = 0$ is shown by a dotted line

rate agrees for both functions. This is of course an essential feature of a fully quantum mechanical theory. As remarked by Drummond [19] in previous calculations on absorptive optical bistability approximations were made which are not consistent with fully quantum mechanical calculations. These calculations lead to tunneling rates which - depending on the representation used - differ by orders of magnitude.

In Fig. 7 the eigenfunction for the Q-function belonging to the lowest nonzero eigenvalue is shown the for parameters of Fig. 5. Near the maximum and minimum the contour lines agree quite accurately with the contour lines near the two maxima in Fig. 5. The nodal line in Fig. 7 separates the two maxima in Fig. 5.

Oscillating Variation of Transition Field

The lowest nonzero eigenvalue has a minimum at a field approximately in the middle of the bistable region. The transition of the average stationary amplitude from the lower branch to the upper branch also occurs approximately at the same field. In the stationary state this transition field $\tilde{F}_{tr}$ may be defined by the condition that the real part of the average amplitude $\langle a \rangle$ vanishes at $\tilde{F}_{tr}$. For small damping constants a strong oscillating variation of this field with respect to the

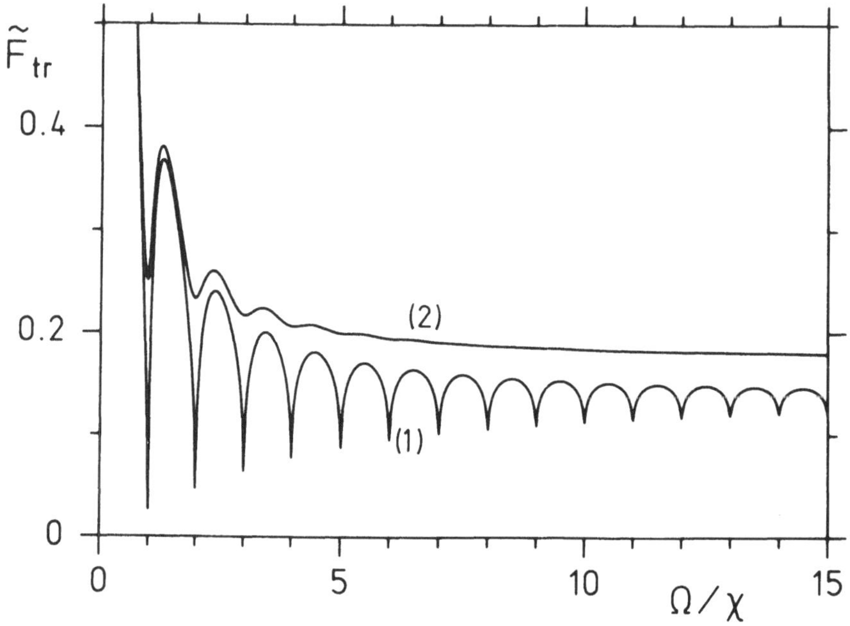

Fig. 8: The field $\tilde{F}_{tr}$ at which the real part of <a> vanishes as a function of Ω/χ for $n_{th} = 0$, $\tilde{\kappa} = 0.001$ (1) and $\tilde{\kappa} = 0.1$ (2)

the parameter Ω/χ was observed, see Fig. 8. (This plot was obtained from Eq. (5.17) of [5]). Near integer values of Ω/χ, $\tilde{F}_{tr}$ reaches very low values for very small $\tilde{\kappa}$. By moving away from these integer values, $\tilde{F}_{tr}$ increases very sharply to values which are not very much affected by changing $\tilde{\kappa}$. It seems that this effect is a typical quantum effect stemming from the discrete energy levels E_m of the Hamilton operator (2.2). By increasing Ω/χ a new eigenstate between the state which evolves from the vacuum state for $F = 0$ and the minimal energy state occurs at integer values of Ω/χ, see also Fig. 2 and Eq. (4.8) of [7]. It was found by the MCF method that the lowest nonzero eigenvalue at the transition field $\tilde{F}_{tr}$ has a larger value at and very near integer values of Ω/χ for small cavity damping. (The method for calculating the eigenvalues described in Sect. 3 and in [7] does not work for integer or near integer Ω/χ values. Here also nondiagonal matrix elements of ρ must be taken into account).

7 Conclusion

In conclusion we have solved the QFPE for the Q- as well as for the P-function with a non positive definite diffusion matrix describing optical bistability with the matrix continued fraction method. Expectation values, the squeezing parameter and the eigenvalues for both functions agree very accurately. The lowest nonzero real eigenvalue has a minimum approximately in the middle of the bistable region. An appreciable oscillating variation of the location of this minimum with the ratio of the detuning to the parameter of the nonlinear susceptibility was found for small cavity damping. Furthermore, the stationary Q-function as well as its eigenfunction belonging to the lowest nonzero eigenvalue have been obtained. The calculation of timedependent correlation functions by the MCF method seems also to be possible.

References

[1] C. M. Bowden, M. Cliftan, H. R. Robl (Eds.): Optical bistability, Plenum Press, New York (1981)

[2] R. Bonifacio (Ed.): Dissipative Systems in Quantum Optics, Topics in Current Physics, Vol. 27, Springer, Berlin (1982)

[3] L. A. Lugiato, Progress in Optics XXI, Ed. E. Wolf, page 69, North-Holland, Amsterdam (1984)

[4] J. C. Englund, R. R. Snapp, W. C. Schieve: Progress in Optics XXI, Ed. E. Wolf, page 355, North-Holland, Amsterdam (1984)

[5] P. D. Drummond, D. F. Walls: J. Phys. **A13**, 725 (1980)

[6] C. W. Gardiner, Handbook of Stochastic Methods, Springer Series in Synergetics, Vol. 13, 2nd ed., Springer, Berlin (1985)

[7] H.Risken, C. Savage, F. Haake, D. F. Walls: Phys. Rev. **A35**,1729 (1987)

[8] K. Vogel, H. Risken: Solution of the Quantum-Fokker-Planck Equation for Dispersive Optical Bistability in Terms of Matrix Continued Fractions, submitted to Opt. Commun.

[9] H. Risken: The Fokker-Planck Equation, Springer Series in Synergetics, Vol. 18, Springer, Berlin (1984)

[10] H. Haken, Laser Theory in Encyclopedia of Physics, Vol. XXV/2c, Ed. S. Flügge, Springer, Berlin (1970)

[11] R. J. Glauber: Phys. Rev. **131**, 2766 (1963)

[12] E. C. G. Sudarshan: Phys. Rev. Lett. **10**, 277 (1963)

[13] D. F. Walls: Nature **306**, 141 (1983)

[14] H. Dörfle, A. Schenzle: Z. Phys. **B65**, 113 (1986)

[15] R. Graham, A. Schenzle: Phys. Rev. **A23**, 1302 (1981)

[16] H. Haug, S. W. Koch, R. Neumann, H. E. Schmidt: Z. Phys. **B49**, 79 (1982)

[17] H. Risken, H. D. Vollmer: Z. Phys. **B39**, 89 and 339 (1980)

[18] H. Risken, H. D. Vollmer: Z. Phys. **B33**, 297 (1979)
H. D. Vollmer, H. Risken: Z. Phys. **B34**, 313 (1979)
H. D. Vollmer, H. Risken: Physica **110A**, 106 (1982)
H. Risken, H. D. Vollmer: Mol. Phys **46**, 555 (1982)

[19] P. D. Drummond: Phys. Rev. **A33**, 4462 (1986)

BERRY'S PHASE AND THE PARALLEL TRANSPORT OF POLARIZATION

Zofia Bialynicka-Birula
Institute of Physics, Polish Academy of Sciences,
Lotnikow 32/46, 02-668 Warsaw, Poland

1. INTRODUCTION

Last year, a simple optical experiment confirmed the existence of the geometrical phase of the wave function predicted by Berry[1] in 1984. This, so called Berry's phase, is a general property of quantum systems evolving in time in adiabatically changing environments.

Originally, Berry considered Hamiltonian systems for which the state vectors evolve according to the Schrödinger equation,

$$i \hbar \partial_t |\psi\rangle = H |\psi\rangle , \tag{1}$$

and the Hamiltonian $H(\vec{R}(t))$ depends on a set of external parameters $\vec{R}(t)$ changing slowly (adiabatically) in time. At any given time the Hamiltonian has a complete set of eigenvectors, denoted by $|n(\vec{R}(t))\rangle$,

$$H(\vec{R}(t)) \; |n(\vec{R}(t))\rangle = E_n(\vec{R}(t)) \; |n(\vec{R}(t))\rangle . \tag{2}$$

In the adiabatic approximation we assume that the system prepared at time t=0 in one of the discrete eigenstate, $|n(\vec{R}(0))\rangle$, of $H(\vec{R}(0))$ will cling to " the same" eigenstate $|n(\vec{R}(t))\rangle$ during its evolution, povided $|n(\vec{R})\rangle$ is a regular and single-valued function of $\vec{R}$ at least in the part of parameter space travelled by the system. It is well known that during such an evolution the wave function will acquire the so called dynamical phase factor, $\exp[-\frac{i}{\hbar}\int_0^t d\tau \, E_n(\vec{R}(\tau))]$, which depends on the time of evolution. Berry discovered an additional phase factor which depends on the path travelled by the system in the space of external parameters but not on the time of the travel.

If the state vector $|\psi(t)\rangle$ at time t>0 is to be directed along $|n(\vec{R}(t))\rangle$ (according to adiabatic approximation) then, in order to satisfy the Schrödinger equation, $|\psi(t)\rangle$ would have to contain a phase factor $\exp(i\gamma_n(t))$ (in addition to the dynamical phase factor) to compensate partly for the change of the vector $|n(\vec{R}(t))\rangle$ resulting from the change of parameters $\vec{R}(t)$. The phase factor $\exp(i\gamma_n(t))$ compensates only that change of $|n(\vec{R}(t))\rangle$ which is directed along $|n(\vec{R}(t))\rangle$,

$$\dot{\gamma}_n(t) = i \langle n(\vec{R}(t))| \nabla_R n(\vec{R}(t))\rangle \cdot\dot{\vec{R}}. \tag{3}$$

If the parameters $\vec{R}$ traverse a closed path C, then the phase γ_n is given by the line integral along C,

$$\gamma_n(C) = i \oint_C \langle n(\vec{R})| \nabla_R n(\vec{R})\rangle \cdot d\vec{R}, \tag{4}$$

and is independent on how this path is traversed in time. Using Stokes theorem one can transform the line integral into the surface integral over the surface S bounded by the contour C,

$$\gamma_n(C) = -\mathrm{Im} \int_S d\vec{\sigma}\cdot\nabla_R \times \langle n(\vec{R})| \nabla_R n(\vec{R})\rangle . \tag{5}$$

The r.h.s. can also be expressed by the matrix elements of the gradiant of the Hamiltonian in the parameter space,

$$\gamma_n(C) = -\mathrm{Im}\int_S d\vec{\sigma}\cdot\sum_{m\neq n} \frac{\langle n(\vec{R})|\nabla_R H(\vec{R})|m(\vec{R})\rangle\times\langle m(\vec{R})|\nabla_R H(\vec{R})|n(\vec{R})\rangle}{[E_m(\vec{R})-E_n(\vec{R})]^2} . \tag{6}$$

This equation shows a somewhat suprising feature of Berry's phase, namely, that it depends on those values of external parameters (inside the contour C) that were not experienced by the system. The dependence on the external parameters is nonlocal, like the nonlocal dependence on the magnetic field in the ordinary space in the Bohm-Aharonov effect. The denominator in Eq.(6) signals also the importance of the degeneracies that may occur for some values of $\vec{R}$ inside the contour C.

The discovery made by Berry is quite general. One can look for his geometrical phase in many areas of physics from gauge-field theories, through the nuclear physics to the molecular physics. Berry, himself, explored several situations in which the geometrical phase could be measured. One of the simplest is the motion of a neutral particle with a magnetic moment, e.g. a neutron, in a magnetic field $\vec{B}(t)$ that slowly changes its direction tracing a closed circuit C. Using Eq.(6) Berry showed that the particle wave function would acquire a phase,

$$\gamma_n(C) = -n \int_S d\vec{\sigma}\cdot\vec{B}/B^3 = -n\ \Omega(C), \tag{7}$$

where n is an eigenvalue of the projection of particle's spin on the direction of the magnetic field, $\vec{s}\cdot\vec{B}$ - the adiabatic invariant; $\Omega(C)$ denotes the solid angle subtended by C at $\vec{B}=0$. ($\vec{B}=0$ is the point of degeneracy.) The phase $\gamma_n(C)$ could be detected in the interference experiment involving two beams of particles: one, which experienced the changing magnetic field and another, which avoided it.

Later, Berry[2] has extended the adiabatic method from Schrödinger theory to Maxwell theory. In this case the role of the changing external parameters is played by the components of the dielectric tensor of a birefringent and gyrotropic medium in which the light beam propagates. Berry's phase resulted in the changes of the polarization of light which, again, could be measured in the interference experiment.

Strictly speaking, in the case of the light beam we have rather to do with the classical analog of Berry's phase. This classical analog was studied by Berry[3] and by Hannay[4] who found in 1985 an additional change of the angle variable resulting from an adiabatic evolution of the classical system governed by an integrable Hamiltonian. Recently, Gozzi and Thacker[5] showed a close connection (a correspondence) between Berry's quantum phase and Hannay's classical angle.

So far, the only experiment in which Berry's phase was measured and its topological properties were verified has been performed on light beams and not on systems described by the Schrödinger equation.

2. OPTICAL MEASUREMENT OF BERRY'S PHASE

The idea of the experiment came from Chiao and Wu[6] who predicted that the linearly polarized light beam travelling through a one-mode, helically wound optical fiber will have its polarization rotated. The angle of rotation is a direct measure of Berry's phase for the photons. Here, like in the case of neutrons in a magnetic field, we again deal with the transport of spin along the particle's trajectory. Helicity (the projection of spin on the wave vector) is an adiabatic invariant. Since the photon is a massless particle, its spin is always directed along its momentum (the wave vector). The helicity quantum number is either +1 or -1. If the fiber is not too sharply kinked, the helicity remains unchanged during the passage of photons through the fiber. For the light propagating along the optical fiber the role of external fields or forces (the magnetic field $\vec{B}$ for neutrons) is played by the changing characteristics of the medium along the trajectory, which change the photon momentum. Thus, the wave vector itself plays the role of the adiabatically changing parameter in Berry's theorem. Chiao and

Wu used the analogy between the projection of spin on the magnetic field, $\vec{s}\cdot\vec{B}$, for neutrons and the photon's helicity, $\vec{s}\cdot\vec{k}$, to predict the change of the phase of the photon wave function. When the optical fiber makes one loop, the photon wave vector traces a circular closed path C in the reciprocal space. The photon wave function for helicity λ acquires a phase

$$\gamma(C) = -\lambda\, \Omega(C), \tag{8}$$

where $\Omega(C)$ is the solid angle subtended by the circuit C at $\vec{k}=0$. ($\vec{k}=0$ is, again, the point of degeneracy.) The changes of the phase will be opposite for opposite helicities; for light that was initially linearly polarized this means a rotation of the plane of polarization.

The experiment proposed by Chiao and Wu was succesfully carried out by Tomita and Chiao[7] at the AT and T Bell Laboratories. They used a 180 cm long, one-mode optical fiber. They inserted it in a Teflon sleeve and they wound it as a helix. At one end the beam of light from a He-Ne laser was injected through a linear polarizer. At the other end the second polarizer detected the rotation of polarization. The experiment was repeated many times for different shapes of regular and irregular helices. It was found that the plane of polarization rotated by the angles that varied from fractions of the radian up to 6 radians. The dependence of this angle of rotation on the solid angle $\Omega(C)$ was found to be strictly linear, as predicted by Eq.(8). Thus, not only Berry's phase for photons was measured, but its topological nature was verified.

3. THEORY OF RELATIVISTIC SPINNING PARTICLES

In the experiment performed by Tomita and Chiao, Berry's phase for photons manifested itself in the rotation of polarization resulting from its parallel transport around a closed circular path in the momentum space. One can understand this phenomenon (as well as the experiment on neutrons in the changing magnetic field, proposed by Berry) on general grounds recalling the action of the Poincaré group in the theory of relativistic spinning particles. This approach is similar in spirit to that of Chiao and Wu. Instead of introducing in the Hamiltonian adiabatically changing external parameters which control the motion of a particle one can study the changes of the particle's momentum which reflect the changes of the environment. The motion of a particle through external fields in the adiabatic limit can be visualized as a series of Poincaré transformations. Therefore, I will now study the properties of such motions to uncover the universal geometrical reason for the appearance of Berry's phase for spinning particles. This Section of my lecture is based on our recent publication.[8]

I will consider a relativistic particle of mass m and spin s. The components of its wave function satisfy a set of differential equations (e.g. the Dirac equation, the Proca equation, the Maxwell equations). Every solution $\psi_a(x)$ of such relativistic wave equations can be expanded into plane waves,

$$\psi_a(x) = \sum_\lambda \int d\Gamma[u_a(\vec{p},\lambda)f_1(\vec{p},\lambda)e^{-ip\cdot x} + v_a(\vec{p},\lambda)f_2(\vec{p},\lambda)e^{ip\cdot x}], \qquad (9)$$

where $x=(\vec{r},t)$, $p\cdot x=E_p t-\vec{p}\cdot\vec{x}$, $E_p=\sqrt{m^2+\vec{p}^2}$, $d\Gamma$ is a Lorentz invariant measure on the mass hyperboloid,

$$d\Gamma = \frac{d^3p}{2E_p(2\pi)^3}, \qquad (10)$$

and $u_a(\vec{p},\lambda)e^{-ip\cdot x}$ and $v_a(\vec{p},\lambda)e^{ip\cdot x}$ are plane-wave solutions of the wave equations with possitive and negative frequencies, respectively, and λ labels different helicity states. In the case of electrons and other spin ½ particles $u_a(\vec{p},\lambda)$ and $v_a(\vec{p},\lambda)$ denote bispinors, for photons - polarization vectors (or polarization tensors). The helicity basis is used because it enables one to treat simultaneously the massive and the massless particles. In the limit $m\to 0$, the spin operator $\vec{s}$ cannot be defined, and only the notion of helicity survives.

The amplitudes $f_1(\vec{p},\lambda)$ and $f_2(\vec{p},\lambda)$ describe the independent degrees of freedom of the wave field $\psi_a(x)$; all the constraints imposed by the wave equations have been fully taken into account.

For photons the role of $\psi_a(x)$ is played by the electromagnetic tensor of the radiation field $f_{\mu\nu}(x)$. Since in this case the solutions with negative frequencies are simply complex conjugate of the respective solutions with positive frequencies, the amplitudes $f_2(\vec{p},\lambda)$ are not independent of $f_1(\vec{p},\lambda)$; one must set $f_2(\vec{p},\lambda)=f_1^*(\vec{p},\lambda)$. That is always the case for particles which have no distinct antiparticles. For simplicity, I will consider now only this case, and I will omit the subscript 1 of $f(\vec{p},\lambda)$. Thus, any wave function in ordinary space-time is fully characterized by one set of functions $f(\vec{p},\lambda)$ in momentum space. The momentum space is restricted to the mass hyperboloid for massive particles, or to the cone for massless particles. The helicity amplitudes $f(\vec{p},\lambda)$ may be called the particle's wave functions in momentum space.

The decomposition of the electromagnetic field tensor into plane waves has the form:

$$f_{\mu\nu}(x) = \int d\Gamma\ [e_{\mu\nu}(\vec{k})f(\vec{k},+1) + e^*_{\mu\nu}(\vec{k})f(\vec{k},-1)]e^{-ik\cdot x} + c.c. \tag{11}$$

The complex antisymmetric polarization tensor $e_{\mu\nu}(\vec{k})$ can be expressed by the complex polarization vector $\vec{e}(k)$,

$$e_{oi}(\vec{k}) = i\omega(\vec{k})\ e_i(\vec{k}) = -\tfrac{1}{2}\ \varepsilon_{ijk}e_{jk}(\vec{k}), \tag{12}$$

where $\omega(k)=|\vec{k}|$.

The complex polarization vector $\vec{e}(\vec{k})$ satisfies a set of algebraic relations which follow from Maxwell equations and the normalization conditions,

$$\vec{k}\cdot\vec{e}(\vec{k}) = 0, \tag{13a}$$

$$\vec{k}\times\vec{e}(\vec{k}) = -i\omega(\vec{k})\ \vec{e}(\vec{k}), \tag{13b}$$

$$\vec{e}(-\vec{k}) = \vec{e}^*(\vec{k}), \tag{13c}$$

$$\vec{e}^*(\vec{k})\cdot\vec{e}(\vec{k}) = 1,\quad \vec{e}(\vec{k})\cdot\vec{e}(\vec{k}) = 0, \tag{13d}$$

$$\vec{e}^*(\vec{k})\times\vec{e}(\vec{k}) = -i\vec{k}/\omega(\vec{k}). \tag{13e}$$

The 10 conserved quantities that serve as the generators of the Poincaré transformations (the energy H, the momentum $\vec{P}$, the angular momentum $\vec{M}$, and the generator of the proper Lorentz transformation $\vec{N}$) can be expressed[9] as the following bilinear combinations of the helicity amplitudes $f(\vec{p},\lambda)$,

$$H = \sum_\lambda \int d\Gamma\ f^*(\vec{p},\lambda)\ E_p\ f(\vec{p},\lambda), \tag{14a}$$

$$\vec{P} = \sum_\lambda \int d\Gamma\ f^*(\vec{p},\lambda)\ \vec{p}\ f(\vec{p},\lambda), \tag{14b}$$

$$\vec{M} = \sum_\lambda \int d\Gamma\ f^*(\vec{p},\lambda)[\ -i\vec{p}\times\vec{D} + \lambda\vec{p}/p\ ]f(\vec{p},\lambda), \tag{14c}$$

$$\vec{N} = \sum_\lambda \int d\Gamma\ f^*(\vec{p},\lambda)[\ i(\ \vec{D}\ E_p + E_p\ \vec{D}\ )/2]\ f(\vec{p},\lambda)$$
$$+\ m \sum_{\lambda\lambda'} \int d\Gamma\ f^*(\vec{p},\lambda)[(\vec{p}\times\vec{s}_{\lambda\lambda'})/p^2]\ f(\vec{p},\lambda'), \tag{14d}$$

where $\vec{s}_{\lambda\lambda'}$ is the set of three spin matrices in the helicity basis.

In Eqs.(14c,d) there appears the "covariant derivative" $\vec{D}$ in momentum space, which has the form:

$$\vec{D} = \nabla + i \lambda \vec{\alpha}(\vec{p}), \tag{15}$$

where ∇ is an ordinary gradient with respect to $\vec{p}$ and the real vector $\vec{\alpha}(\vec{p})$ is an analog of the vector potential. The helicity amplitudes $f(\vec{p},\lambda)$ are determined up to a phase factor, since the polarization functions $u_a(\vec{p},\lambda)$, $v_a(\vec{p},\lambda)$ or $e_{\mu\nu}(\vec{k})$ are also determined up to a phase factor. The presence of the covariant derivative $\vec{D}$ in the expressions (14) for generators makes these expressions explicitely invariant under the change of phase of $f(\vec{p},\lambda)$. It also helps to split the angular momentum $\vec{M}$ into two parts: one orthogonal to the momentum $\vec{p}$ (orbital angular momentum) and one parallel to $\vec{p}$ (helicity).

Our calculations of generators yielded the following expressions for $\vec{\alpha}(\vec{p})$ for spin ½ and spin 1 particles:

$$\vec{\alpha}(\vec{p}) = \frac{1}{p(p+p_3)}(p_2, -p_1, 0) \quad \text{for spin } \tfrac{1}{2}, \tag{16a}$$

$$\vec{\alpha}(\vec{p}) = \frac{p_3}{p(p_1^2+p_2^2)}(p_2, -p_1, 0) \quad \text{for spin } 1, \tag{16b}$$

which are singular along a line. For photons the vector $\vec{\alpha}(\vec{k})$ was calculated from the equation:

$$\vec{\alpha}(\vec{k}) = i\, e_i^*(\vec{k}) \nabla e_i(\vec{k}). \tag{17}$$

The curl of $\vec{\alpha}(\vec{p})$ has the form of the magnetic field of a monopole of a unit strength located at the origin in momentum space,

$$\nabla \times \vec{\alpha}(\vec{p}) = \vec{p}/p^3. \tag{18}$$

The motions of a particle in space-time,viewed as Poincaré transformations, result in the action of the appropriate generators on the helicity amplitude $f(\vec{p},\lambda)$. In the case of rotations it is the action of the angular momentum operator. For rotations around the z-axis, we have:

$$M_z\, f(\vec{p},\lambda) = [\, -i\vec{p} \times \vec{D} + \lambda\vec{p}/p\,]_z\, f(\vec{p},\lambda). \tag{19}$$

The rotation around a fixed axis will be considered in the next Section in relation to Berry's phase.

4. THE PARALLEL TRANSPORT OF POLARIZATION

Let us consider now a particle moving with a constant energy and a slowly varying momentum. In this case the changes of the momentum are due to pure rotations. To make contact with the optical experiment on Berry's phase I will calculate the change of the phase of the wave function $f(\vec{p},\lambda)$ for the 2π rotation around a fixed axis, i.e. for a circular closed path in momentum space. One can view this problem as a parallel transport of a vector along a closed path - a classical problem of differential geometry.

In order to directly apply the methods of differential geometry, the function $f(\vec{p},\lambda)$ will be represented by a real 2-component vector $F^a(\vec{p},\lambda)$ whose components are the real and the imaginary parts of $f(\vec{p},\lambda)$.

The covariant derivative of the vector field $F^a(\vec{p},\lambda)$ is:

$$D_i F^a(\vec{p},\lambda) = \left[\nabla_i + \Gamma_i{}^a{}_b(\vec{p})\right] F^b(\vec{p},\lambda). \tag{20}$$

According to Eq.(15), the affine connection $\Gamma_i{}^a{}_b(\vec{p})$ has the form:

$$\Gamma_i{}^a{}_b(\vec{p}) = \lambda\ \alpha_i(\vec{p})\ \varepsilon^a{}_b, \tag{21}$$

where $\varepsilon^a{}_b$ is the antisymmetric matrix: $\varepsilon^2{}_1 = 1 = -\varepsilon^1{}_2$, $\varepsilon^1{}_1 = 0$, $\varepsilon^2{}_2 = 0$.

The curvature tensor $R_{ij}{}^a{}_b$ derived from this connection is:

$$R_{ij}{}^a{}_b = \partial_i\, \Gamma_j{}^a{}_b - \partial_j\, \Gamma_i{}^a{}_b + \Gamma_i{}^a{}_c\, \Gamma_j{}^c{}_b - \Gamma_j{}^a{}_c\, \Gamma_i{}^c{}_b$$

$$= \lambda\ (\nabla_i\, \alpha_j \times \nabla_j\, \alpha_i)\ \varepsilon^a{}_b\ . \tag{22}$$

The nonvanishing of the curvature tensor in the momentum space is the reason for Berry's phase to appear; the curvature causes the rotation of the vector $F^a(\vec{p},\lambda)$ when it is transported along a closed path C in momentum space. The angle of rotation is given by the integral of the curvature tensor over the surface spanned by the contour C,

$$\Delta\phi = \pm\ \lambda \int_S d\vec{\sigma}\cdot(\nabla\times\vec{\alpha}) = \pm\ \lambda \int_S d\vec{\sigma}\cdot\vec{p}/p^3 = \pm\ \lambda\ \Omega(C), \tag{23}$$

where $\Omega(C)$ is a solid angle subtended by the contour C at $\vec{p}=0$; the sign depends on the orientation of the contour.

The rotation of the real vector $F^a(\vec{p},\lambda)$ means a change of the phase of the complex function $f(\vec{p},\lambda)$ - Berry's phase.

For photons the angle of rotation (23) is exactly the angle of the rotation of the linear polarization.

5. FINAL REMARKS

I have shown that the curvature associated with the transport of the polarization in momentum space is the reason for the appearance of Berry's phase in the case of relativistic spinning particles moving in a slowly changing environment.

However, it is not the only situation in which Berry's phase can manifest itself. Another field, in which one can expect Berry's theorem to work, is molecular physics. Very often one can use the Born-Oppenheimer approximation and one can treat internal coordinates of nuclei as slowly varying parameters, to which the electronic wave functions can adjust at any given time. Those internal coordinates of nuclei may play a role of the adiabatically changing external parameters in Berry's theorem. Recently, an experiment on the two-photon resonant ionization of Na_3 clusters was reported[10] which indirectly proved the existence of Berry's phase for molecules.

REFERENCES AND NOTES

1. M.V.Berry, Proc.Roy.Soc.(London) A 392, 45 (1984).
2. M.V.Berry, in *Fundamental Aspects of Quantum Theory*, V.Gorini and A.Frigerio, Eds. (Plenum, New York, 1986).
3. M.V.Berry, J.Phys.A 18, 15 (1985).
4. J.H.Hannay, J.Phys.A 18, 221 (1985).
5. E.Gozzi and W.D.Thacker, Max-Plack-Institut Preprint MPI-PAE/PTh 57/86.
6. R.Y.Chiao and Yong-Shi Wu, Phys.Rev.Lett. 57, 933 (1986).
7. A.Tomita and R.Y.Chiao, Phys.Rev.Lett. 57, 937 (1986).
8. I.Bialynicki-Birula and Z.Bialynicka-Birula, to be published in Phys.Rev.D.
9. These formulas were derived earlier (I.Bialynicki-Birula and Z.Bialynicka-Birula, *Quantum Electrodynamics*, (Pergamon, Oxford, 1975)) for spin 1 particles by inserting the plane wave expansions of the free fields into the formulas for generators expressed by the components of the energy-momentum tensor.
10. G.Delacrétaz, E.R.Grant, R.L.Whetten, L.Wöste, and J.W.Zwanziger, Phys.Rev.Lett. 56, 2598 (1986).

RAMAN HETERODYNE RAMSEY SPECTROSCOPY IN LOCAL SPACE AND VELOCITY SPACE

Jürgen Mlynek and Rudolf Grimm
Institute of Quantum Electronics
Swiss Federal Institute of Technology (ETH) Zürich
CH-8093 Zürich, Switzerland

Egbert Buhr and Volker Jordan
Institute of Quantum Optics, University of Hannover
D-3000 Hannover 1, Welfengarten 1, Fed. Rep. Germany

Abstract

A novel optical heterodyne technique for Raman Ramsey spectroscopy of atomic radio-frequency resonances is reported. The method allows for high-resolution studies in an atomic beam as well as for the study of collisional velocity diffusion of atoms within an optical Doppler distribution. Our experiments are performed on Zeeman sublevels in the Samarium $\lambda = 570.68$ nm (J=1)-(J'=0) transition both in an atomic beam and in a vapor; in the latter case rare gas perturbers are added as collision partners. All our experimental findings are in satisfactory agreement with theoretical predictions.

I. Introduction

Ramsey's method of separated fields for the observation of narrow radio-frequency (rf) resonances is well known from atomic and molecular beam experiments /1/; in general, these Ramsey fringes are induced by use of two spatially separated rf fields. More recently, Ramsey resonances have been observed also for a resonance Raman transition in an atomic beam /2/: Here, the two rf fields simply were replaced by two modulated laser fields which can create sublevel coherence with high efficiency; the Ramsey fringes were detected via fluorescence from the optically excited state that couples the light fields to the ground state sublevels.

In this contribution we report on a novel *coherent* optical technique for Ramsey spectroscopy of rf transitions in free atoms. The method relies on resonance Raman transitions to optically excite *and probe* sublevel coherence in atomic ground states or optically excited states using two separated atom-field interaction regions. In contrast to previous work, *no second "oscillatory" field* is needed: Again a modulated laser field is used for coherent Raman excitation of sublevel coherence; the required phase sensitive detection, however, is achieved by *Raman heterodyne detection* /3,4 / of the atomic coherence using an unmodulated probe laser field in the second interaction region. This coherent detection scheme is exclusively sensitive on oscillating sublevel coherence; thus an essentially background-free observation of the Ramsey resonances can be obtained. Let us point out that the Raman Ramsey method is based on the creation and detection of *oscillating sublevel coherence;* in this respect, it is far different from other previous observations of Ramsey fringes using lasers in connection with atomic coherences at optical frequencies /5/.

The coherent Ramsey technique reported here can be used to investigate rf resonances in atomic beams or in atomic vapors: (i) In atomic beam experiments using spatially separated fields, the new technique might be an interesting alternative for ultra-high resolution spectroscopy in the rf range (see Sec. II). (ii) When being applied to an atomic vapor, the method allows the study of Ramsey resonances in velocity space /6,7/. In this case, counterpropagating excitation and probe fields are used; Ramsey interference patterns then show up due to collisional velocity diffusion of atoms with sublevel coherence within an optical Doppler distribution thereby allowing detailed collision studies (see Sec. III).

II. Raman Heterodyne Detection of Ramsey Resonances in an Atomic Beam

The principle of our atomic beam experiment is illustrated in Fig.1. The atoms interact successively with two laser fields; in the first interaction zone a modulated laser field consisting of the carrier with frequency ω_E and the sidebands with frequencies $\omega_E \pm \omega_M$ drives a Zeeman split J=1 to J'=0 transition (see Fig.1a). In a resonant Raman process a coherence between the Zeeman sublevels is induced; here, we consider the creation of $|\Delta m| = 1$ coherence only. After leaving the laser beam the sublevel coherence simply oscillates at its eigenfrequency Ω (Fig.1b); thereby a phase shift between the sublevel coherence (Ω) and the oscillatory laser modulation (ω_M) is built up. When, after a time τ , the atoms arrive at the second interaction zone, this phase shift accumulates to $\Delta_M \tau$ with $\Delta_M = \omega_M - \Omega$ being the detuning of the modulation frequency ω_M from the sublevel splitting frequency Ω. In the second interaction region the sublevel coherence is probed by an unmodulated laser field of frequency ω_E and, via a coherent scattering process, copropagating coherent Raman fields with frequencies $\omega_E \pm \Omega$ are produced (Fig.1c). On a fast photodiode the Raman fields together with the probe laser field yield a heterodyne beat signal of frequency $\Omega = | \omega_E - (\omega_E \pm \Omega) |$. The amplitude and phase of this signal directly reveal the amplitude and phase of the sublevel coherence; thus, using a phase sensitive detection scheme like a double balanced mixer or a rf lock-in amplifier, a phase sensitive detection of the sublevel coherence can be achieved (see Fig.2).

The phase shift $\Delta_M \tau$ resulting from the free evolution of the sublevel coherence in the field-free zone now shows up in the output signal of the lock-in amplifier; more precisely, a Ramsey type interference pattern proportional to $\cos(\Delta_M \tau)$ appears in the signal lineshape. This interference structure can be observed by sweeping the sublevel detuning; it becomes increasingly narrow for a larger time of flight τ , i.e. for a larger distance L between both interaction zones. Due to the velocity distribution of the atoms in the atomic beam, higher order Ramsey fringes (large Δ_M) will be smeared out but will essentially leave the central fringe (small Δ_M).

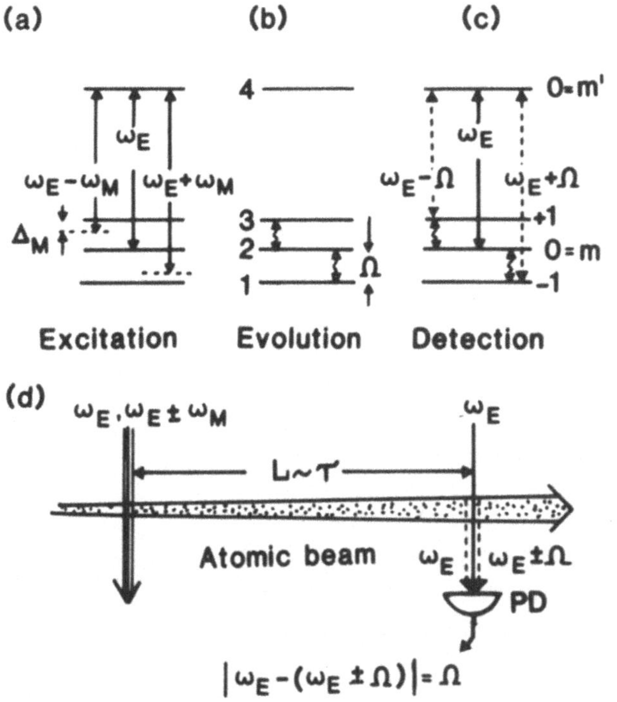

Fig.1: (a) Modulated excitation process of Zeeman coherence for a Zeeman-split (J=1)-(J'=0) transition. (b) Free evolution of sublevel coherence during spatial motion in between the two interaction zones.(c) Detection process showing the induced Raman sidebands. (d) Schematic of the experimental setup for the observation of Ramsey resonances. The photodiode PD is used for optical heterodyne detection of the Raman sidebands.

A schematic of our experimental arrangement is shown in Fig.2. The beam of a cw dye ring laser was split into two parts to yield the spatially separated pump- and probe beams; the pump beam was phase modulated by means of an electrooptic modulation system MS in order to generate the modulation sidebands with frequenies $\omega_E \pm \omega_M$. The carrier and the sidebands were orthogonally polarized with respect to each other; the polarization direction of the carrier was chosen to be parallel to the transverse static magnetic field B that lifted the ground state Zeeman level degeneracy. In this way (see Fig.1) the carrier drives the optical π -transition ($\Delta m = 0$), whereas the sidebands solely couple the σ-transitions ($\Delta m = \pm 1$).

The unmodulated probe beam was polarized parallel to the static magnetic field B and therefore only drives the optical π -transition. The simultaneous presence of oscillating $|\Delta m| = 1$ Zeeman coherence in the second interaction zone gives rise to coherent Stokes and anti-Stokes Raman sidebands of frequencies $\omega_E \pm \Omega$ that propagate in the same direction as the probe field. As a consequence of the selection rules, the Raman sidebands are linearly polarized in a direction that is perpendicular to the static magnetic field B. The orthogonally polarized probe field and Raman sideband fields were then sent through a $\lambda/4$-plate which introduces an optical phase shift of $\pi/2$ between the carrier and the Raman sidebands; this optical phase shift ensures that

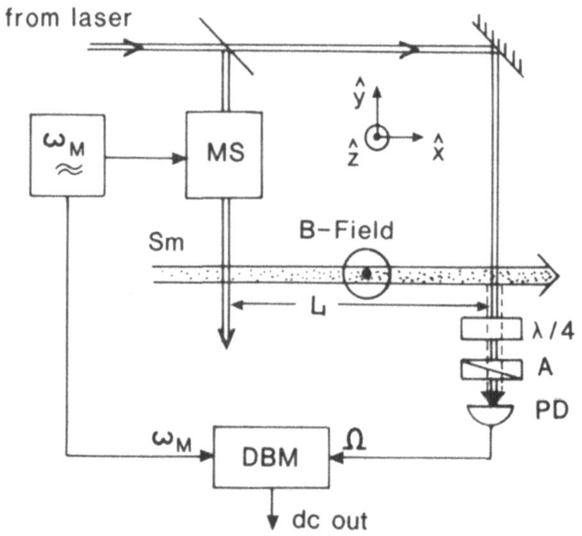

Fig.2: Experimental scheme: MS, modulation system; $B\hat{z}$, static magnetic field; λ/4, retardation plate; A, polarization analyzer; PD, photodetector; DBM, double balanced mixer. The two principal axes of the λ/4-plate are oriented parallel to the $\hat{z}$ and $\hat{x}$ directions, respectively.

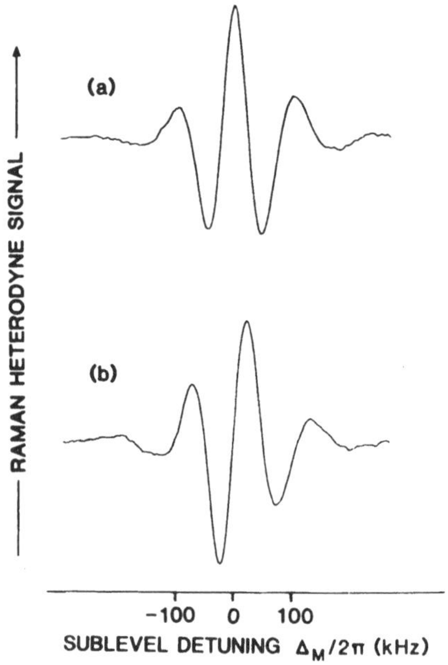

Fig.3: Measured Raman heterodyne signals of the detection beam as a function of sublevel detuning Δ_M for a beam separation of L=7 mm. Depending on the phase setting of the rf lock-in, the in-phase (a) and quadrature (b) component of the beat signal can be monitored.

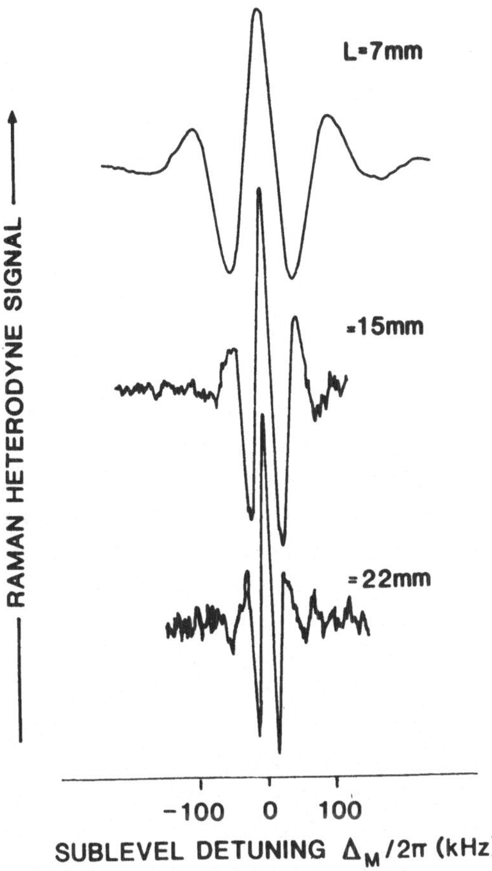

Fig.4: Measured in-phase Raman heterodyne signals as a function of sublevel detuning for three values of beam separation L.

we detect those Raman signal contributions that result from atoms being optically resonant with the probe beam /7/. The polarization analyzer A behind the $\lambda/4$-plate finally projects the carrier and Raman sidebands along a common direction in order to enable an optical interference at the photodiode PD. The heterodyne beat signal can then be detected by a phase sensitive electronic device; in our experiments we used a rf lock-in amplifier for this purpose.

Preliminary experiments were carried out in a Sm atomic beam; here the ^{154} Sm λ = 570.68 nm ($4f^6\ 6s^2\ {}^7F_1$ - $4f^6\ 6s6p\,{}^7F_0$) line was used for optical excitation. Typically, laser powers in the mW-range were used with beam diameters of about 500 µm. According to the Lande factor of g_J =1.5 for the J=1 ground state of the even Sm isotopes, the Bohr frequency of the Zeeman splitting between the m-sublevels is given by $\Omega = 2\pi * 21.0$ MHz (B/mT). Raman heterodyne signals of the Zeeman resonances were observed at a fixed modulation frequency $\omega_M = 2\pi *$ 10.0 MHz by variation of the Zeeman splitting Ω via the external field B. In Fig.3 typical Raman heterodyne signals are shown for a separation of 7 mm between the interaction zones of pump-

and probe beam; depending on the phase setting of the rf lock-in amplifier with respect to the phase of the driving modulation field ω_M , the "in-phase" and "out-of- phase" components of the heterodyne beat signal could be monitored. When the spatial separation between the two laser beams was increased, the linewidth of the Ramsey resonances decreased as displayed in Fig.4. The data also show that the signal-to-noise ratio strongly decreases for larger beam separations; this effect is mainly due to the bad quality of our present atomic beam and due to magnetic field inhomogeneties along the beam axis. More systematic studies on the signal lineshapes are in preparation using an improved atomic beam apparatus.

As in the case of the Raman Ramsey experiments performed by Ezekiel and coworkers /2/, various frequency error sources may be present and also have to be studied in detail with respect to our method: these include, e.g., atomic beam misalignment, laser beam misalignement, optically-induced level shifts and magnetic field inhomogeneties. In this context let us note that no frequency error should occur due to optical path differences from the beam splitter to each of the two interaction regions shown in Fig.2 as only the excitation field is modulated; in principle, even two independent lasers can be used to induce and detect the atomic sublevel coherence.

Our theoretical treatment is based on a perturbation solution for the Raman heterodyne signal. The medium is modeled as an ensemble of 4- level atoms as shown in Fig.1. We are using the time domain for description: the atoms first experience a pump pulse and, after some time τ, another probe pulse that reads out the oscillating sublevel coherence. The solution of the time-dependent equations of motion can then be transformed easily into the local space. Special care is taken with respect to the integration over the atomic velocity distribution within the beam. First results show that the calculated signal lineshapes are in satisfactory qualitative agreement with the measurements; details of the experiment together with the theory will be published elsewhere /8/. In this respect let us mention that Dalton et al. recently gave a non-perturbative theoretical analysis of Ramsey interference lineshapes in three-level lambda systems excited by laser fields /9/; their description, however, is closely related to the experiments of Ezekiel and coworkers /2/ and does not apply to our experimental situation.

In contrast to previous experiments using Raman Ramsey spectroscopy /2/, our technique does not need a second oscillatory field to induce Ramsey resonances: The Raman heterodyne method yields phase-sensitive information on the sublevel coherence in a direct way and the Ramsey interference pattern is obtained subsequently by demodulation of the beat signal in a phase-sensitive electronic device. Moreover, the technique presented here is essentially background-free; it is a simple transmission method with high sensitivity through optical heterodyning.

Obviously Raman Ramsey experiments are of interest because they have possible applications both in spectroscopy and in the development of new time and frequency standards, especially in the rf- and microwave range of the spectrum. This work clearly shows that very narrow optical features can be obtained by using coherent resonance Raman processes to induce and monitor rf resonances between two long- lived ground sublevels in a samarium atomic beam. The technique itself should be also applicable to rubidium or cesium; in this case commercially available semiconductor lasers may be used for excitation and detection of the hyperfine structure resonances of interest.

We also note that our technique may be applied to study sublevel resonances in optically excited states. Here, of course, the upper state lifetime has to be taken into account; for a reasonable time delay between excitation and detection, however, subnatural linewidth features should be observable with acceptable signal-to-noise ratio. Moreover, the Raman heterodyne detection technique may be directly applied to trapped ions /10/; in this case Ramsey resonances may be obtained by using a time-separated pulsed excitation and probe field. Thus the Raman heterodyne detection of Ramsey resonances as reported here can find various useful applications in high-resolution sublevel spectroscopy.

Finally let us briefly discuss a somewhat different idea of Raman heterodyne detection of Ramsey resonances in an atomic beam. Over the last few years several projects are under way to study and demonstrate the potential performance achievable in cesium beam frequency standards in which laser driven optical pumping is used for the atomic state selection and state detection in place of the conventional magnetic state selection /11/. Our proposal combines this approach with the Raman heterodyne detection technique: its main idea is outlined in Fig. 5. In a first step, a sublevel population difference is created by optical pumping (Fig.5a). Then a rf field (ω_H) resonantly excites the sublevel coherence (Fig.5b); this corresponds to the first atom-field interaction region. After leaving this rf field region, the sublevel coherence freely evolves in time with frequency Ω as the atoms propagate along the atomic beam (Fig.5c). Now, in a second

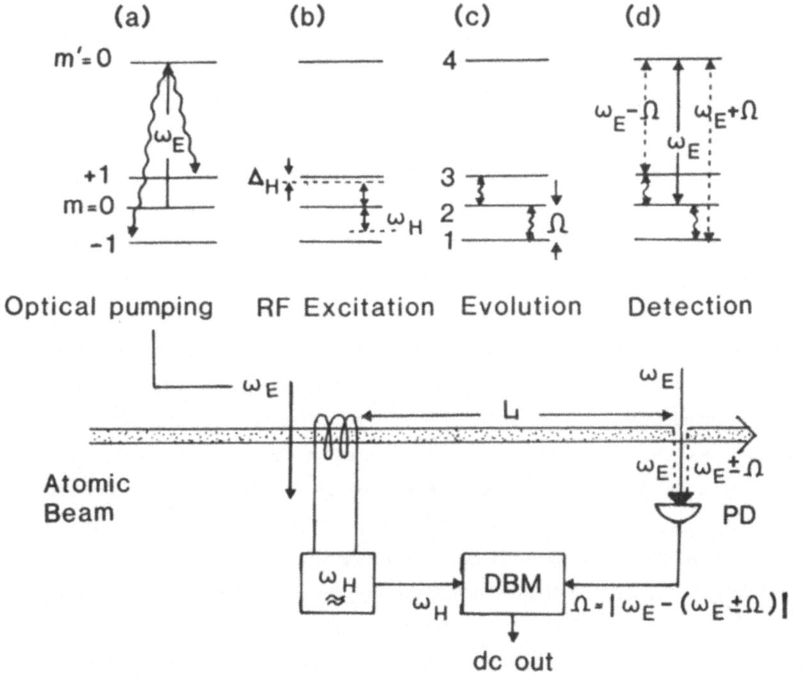

Fig.5: Outline of a proposal for Raman heterodyne detection of rf- induced resonances in a Ramsey-type atomic beam experiment. For details see text.

atom-field interaction region, the sublevel coherence is read out by an optical probe beam yielding coherent Raman sidebands (Fig.5d); these Raman sidebands can be monitored using optical heterodyne detection in the same way as discussed above. The rf beat signal from the photodetector can then be demodulated using , e.g., a bouble balanced mixer as shown in Fig.5; obviously Ramsey interference patterns will appear in the output of the electronic device for a sufficiently large separation L between the two interaction regions.

Details of this new proposal for Ramsey spectroscopy using Raman heterodyne detection will be published elsewhere /12/; here we only mention that this technique is closely related to previous work on optical heterodyne detection of rf resonances in atomic vapors /13/. In principle, the proposed detection scheme may easily be incorporated into atomic beam setups that already exist for the study of Ramsey resonances using rf fields in combination with optical pumping techniques.

III. Collision-Induced Ramsey Resonances in Velocity Space

The Ramsey resonance experiment discussed in this section utilizes the same technique for the coherent optical excitation and phase sensitive detection of sublevel coherence as described above; it is, however, performed in an atomic vapor, where the two laser fields spatially overlap (see Fig.6e). Here, the modulated pump field and the unmodulated probe field propagate in opposite directions within the sample cell. Consequently, for a nonzero laser frequency detuning with respect to the Doppler broadened optical transition ($\Delta_E \neq 0$), the two fields interact with different atomic subgroups having opposite velocities (Fig.6d). The width of the interaction zones in velocity space is determined by the homogeneous optical linewidth Γ . Thus, a sublevel coherence is created by the modulated pump field in atoms with a velocity centered around $-v_o$, whereas the probe field is capable to detect the sublevel coherence in atoms with velocity $+v_o$. Hence, the probe beam can create coherent Raman sidebands only if the active atoms change their velocity from $-v_o$ to $+v_o$, i.e. if they "move" in velocity space.

One well known mechanism that provides velocity changes in a vapor is due to collisions that preserve the internal atomic structure /14/. Here we consider velocity changing collisions of the active atoms with added rare gas perturbers. During this collisional redistribution from the velocity $-v_o$ to other velocities the active atoms are not affected by the laser fields due to the Doppler effect; as a consequence, the sublevel coherence evolves at its eigenfrequency Ω (see Fig.6b). If, after a certain time τ , the active atoms accidentally "arrive" in the opposite velocity subgroup $+v_o$, the interaction with the unmodulated probe field leads to the generation of coherent Raman sidebands (Fig.6c); of course, the Raman sidebands can be created only if the sublevel coherence has not been destroyed during the velocity diffusion process. Then, quite similar to the more conventional atomic beam Ramsey resonance experiment (see Sec.II), these Raman sidebands can easily be monitored using the Raman heterodyne technique (Fig.7). Most importantly, the velocity diffusion time τ leads to a phase shift $\Delta_M \tau$ of the photodetector signal (Ω) with respect to the reference signal (ω_M), thereby introducing Ramsey type interference patterns proportional to

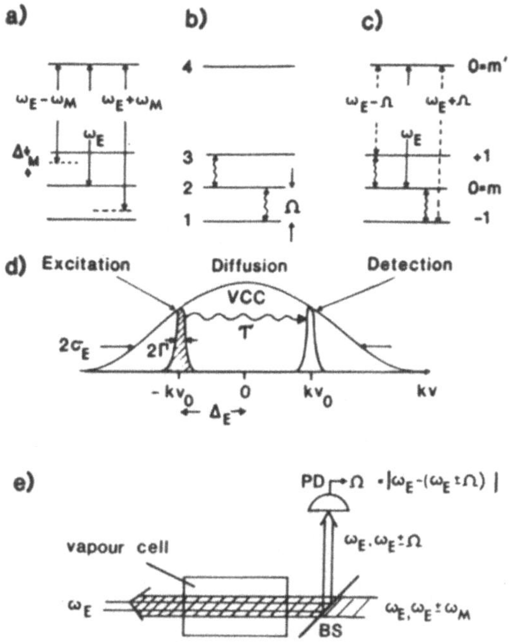

Fig.6: (a) Modulated excitation process of Zeeman coherence for a nondegenerate (J=1)-(J'=0) transition. (b) Free evolution of sublevel coherence during velocity diffusion. (c) Detection process showing the induced Raman sidebands. (d) Outline of the velocity selectivity of the excitation and detection of sublevel coherence in a Doppler- broadened optical transition. (e) Experimental scheme for the observation of collision-induced Ramsey resonances in Sm vapor.

$\cos(\Delta_M \tau)$ in the detected signal lineshapes.

In this Ramsey resonance experiment the "motion" of the active atoms in velocity space due to VCC plays a role similar to the spatial motion of the atoms in the atomic beam experiment. Thus Ramsey-type fringes are expected to occur, if there is a separation between both interaction regions in velocity space, i.e. if the laser frequency is detuned from the center of the Doppler profile ($\Delta_E \neq 0$).

A more detailed schematic of our experimental arrangement is shown in Fig.7. The pump beam of frequency ω_E was phase modulated by means of an electrooptic modulation system to yield the modulation sidebands with frequencies $\omega_E \pm \omega_M$. The amplitudes of the orthogonally polarized carrier and sidebands were adjusted to yield equal dipole coupling strengths for the three Sm transitions in order to avoid optical pumping between the m-sublevels of the ground state /7/. This modulation and polarization scheme permits an excitation of Zeeman coherence in a single velocity subgroup only /6,7/. The counterpropagating probe beam was polarized parallel to the static magnetic field B. The total field behind the sample cell, consisting of the probe field and the Raman sidebands, passed through a $\lambda/4$-plate and a polarization analyzer A for Raman heterodyne detection of the sublevel coherence. The heterodyne beat signal then was monitored by

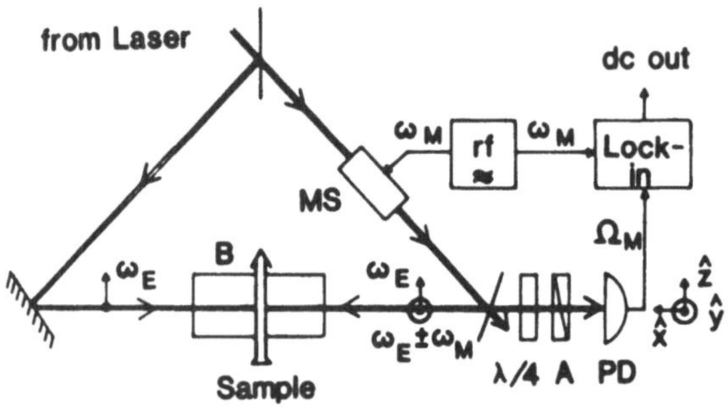

Fig.7: Experimental scheme: MS, modulation system; B, static magnetic field; λ/4, retardation plate; A, polarization analyzer; PD, photodetector. The two principal axes of the λ/4-plate are oriented parallel to the $\hat{y}$ and $\hat{z}$ directions, respectively.

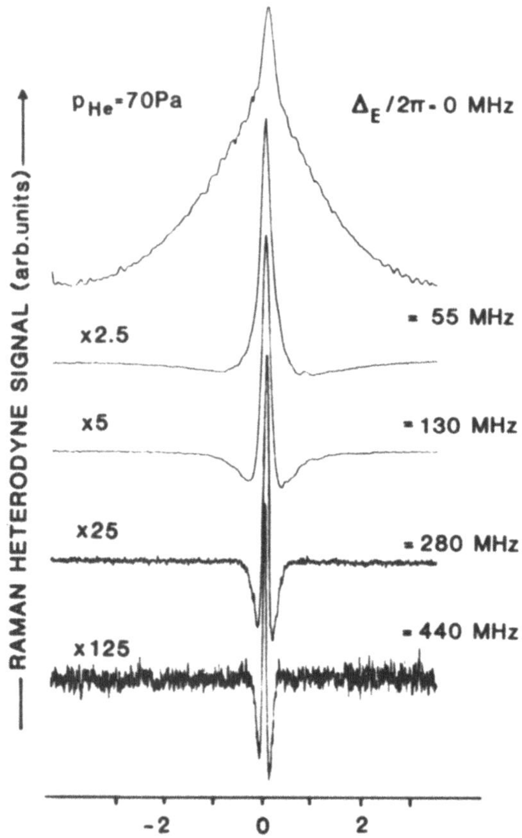

Fig.8: Measured RHS in the case of He perturbers as a function of the sublevel detuning Δ_M for different laser detunings Δ_E , showing narrow Ramsey-type interference patterns.

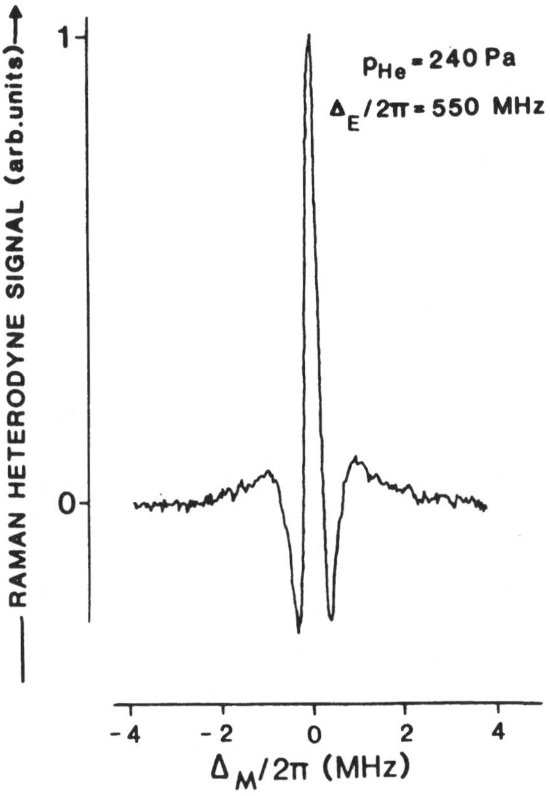

Fig.9: Measured RHS for a relatively large laser detuning (Δ_E /2π =550 Mhz) and a long signal averaging time, showing a pronounced Ramsey-type resonance.

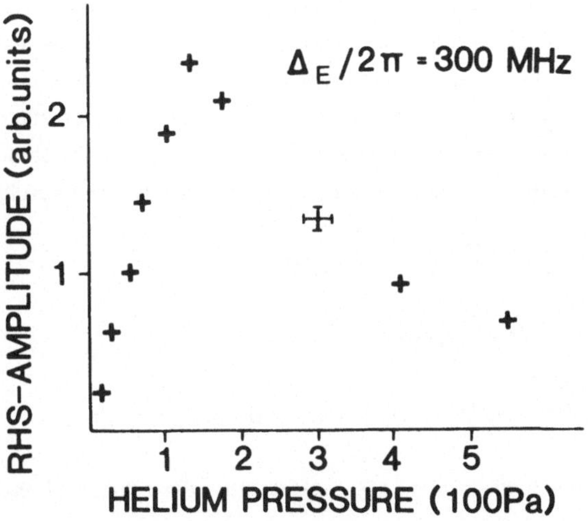

Fig.10: He pressure dependence of the RHS amplitude for a laser detuning Δ_E /2π = 300 Mhz.

means of a rf lock-in amplifier as discussed in Sec.II.

The experiments were performed in atomic samarium vapor contained in an aluminum ceramic tube. The ^{154}Sm λ = 570.68 nm ($4f^6\ 6s^2\ {}^7F_1 - 4f^6\ 6s6p\ {}^7F_0$) line was used for optical excitation. Natural Sm consists of seven isotopes; amoung these isotopes the ^{154} Sm and the ^{152} Sm are the most abundant ones. Corresponding to a cell temperature of about 1000 K, the width of the Doppler broadened transition amounts to ku = $2\pi *$ 580 MHz; the homogeneous optical linewidth (HWHM) is $2\pi *$ 230 kHz according to the lifetime (340 ns) of the excited $^7F_0$ state.

For the measurements of the sublevel resonance signals, the modulation frequency ω_M was kept fixed ($\omega_M = 2\pi *$ 10.0 MHz) and the Zeeman splitting was varied by sweeping the external magnetic field B. In Fig.8 typical in-phase Raman heterodyne signals (RHS) are shown as a function of the sublevel detuning with the laser frequency detuning as a parameter; here He buffer gas was added at a pressure of 70 Pa. The sublevel detuning Δ_M is given in frequency units according to the magnetic field dependence of the Zeeman splitting frequency. It can be seen clearly that the lineshape of the Zeeman resonances changes drastically for increasing laser detunings Δ_E . In the case of $\Delta_E = 0$ the observed RHS display a very broad resonance structure. However, if the laser is slightly detuned from the center of the Doppler profile (e.g. for $\Delta_E/2\pi$ = 55 MHz), the width of the resonance decreases by almost one order of magnitude and a pair of symmetrical sidelopes begins to develop. For larger laser frequency detunings this characteristic lineshape patterns becomes more and more pronounced; it is accompanied by a further decrease of the resonance linewidth and of the signal amplitude. The observed lineshapes thus show the typical interference pattern of Ramsey resonances. This similarity is seen most obviously in Fig.9, which shows a particular nice RHS in the case of a very large laser frequency detuning; here again, He was used as a buffer gas. Within the investigated pressure range, the basic RHS lineshape features remain qualitatively the same for the different rare gas species. The Ramsey-type interference pattern, however, is less pronounced for the heavier rare gases /7/.

The important role of collisions for the generation of the RHS becomes evident by considering the pressure dependence of the RHS amplitude. A typical result for He perturbers is shown in Fig.10; during this measurement the laser detuning was kept fixed at $\Delta_E/2\pi$ = 300 MHz. At first, the RHS amplitude increases with increasing buffer gas pressure; it subsequently approaches a maximum value at a certain pressure and finally decreases again.

Our theoretical treatment relies on a steady state perturbative solution of the density matrix equations for the coherent excitation and detection processes shown in Figs. 6a and 6c, respectively; the effects of VCC on the ground state sublevel coherence are taken into account by means of a semiclassical transport equation. The resulting RHS can be written in the form /6,7/

$$S(\Delta_M) = \int_0^\infty F(t) \cos(\Delta_M t)\, dt \quad , \tag{1}$$

with F(t) given by

$$F(t) \sim \exp[(-v_0/u)^2] \ \exp(-\gamma t) \ \ p(-v_0 \rightarrow v_0, t) \ . \tag{2}$$

The function $p(-v_0 \rightarrow v_0,\ t)$ denotes the solution of the Boltzmann transport equation

$$\frac{\partial}{\partial t} p(-v_o \rightarrow v_o, t) = -\gamma_{VCC}\, p(-v_o \rightarrow v_o, t) + \int dv\, p(-v_o \rightarrow v, t)\, W(v \rightarrow v_o)\,; \qquad (3)$$

here, $W(v \rightarrow v_o)$ is the one-dimensional collision kernel and γ_{VCC} denotes the collision rate for VCC. $p(-v_o \rightarrow v_o, t)$ describes the probability to find an atom at time t with velocity v_o, if at time t = 0 its velocity is $-v_o$. The Gaussian in Eq.(2) describes the usual Doppler profile that yields information on the number of the initially excited atoms; the exponential $\exp(-\gamma t)$ describes the decay of sublevel coherence during the velocity diffusion process. The loss rate $\gamma = \gamma_{col} + \gamma_{tr}$ results from depolarizing collisions (γ_{col}) and spatial diffusion (γ_{tr}) out of the laser beam.

The experimental curves can be analyzed by means of the theoretical result Eq.(1), which essentially predicts that the sublevel resonance lineshape generally can be written as a Fourier transform of a certain function F(t). This function F(t) describes the distribution of diffusion times of atoms with Zeeman coherence moving from $-v_o$ to $+v_o$; it can obviously be obtained from the measured signals by a simple Fourier transformation according to

$$F(t) = \frac{1}{\pi} \int_{-\infty}^{\infty} S(\Delta_M) \cos(\Delta_M t)\, d\Delta_M \quad . \qquad (4)$$

In this way a temporal resolution of the velocity diffusion process is achieved. When the Fourier transformation (4) is applied to the measured curves, the data points (+++) shown in Fig.11 are obtained. These Fourier transforms obviously demonstrate that the most probable as well as the mean diffusion time get larger for an increasing laser detuning, i.e. for a larger separation between the interaction zones of the pump- and probe fields in velocity space. Particularly for large laser detunings the Fourier transforms show that there are barely any signal contributions

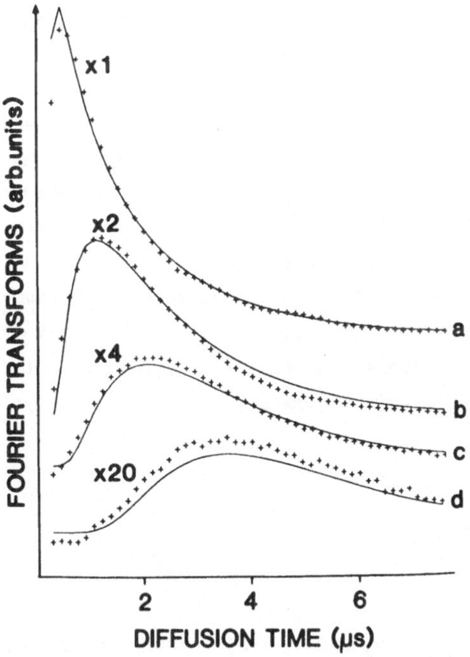

Fig.11: Typical distributions of diffusion times of Zeeman coherence in a He atmosphere (p_{He} = 70 Pa) for different laser detunings (a: $\Delta_E/2\pi$ = 55 MHz; b: 130 MHz; c: 220 MHz; d: 360 MHz). The data points are obtained by a Fourier transformation of the measured RHS; the solid lines correspond to an overall fit based on the FP-model: The values for the fit parameters are $\gamma = 6.0 * 10^5$ /s and $\gamma_{th} = 1.4 * 10^5$ /s.

arising a short time after the excitation: it takes a certain time until the first Sm atoms "arrive" at the opposite velocity subgroup $+v_o$. Due to this "intrinsic" time delay between the excitation and detection process and due to the phase sensitive detection scheme, the resonance lineshapes show the observed Ramsey -type interference structure that is connected with the linewidth narrowing. The relative heights of the Fourier transforms (or, equivalently of the measured RHS) indicate that the number of atoms contributing to the RHS is strongly reduced at large detunings Δ_M : First, the number of atoms which initially were excited is reduced according to the Doppler factor and, more important, the amount of detectable sublevel coherence is decreased due to losses during the larger diffusion time.

In order to analyse our experimental data on a quantitative base, we tried to fit the Fourier transforms of the measured RHS. The calculation of F(t) requires a solution of the Boltzmann transport equation (3); if we assume a Brownian motion-type collision model, i.e. the "weak" collision case, the Boltzmann equation (3) reduces to a Fokker-Planck (FP) equation /15/. The FP-equation contains only one time constant τ_{th} that characterizes the approach of a given velocity anisotropy towards thermal equilibrium. With the use of the well known solution of the FP-equation /6,7/ we have fitted the distribution functions F(t) to the measured curves using only two free parameters, the FP-parameter $\gamma_{th} = 1/\tau_{th}$ and the loss rate γ. An example of such an overall fit is shown in Fig.11; obviously, the FP-approach yields a good description of the experimental data. From these fits the thermalization time τ_{th} and the loss rate γ can be determined; for Sm-He collisions this time constant yields τ_{th} = (570 ± 100 µs Pa)/p, and the pressure broadening $\gamma_{col} = c_{col} * p$ for depolarizing collisions yields the coefficient c_{col} = $(2\pi * 0.48 \pm 0.05$ kHz)/Pa. In the case of heavier rare gas perturbers like Ar or Xe, however, the FP-model fails to describe the collisional velocity diffusion process; here, a more complex description of VCC based on the phenomenological Keilson- Storer collision kernel yields a better agreement with the measured collision data /7/.

In conclusion, we have reported on the observation of collision- induced Ramsey resonances in atomic Sm vapor in the presence of rare gas perturbers. The formation of these resonances is due to the collisional diffusion of Sm atoms in velocity space; here, in the creation of sublevel coherence, collisions are of no importance, in contrast to recent studies by Zou and Bloembergen on collision- assisted Zeeman coherence /16/. Moreover, our measurements have demonstrated that the observed linewidths of the Zeeman resonances can assume values lying below the limit given by the time rate of depolarizing collisions and by the transit time broadening /6,7/. Finally, we have shown that the time constants for the collisional thermalization of the active atoms can be derived from our data. Thus our experimental technique may find useful applications in high resolution sublevel spectroscopy in velocity space and in collision studies.

IV. Acknowledgements

One of us (J.M.) acknowledges partial support by the Heisenberg-Programm of the Deutsche Forschungsgemeinschaft and another of us (E.B.) by the Niedersächsische Graduiertenförderung.

V. References

/1/ N.F. Ramsey, "Molecular Beams" (Oxford Univ. Press, London, 1963).

/2/ J.E. Thomas, S. Ezekiel, C.C. Leiby,Jr., R.H. Picard, and C.R. Willis, Opt. Lett. 6, 298 (1981); J.E. Thomas, P.R. Hemmer, S. Ezekiel, C.C. Leiby,Jr., R.H. Picard, and C.R. Willis, Phys. Rev. Lett. 48, 867 (1982); P.R. Hemmer, S. Ezekiel, and C.C. Leiby, Jr., Opt. Lett. 8, 440 (1983).

/3/ J. Mlynek, K.H. Drake, G. Kersten, D. Frölich, and W. Lange, Opt. Lett. 6, 87 (1981).

/4/ J. Mlynek, N.C. Wong, R.G. deVoe, E.S. Kintzer, and R.G. Brewer, Phys. Rev. Lett. 50, 993 (1983).

/5/ See, e.g., M.D. Levenson, "Introduction to Nonlinear Laser Spectroscopy" (Academic Press, New York, 1982).

/6/ E. Buhr and J. Mlynek, Phys. Rev. Lett. 57, 1300 (1986).

/7/ E. Buhr and J. Mlynek, submitted to Phys. Rev. A.

/8/ V. Jordan, E. Buhr, R. Grimm, and J. Mlynek, to be published.

/9/ B.J. Dalton, T.D. Kieu and P.L. Knight, Opt. Acta 33, 459 (1986).

/10/ D.J. Wineland, J.J. Bollinger, and W.M. Itano, Phys. Rev. Lett. 50, 628 (1983).

/11/ A. Derbyshire et al, Proceedings of the 39th Annual Frequency Symposium 1985 (IEEE Catalog No. 85CH2186-5) p. 18; M. Arditi and J.L. Picque, J. Phys. (Paris) 41, L-379 (1980).

/12/ R. Grimm and J. Mlynek, to be published.

/13/ J. Mlynek, Chr. Tamm, E. Buhr, and N.C. Wong, Phys. Rev. Lett. 53, 1814 (1984); Chr. Tamm, E. Buhr, and J. Mlynek, Phys. Rev. A34, 1977(1986).

/14/ See, e.g., P.R. Berman, in "New Trends in Atomic Physics", Eds.: G. Grynberg and R. Stora (North Holland, Amsterdam,1984) Vol.1, p. 451.

/15/ P.R. Berman, Phys. Rev. A9, 2170 (1974).

/16/ Y.H. Zou and N. Bloembergen, Phys. Rev. A33, 1730 (1986).

CONTRIBUTED PAPERS

(Abstracts)

DISSIPATIVE DEATH OF QUANTUM EFFECTS IN A SPIN SYSTEM

Rainer Grobe and Fritz Haake
Fachbereich Physik, Universität-GHS Essen
Postfach 103 764, D-4300 Essen

We investigate the influence of very weak dissipation on the dynamics of a kicked spin system whose nondamped part [1] is described by the Hamiltonian

$$H = p\, J_y + \frac{K}{2j}\, J_z^2 \sum_{n=-\infty}^{n=\infty} \delta(t-n) \quad .$$

The timedependent Hamiltonian H commutes with the angular momentum vector $\vec{J}^2$ so that we can describe the system in a(2j+1) dimensional Hilbert space. The classical limit can be thought of as putting the angular momentum quantum number j to infinity.

Depending on the strength of the nonlinearity K we find classically regular or chaotic dynamics. We have compared the evolution of the quantum expectation value of a scaled angular momentum component $\frac{J_z}{j}$ with an ensemble average of many classical trajectories. We found that in the classically regular domain the corresponding quantum expectation value reveals a rather regular sequence of collapses and revivals with a period proportional to j. In the chaotic regime, however, we found erratic recurrencies.

The dissipation mechanism chosen in this model also leaves the "length" of our angular momentum vector $\vec{J}$ invariant and is described by the superradiance master equation [2] for the reduced density operator. Classically it corresponds to a relaxation of the vector $\vec{J}$ to the pole $J_z = -j$ of the sphere with radius j.

Surprisingly we found [3] in the quantum system dissipation manifests itself already for times much smaller than the inverse damping constant. Typical quantum phenomena like recurrencies are exponentially damped at a rate which is proportional to j. After a time proportional to the inverse spin length all coherencies are dead.

In the classically regular domain dissipation acts quite selectively: Classically important eigenstates of the nondamped dynamics are affected only weakly whereas the modes contributing only to quantum fluctuations are very sensitive to dissipation.

As a second phenomena we have investigated the influence of dissipation on coherent tunneling. A perturbative analysis reveals that the coherent tunneling frequency is decreased by the same amount by which a j-proportional damping would shift the natural frequency of the classical harmonic oscillator. These findings are in excellent agreement with our numerical data.

References

[1]: F. Haake, M. Kuś, and R. Scharf, Z. Phys. B65, 381 (1987).
[2]: R. Bonifacio, P. Schwendimann, and F. Haake, Phys. Rev. A4, 302 (1971); Phys. Rev. A4, 854 (1971).
[3]: R. Grobe and F. Haake, to be published.

PERIOD DOUBLING IN A QUANTIZED VERSION OF HENON'S MAP

R. Graham, S. Isermann
Fachbereich Physik, Universität Essen GHS

We consider a renormalization approach under period doubling of a quantized version of Hénon's map, which is known to exhibit chaos and to reach the chaotic state via the period doubling route. The influence of quantum noise on the period doubling is investigated. Writing Hénons map in the form

$$x_{n+1} = r(x_n, y_n) = 1 - ax_n^2 - Ey_n \qquad (1)$$
$$y_{n+1} = s(x_n, y_n) = Ey_n$$

the quantized version can be deduced from a damped, kicked oscillator [1] and the quantized map can be rewritten in the form of a simple c-number-map in which non-deterministic c-number-quantities appear, which are distributed according to some quasi-propability densities [2]

$$x_{n+1} = r(x_n, y_n) + \xi_n \qquad (2)$$
$$y_{n+1} = s(x_n, y_n) + \eta_n$$

Here ξ_n and η_n are stochastic c-numbers with *non-classical* statistical properties. Combining these equations to a single two-step recursive map in standard form [2] the latter reads

$$z_{n+1} + Bz_{n-1} = 2Cz_n + 2z_n^2 + \varsigma_n \quad , \; B = E^2 \qquad (3)$$

after appropriate redefinitions [2]. The special case $\varsigma_n=0$ will be referred to as the deterministic case. For the nonvanishing cumulants of the non-classical force ς we obtain

$$\langle \varsigma_n^2 \rangle \equiv \hbar Q_0 = \left[\frac{\hbar a^2}{\omega}\right] \frac{1+B}{4} \frac{1-B}{2} \coth\left[\frac{1}{2}\frac{\hbar\omega}{KT}\right] \qquad (4)$$

$$\langle \varsigma_n^3 \rangle \equiv \hbar R_0 = \left[\frac{\hbar a^2}{\omega}\right]^2 \frac{B}{16}$$

where ω is the frequency of the kicked oscillator and T is the temperature of the heat bath in which energy is dissipated. Renormalization transformations will later generate the following cumulants

$$\langle \varsigma_n \varsigma_{n\pm 1} \rangle \equiv Q_1 \qquad (5)$$

$$\langle \varsigma_n^2 \varsigma_{n\pm 1} \rangle \equiv R_{\pm 1}$$

An approximate renormalization scheme for the deterministic map has been developed by Helleman and McKay [3,4]. After each period doubling the map (3) can be rewritten after an appropriate redefinition of the variable z_n in its original form with well known recursion relations for the coefficients B and C [3,4]. In addition to these relations we get recursion relations for the non-classical stochastic force

$$\varsigma_n' = \alpha(B\varsigma_{2n-1} + e\varsigma_{2n} + \varsigma_{2n+1}) \qquad (6)$$

with

$$\alpha = e + \frac{(2C + 4z_{2+})^2}{1 + B} \qquad (7)$$
$$e = 2C + 4z_{2-}$$

where z_{2+}, z_{2-} denote the stable two-cycle of the deterministic map (3). The following recursion relations for the cumulants are generated from eqs. (4)-(6) [2,5].

$$\begin{bmatrix} Q_0' \\ Q_1' \end{bmatrix} = \alpha^2 \begin{bmatrix} (1+e^2+B) & 2e(1+B) \\ B & e(1+B) \end{bmatrix} \begin{bmatrix} Q_0' \\ Q_1' \end{bmatrix} \qquad \begin{bmatrix} R_0' \\ R_{+1}' \\ R_{-1}' \end{bmatrix} = \alpha^3 \begin{bmatrix} (1+e^3+B) & 3e(e+B^2) & 3e(1+Be) \\ B & e(1+Be) & 2Be \\ B^2 & 2Be & e(e+B^2) \end{bmatrix} \begin{bmatrix} R_0 \\ R_{+1} \\ R_{-1} \end{bmatrix} \qquad (8)$$

These recursion relations determine the scaling behavior of the quantum fluctuations. In the conservative case ($B=1$) and C close to $C_\infty(1)$ the second order cumulants vanish. The largest eigenvalue of the matrix renormalizing the cubic cumulants leads to an increase of the effective $\hbar$ after each period doubling according to

$$\hbar' \simeq 69.90\ \hbar \qquad (9)$$

A similar numerical result is obtained by Grempel et al. [6] by a different method ($\hbar' = |\alpha^2 e| \hbar \simeq 70.83\hbar$) which is only applicable to conservative maps however. In the case of very weak dissipation the two distinct types of quantum fluctuations, the cubic cumulants associated with the nonlinearity of the classical map and the quadratic cumulants with the linear mechanism of dissipation, compete with each other. Since the scaling of $\hbar$ is already fixed we get a rescaling of the effective temperature by the matrix, which renormalizes the second order cumulants. We find from its largest eigenvalue

$$\coth\left[\frac{1}{2}\frac{\hbar\omega}{kT}\right]' \simeq 2.247 \coth\left[\frac{1}{2}\frac{\hbar\omega}{kT}\right] \qquad (10)$$

This shows, that despite of the effective increase of $\hbar$ by renormalization the ratio $\hbar\omega/kT$ effectively decreases. For the strongly dissipative case $B \to 0$ the dimensionless ratio $\langle\zeta_n^3\rangle^2/\langle\zeta_n^2\rangle^3$, which measures the relative importance of the cubic cumulant, decreases rather quickly under renormalization by a factor 0.132.

In conclusion we can say that quantum fluctuations in a system, which is not strictly conservative, on the period doubling route to chaos very rapidly act like classical fluctuations.

[1] R. Graham, T.Tél, Z. Phys. B60, 127 (1985)

[2] R. Graham, Period Doubling in Dissipative Quantum Systems, preprint, Essen 1986

[3] A.J. Lichtenberg, M.A. Lieberman, "Regular and Stochastic Motion", Springer, Berlin 1986

[4] R.H.G. Helleman in "Fundamental Problems in Statistical Mechanics", Vol. 5, E.G.D. Cohen ed. North Holland, Amsterdam, p. 165

[5] S. Isermann, thesis, Universität Essen GHS, 1987, unpublished

[6] D.R. Grempel, S. Fishman, R.E. Prange, Phys. Rev. Lett. 53, 1212 (1984)

RESONANCE OVERLAP AND DIFFUSION OF THE ACTION VARIABLE IN THE LASEREXCITATION OF MOLECULAR VIBRATIONS

M. Höhnerbach, R. Graham
Fachbereich Physik, Universität Essen

Experimentally there are strong indications that chaotic dynamics is involved in the multi-photon dissociation of molecules in strong infrared laserfields. We discuss an appropriate anharmonic oscillator model classically and apply arguments from the theory of quantum localization to make qualitative predictions of quantum effects. The model we investigate is

$$H(p,x,t) = \frac{p^2}{2\mu} + D\left[1 - \exp(-\alpha x)\right]^2 - E_0\, x \cos\omega t \tag{1}$$

The anharmonic oscillator is of the Morse type, which is a suitable model to describe vibrational excitation of a diatomic molecule. In order to discuss classical chaotic dynamics it is useful to transform the Hamiltonian (1) into action and angle variables [1].

$$H(I,\theta,t) = 2I - I^2 + g\cos\omega t\,\ln\left[\frac{1+\sqrt{2I-I^2}\cos\theta}{(1-I)^2}\right] \tag{2}$$

Here we made the following rescaling of variables.

$$g = \frac{\mu E_0}{\alpha D}\ ,\quad \frac{\alpha}{\sqrt{2\mu D}}\, I \to I\ ,\quad \alpha\sqrt{\frac{D}{2\mu}}\, t \to t\ ,\quad \frac{1}{\alpha}\sqrt{\frac{2\mu}{D}}\,\omega \to \omega\ ,\quad \theta \to \theta - \pi/2 \tag{3}$$

As a criterion for the onset of chaotic behaviour we use Chirikov's criterion of resonace overlap [2]. A similar analysis was performed by Jensen [3] for a classical electron in a 1-dimensional Coulomb potential. The criterion requires the Fourier series expansion of the perturbation.

$$H(I,\theta,t) = 2I - I^2 + g\sum_{m=-\infty}^{\infty} V_m(I)\cos(m\theta-\omega t) \tag{4}$$

The maximum distortion of orbits in action - angle space will occur at resonances, where the phase , $m\theta - \omega t$, is stationary. The values for the resonant action variables are

$$I_m = 1 - \frac{\omega}{2m} \tag{5}$$

To consider the width of the resonances, we made an approximate transformation in the vicinity of the resonant islands. The width W_m is when be given by

$$W_m = 4\left[\frac{g\,|V_m(I)|}{2}\right]^{1/2}\Bigg|_{I=I_m} \tag{6}$$

The Fourier amplitudes $V_m(I)$ can be found exatly

$$V_m(I) = \frac{(-1)^{m+1}}{m} \left[\frac{I}{2-I} \right]^{m/2} - 2 \ln (1-I) \, \delta_{m,0} \tag{7}$$

The result for the width W_m is then

$$W_m = 2 \sqrt{2g/m} \left[\frac{1 - \frac{\omega}{2m}}{1 + \frac{\omega}{2m}} \right]^{m/4} \tag{8}$$

Resonance overlap occurs when the width of the islands is greater then the distance between two neighbouring islands, i.e.

$$\frac{W_m + W_{m+1}}{2} > I_{m+1} - I_m \tag{9}$$

This inequality defines an approximate criterion for the critical coupling g_c required to destroy all KAM surfaces between the m and m+1 island chains. For m » 1 we find

$$g_c(m) = \frac{\omega^2}{32m^3} e^{\frac{\omega}{2}} \left[1 + \mathcal{O}\left(\frac{1}{m} \right) \right] \tag{10}$$

The comparision of eq. (10) with the numerical calculations shows good agreement. For $g > g_c(m)$ all resonances (5) of the order larger than m can be reached from the m'th resonance by a classical diffusion process of the action variable. The diffusion constant D(I) of this process can be calculated in the quasi-linear approximation [3,4] and we find

$$D(I) = \frac{g^2 \pi}{4} \frac{1}{(1-I)} \left[\frac{I}{2-I} \right]^{\omega/2(1-I)} \tag{11}$$

The mean first passage time τ for the action I(t) to diffuse across the border of dissociation at I=1 is then

$$\tau = \frac{4 \, (1-I(0))^3}{3g^2 \, \pi \, e^{-\omega}} \left[\frac{3}{2} \frac{(4\omega g)^{1/3} \, e^{-\omega/6}}{(1-I(0))} - 1 \right] \tag{12}$$

in good qualitative agreement with numerical results. Quantum mechanically classical chaos may induce Anderson-type localization of the quasi-energy states with respect to the action variable. Preliminary investigations indicate that this does not happen in the present case due to the fact that D(I) diverges for $I \to 1$, which would imply a large localization length, while at the same time the size of the interval (1-I) where localization could occur shrinks to zero. However, a different type of localization due to Cantori (= fractal remnants of KAM tori) exists in this system, as has recently been shown in [7].

[1] C. C. Rankin, W. H. Miller, J. Chem. Phys. 55, 3150 (71)
[2] B. V. Chirikov, Phys. Preorts 52, 265 (79)
[3] R. V. Jensen, Phys. Rev A, 30, 386 (84)
[4] G. M. Zaslavski, N. N. Filonenko, Sov. Phys. JETP 27, 851 (68)
[5] G. Casati, B.V.Chirikov, I. Guarneri, D.L.Shepelyansky, Phys. Rev. Lett 56, 2437 (86)
[6] R. Blümel, U. Smilansky, preprint WIS-86/48-Ph Sept, (86)
[7] R. C. Brown, R. E. Wyatt, Phys. Rev. Lett. 57, 1 (86)

WINDING NUMBERS AND COLLISIONS BETWEEN ATTRACTORS IN A LASER SYSTEM

J. Tredicce, R. Gilmore, H. G. Solari and E. Eschenazi.
Department of Physics and Atmospheric Science, Drexel University,
Philadelphia, PA 19104, U.S.A. .

We consider the standard model for a CO_2 laser based on the rate equations for the intensity, u, the population invertion, z, and modulated loses, in the (normalized) form

$$du/dt = (z - R\cos(\Omega t))\,u$$
$$dz/dt = 1 - \varepsilon_1 z - (1 + \varepsilon_2 z)\,u \qquad (1)$$

It has been shown experimentally[1] that the CO_2 laser with modulated loses displays chaotic behavior even when forced at moderate amplitude.

A detailed[2] study of equation (1) is able to explain most (if not all) the observed features, in terms of the coexistence of several attractors with different basins of attraction for a given set of parameters.

The general description of the dynamics through the bifurcation diagram is as follows:

For no force (R = 0) there are two fixed points; a saddle at $(u,z) = (0, 1/\varepsilon_1)$ and a node at (1,0). This last one is the unperturbed mode of operation of the laser.

The saddle point is invariant under changes of the strengh R and remains at $(0, 1/\varepsilon_1)$ for all R. The node initiates a period doubling cascade which back bends (at R_{in}) in an inverse saddle-node bifurcation. This bifurcation involves the period two orbit born out of the period one in a pitchfork bifurcation together with an unstable period two orbit born at a saddle-node bifurcation at R_2. There is a coexistence of attractors, one belonging to the unperturbed state of the laser and the other formed by the node created at R_2 and its cascade.

The succession of pitchfork bifurcations reachs the acumulation point and subsequently undergoes a period halving (inverse) cascade.

There is simultaneously a series of saddle-node bifurcations which creates periodic orbits of periods 3,4,...,n,... for increasing values of R, (R_n). The nodes of these orbits undergo complete cascades. We call a branch the series of events starting in a saddle-node bifurcation, including the noisy periodic attractors of the inverse cascade.

The orbits created at R_n are characterized by one peak an a charge-time of n-1 periods approximately. In addition to this serie, there exist others branches which create more coexisting attractors, although these orbits are not easy to detect due to the smallness of their basins of attraction. In particular, we have found another branch of period 5 coexisting with members of the main branch and the branches born at R_3 and R_4.

Different coexisting attractors can interact through external and boundary crisis resulting in the fusion of their respective basins of attraction.

For a fixed value of the parameters ε_1, ε_2 and Ω in (1), those external and boundary crisis came in a given order with respect to R but, would they came in the same order if we change ε_1, ε_2 or Ω ?.

If the dynamics is higly sensitive to changes in the parameters we lose all possibilities of comparision between the theory (or models) and the experiments and even the reproductiveness of the experiments.

Fortunately enough, there are constraints impossed on the dynamics by the way in which the oribts wrap around each other. Those constraints are well represented (for forced damped two dimensional systems) by the winding numbers and the collection of them, the intertwining matrix[3].

The relative winding number between the orbits γ, $(x_\gamma=(u_\gamma,z_\gamma)(t))$, and β, $(x_\beta=(u_\beta,z_\beta)(t))$ of an arbitrary forced two dimensional system is defined by:

$$n^{ij} = 1/(\tau\, p_\gamma\, p_\beta)\int_{[0,\,\tau\, p_\gamma\, p_\beta]} dt\, \{[(x_\gamma{}^{i} - x_\beta{}^{j}) \wedge d(x_\gamma{}^{i} - x_\beta{}^{j})/dt\,]/\,(x_\gamma{}^{i} - x_\beta{}^{j})^2\} \qquad (2)$$

where $\wedge$ means external product, and $p_\gamma\, p_\beta$ are the respective periods of γ and β. The index i j in (2) stand for the several different pairs of initial conditions possible for a fixed phase of the forcing term, i. e.

$$x^i = x(i\,\tau)\ i=0\ldots p_\gamma \quad , \quad x^j = x(j\,\tau)\ j=0\ldots p_\beta\,.$$

It has been shown[3] that the winding numbers (W.N.) are invariant under continuous deformations if both orbits γ and β continue to exist during the process.

The W.N. can be generalized to pairs composed by a chaotic localized attractor and a periodic orbit and even two chaotic attractors.

One of the most useful properties of the W.N. is that if two orbits (chaotic or not) are involved in a collision, then they have the same W.N. with respect to all the other orbits (not necesarily attractors) present at the parameter value of the collision and which are not in the closure of the attractor.

Based on this property and the intertwining matrix for the system (in our case eq. (1)), it is possible to understand what will be the order of the succesive crisis.

Predictions based on this kind of analysis of eq. (1) have been recently confirmed in experimental observations[4].

We finally stress that W.N. can be measured directly from the experimental data, even for orbits which do not coexist simultaneously.

H. G. Solari is a fellow of the Consejo Nacional de Investigaciones Científicas y Técnicas of Argentina. J. R. Tredicce acknowledges a "Josep H. DeFrees" grant of Research Corporation.

1. J. R. Tredicce, F. T. Arecchi, G. P. Puccioni, A. Poggi and W. Gadowski. Phys. Rev. **A34**, 2073 (1986).
2. H. G. Solari, R. Gilmore, E. Schenazi and J. R. Tredicce. (To be published).
3. R. Gilmore and H. G. Solari. (To be published).
4. F. T. Arecchi, R. Meucci, A. Poggi and J. R. Tredicce. (To be published).

SQUEEZED QUANTUM FLUCTUATIONS AND NOISE LIMITS IN AMPLIFIERS AND ATTENUATORS

S M Barnett[1], C R Gilson[2], S Stenholm[3] and M A Dupertuis[3].
[1]Optics Section, Blackett Laboratory, Imperial College, London SW7 2BZ, England.
[2]Kings College London. [3]University of Helsinki.

Conventional linear amplifiers and attenuators add quantum fluctuations to a signal (1). This additional noise is an unavoidable consequence of quantum mechanics and is responsible for a reduction of the signal to noise ratio during processing. Moreover, these fluctuations mean that even strongly squeezed input light leads to an unsqueezed output if the amplifier gain exceeds a factor of two (2). However, devices exhibiting squeezed fluctuations in the gain or loss medium may add squeezed quantum noise to the signal. This squeezing of the added noise leads to an improved signal to noise ratio, in one quadrature of the output, compared to the limits associated with conventional devices (3-6).

An ideal, quantum mechanical, linear amplifier or attenuator multiplies an input signal by a gain factor $G^{\frac{1}{2}}$ and adds Langevin-type quantum noise to the signal. In the Heisenberg interaction picture the annihilation operator for the output, a^{out}, is related to the input operator, a^{in}, by the expression

$$a^{out} = G^{\frac{1}{2}}a^{in} + R$$

Here R is the result of quantum fluctuations in the gain or loss medium. For truly linear operation we assume the added noise operators to commute with the input-field operators. The unitarity of quantum mechanics requires the conservation of the commutator (a,a^{+}). Therefore, the added noise operators are constrained to obey the commutator $(R,R^{+}) = 1-G$.

We define the hermitian quadrature phases by relation to the annihilation operator $a = a_1 + ia_2$, with a similar expression for the reservoir noise operator $R = R_1 + iR_2$. If we assume that the initial states of the signal mode and gain or loss medium are decorrelated, then the variances in the output quadratures are

$$(\Delta a^{out}_{1,2})^2 = G(\Delta a^{in}_{1,2})^2 + \Delta R^2_{1,2}$$

Clearly, the fluctuations in the output are supplemented by the additive contributions $\Delta R^2_{1,2}$. The additional fluctuations are limited by the Heisenberg uncertainty principle.

$$\Delta R_1 \Delta R_2 > 1/4|1 - G|.$$

Conventional devices have phase-insensitive noise associated with the gain or loss medium and so produce equal added noise variances $\Delta R_1^2 = \Delta R_2^2$. If the noise associated with the gain or loss medium is squeezed, then the noise added to one of the output quadratures, during processing, is less than that added by any conventional device. Naturally, the conjugate quadrature must become very noisy for the uncertainty principle to hold.

The principle of low noise signal processing may be demonstrated by considering a specific amplifier or attenuator model. We have specialised our general discussion to a simple model in which the gain or loss medium takes the form of a reservoir of inverted or conventional harmonic oscillators respectively (3,4). In this model, the noise added to the signal during processing is dominated by those elements of the gain or loss medium with transition frequencies that are near to resonance with the signal mode. The signal and its associated fluctuations can be manipulated by specially preparing or 'rigging' the reservoir state. Three classes of reservoir state are of particular interest. If the reservoir oscillators are prepared in their vacuum states, then the amplifier and attenuator reproduce the limiting behaviour of conventional devices. Thermal reservoirs, in which the loss medium is prepared at a finite temperature or the gain medium is prepared at a finite negative temperature, degrade the output signal to noise ratio below this conventional limit. If the reservoir is prepared in a multimode squeezed state, then amplification or attenuation with reduced quantum noise is possible for one quadrature of the output. Squeezed reservoirs may also reproduce noisy thermal behaviour.

REFERENCES

(1) C M Caves; Phys Rev D 26,1817 (1982).
(2) R Loudon and T J Shepherd; Optica Acta 31,1243 (1984).
(3) M A Dupertuis and S Stenholm; J Opt Soc Am B, in press (1987).
M A Dupertuis, S M Barnett and S Stenholm; J Opt Soc Am B, in press (1987).
(4) C R Gilson, S M Barnett and S Stenholm; J Mod Opt, in press (1987).
(5) J A Vaccaro and D T Pegg; Optica Acta, 33,1141 (1986).
(6) G J Milburn, M L Steyn-Moss and D F Walls, to be published.

ORDERED STRUCTURES OF IONS STORED IN A RF-TRAP

R.Casdorff, R.Blatt and P.E.Toschek
I.Institut für Experimentalphysik, Universität Hamburg
Jungiusstraße 9, 2000 Hamburg 36

Laser cooling of ions /1/ and atoms /2/ and subsequent realization of magnetic /3/ and optical traps /4/ for neutral atoms have lead to the consideration of crystallization of ions /5/ and possible Bose condensation in light pressure traps /6/. In particular, ordered structures have been found to occur /7/ for ions confined in a rf-trap, and various spatial configurations have been predicted. However, experimental verification is still lacking. On the other hand, many years ago, Wuerker et al. /8/ performed an experiment with charged aluminum particles stored in a rf-trap almost cooled to rest of the viscous drag of the atmospheric background pressure,where ordered structures were indeed observed, however in patterns different from the results of Ref. 7. Numerical calculations based on a Monte-Carlo-Simulation enable us to predict the shape of various ordered structures of few trapped ions under the irradiation by the cooling light of a cw dye laser. The ions are treated as classical particles, their motion being determined by the time dependent trap potential and the mutual Coulomb repulsion. Laser cooling is included by random spontaneous emission as in Ref. /9/ where the simulation procedure has been derived for free particles. Comparison of the simulation results with the experimentally observed structures /8/ and with experimental as well as theoretical distribution functions /10/ of ion clouds shows excellent agreement. For these reasons we believe our simulation procedure to be correct; we are able to predict reliable ordered structures.

The calculation reveals the following features:
(1) Ordered structures are readily obtained in the process of laser cooling. In contrast with Ref. /7/, the structural arrangement of up to 10 ions is in the X-Y plane of a rf-trap, provided no dc-potential is applied (trap geometry: $Z_0 = X_0 / \sqrt{2}$). If more than 10 ions are loaded to the trap the structures develop additional layers parallel

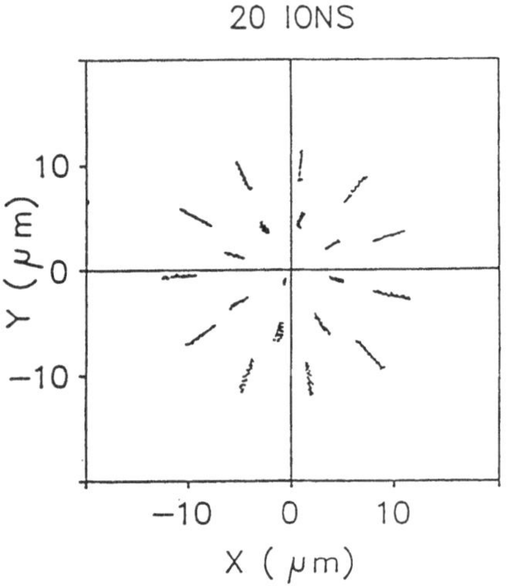

Fig. 1
Structure of 20 ions in a re-trap X-Y cut, a = 0, q = 0.4, Ω/2π = 6.4 MHz

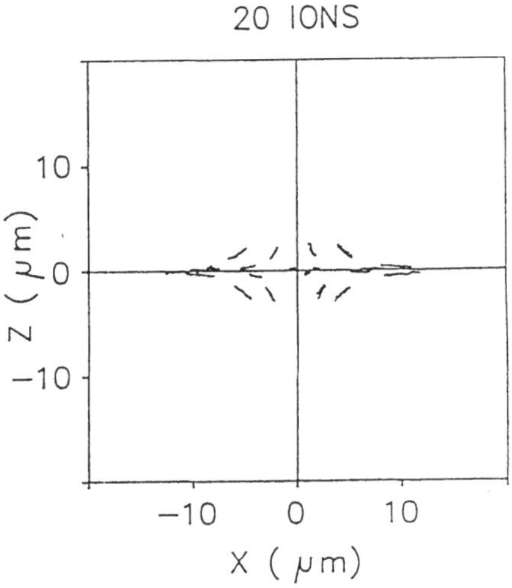

Fig. 2
X-Z cut of structure of Fig. 1. This reproduces essentially the observed structures of Ref. 8

to the X-Y plane, the ions generally being aligned along the field lines of the trapping field (cf. Fig. 2).
(2) Small imperfections in the trap geometry or slight asymmetries of the cooling laser with respect to the center of gravity of the ion cloud lead to rotation of the structures.
(3) The ratio of Coulomb-energy over kinetic energy which determines the cooperative behaviour of such a dilute plasma /11/ is periodic in the drive frequency and mean values ∿ 100 can be achieved. At values ∿ 170 crystallization has been predicted /12/.

The numerical simulation of ordered structures of trapped ions in a rf-trap under the influence of laser cooling is a powerful tool for the clarification of the dynamics of few-body systems.

/1/ W.Neuhauser, M.Hohenstatt, P.E.Toschek and H.G.Dehmelt Phys.Rev. A22, 1137 (1980)

/2/ J.Prodan, A.Migdall, W.D.Phillips, I.So, H.Metcalf and J.Dalibard Phys.Rev.Lett. 54, 992 (1985), W.Ertmer, R.Blatt, J.L.Hall and M.Zhu Phys.Rev.Lett. 54, 996 (1985)

/3/ A.Migdall, J.V.Prodan, W.D.Phillips, T.H.Bergeman and H.J. Metcalf Phys. Rev.Lett. 54, 2596 (1985)

/4/ S.Chu, J.E.Bjorkholm, A.Ashkin and A.Cable Phys.Rev.Lett. 57, 314 (1986)

/5/ J.J.Bollinger and D.J.Wineland Phys.Rev.Lett. 53, 348 (1984)

/6/ J.Javanainen, cf. these Proceedings

/7/ E.V.Baklanov and V.P.Chebotayev Appl.Phys. B39, 179 (1986)

/8/ R.F.Wuerker, H.Shelton, R.V. Langmuir J.Appl.Phys. 30, 342 (1959)

/9/ R.Blatt, W.Ertmer, P.Zoller and J.L.Hall Phys.Rev. A34, 3022 (1986)

/10/ R.Blatt, P.Zoller, G.Holzmüller and I.Siemers Z.Phys. D4, 121 (1986)

/11/ S.Ichimaru, Rev.Mod.Phys. 54, 1017 (1982)

/12/ W.L.Slattery, G.D.Dooley and H.E. de Witt Phys.Rev. A21, 2087 (1980)

QUENCHING OF QUANTUM NOISE AND DETECTION OF WEAK OPTICAL SIGNALS IN THE QUANTUM BEAT LASER

János A. Bergou
Center for Advanced Studies, Department of Physics and Astronomy, University of New Mexico, Albuquerque, NM 87131
and
Central Research Institute for Physics, H-1525 Budapest, P.O.Box 49, Hungary

Abstract: Starting from a Hamiltonian model of the coupled three-level/two-mode system, a nonlinear quantum theory of the quantum beat laser is developed. The theory is valid under the special conditions that lead to correlated spontaneous emission laser (CEL) operation in the linear theory. It is shown that vanishing of the diffusion constant for the relative phase persists in the nonlinear theory and this operation is stable above threshold. We show, on the example of the CEL gyro/gravity wave detector, how this effect can advantageously be applied to the detection of weak optical signals. A detailed analysis of the noise performance suggests that the "in principle" sensitivity of CEL detectors might exceed the standard quantum limit. A theoretical upper limit for the sensitivity increase is established.

In the optical detection of small changes of a given physical quantity the change is converted into a phase shift (in the passive scheme) or frequency shift (active scheme) of a laser field. This is accomplished by sending the laser light through or generating it in a cavity whose optical path length is sensitive to the physical effect to be detected. The shift is measured by beating the output light with that from a reference laser.

In this paper we deal with active detectors. In the active detection scheme the limiting noise source is the fluctuation, caused by independent spontaneous emission events, in the relative phase between the signal and reference lasers. It has been shown in a recent paper that the linewidth and the associated uncertainty in the relative phase may be eliminated by preparing the laser medium in a coherent superposition of two upper states [1], as e.g. in the quantum beat laser and Hanle laser.

In the present paper we develop a nonlinear theory of the CEL quantum beat laser. The key feature of our approach is that, by introducing appropriately defined "dressed states", the strong classical interaction of upper two levels is eliminated and, in terms of these dressed states, we have two uncoupled upper states. Then we show that the particular detuning conditions of Ref. 1 amount to selecting one of the dressed states as the upper state for the laser transitions. In fact, one can, at this point, introduce two orthogonal dressed modes and show that, when the above conditions are met, only one of them will lase [2]. A similar "dressed atom-dressed mode" picture applies to the case of a Hanle laser, as well [3]. The CEL operation can now be interpreted as follows. The original two bare modes can both be expressed in terms of the single lasing "dressed mode". In particular, the spontaneous emission contribution into both bare modes will be common and they cancel from the beat signal. This is the physical origin of the quenching of quantum noise.

In the next step we show that, if there is an additional frequency shift between the two modes due to some physical effect (change of arm length due to gravity waves or change of path lengths between two counterpropagating waves in a ring interferometer like in the Sagnac effect) then, the minimum detactable frequency might in principle be by a factor

$$\varepsilon = \sqrt{\frac{\Omega}{\gamma_c}} \tag{1}$$

smaller than the standard quantum limit. Here γ_c is the cavity linewidth (inverse of photon lifetime in the cavity) and Ω is a characteristic frequency of the effect to be detected (Ω is the frequency of gravity waves for gravity wave detectors and the rotation for laser gyros). However, when CEL operation occurs, for a large range of parameters also frequency locking occurs and in this case ε should simply be replaced by 1. In this case the CEL detector operates on the standard quantum limit. It still has the advantage over the usual active systems that the signal appears directly as a phase shift and not as a frequency shift and the system is free of the so-called "dead band" associated with active systems of detection.

The most interesting aspect of CEL detectors is that, by simply increasing its geometrical dimensions, it can be unlocked. This leads to the following requirement

$$S \frac{\Omega}{\gamma_c} > 1 \qquad (1)$$

where S is a characteristic scaling factor of the effect to be measured. Its explicit expression for the Sagnac effect is

$$S = \frac{4A}{\lambda\!\!\!^{-} p} \qquad (3)$$

where A is the area, p the perimeter of the ring cavity and $\lambda\!\!\!^{-}$ is the reduced wavelength. Using this in (2) we find

$$r^2 > \frac{c\ t^2}{16\ \pi^2} \frac{\lambda}{\Omega} \qquad (4)$$

as the unlocking condition. Here c is the velocity of light and t is the transmission of mirrors. We also used the standard expression of γ_c via c, p and t . If we are in the unlocked regime the sensitivity improvement is not quite as large as one might expect it from (1) because γ_c decreases with the increase of the geometrical dimensions but it can still be quite significant. For example, for case of detecting a rotation rate equal to that of the earth ($\sim 10^{-4}$ Hc) with the help of an unlocked CEL Sagnac detector $\varepsilon \sim 10^{-3}$. This is a significant effect and might render some of the predicted effects of general relativity observable.

The author is indebted for several stimulating discussions on different aspects of the problem to M.O. Scully, M. Orszag and L.M. Pedrotti. This work was supported by the ONR.

References

[1] M.O. Scully, Phys. Rev. Lett. 55, 2802 (1985)

[2] J.A. Bergou, M. Orszag, and M.O. Scully, Nonlinear theory of the quantum beat laser, submitted to Phys. Rev. A.

[3] J.A. Bergou and M.O. Scully, Quantum theory of the Hanle laser and its use as a metric gravity probe, invited paper in 'Hanle probes of the micro-and macrocosmos', eds. R. Stenflo and R. Strumia (honorary volume dedicated to the 85th anniversary of W. Hanle, Springer, 1987 to appear).

LIGHT PRESSURE INDUCED NONLINEAR DISPERSION IN A DOPPLER-BROADENED MEDIUM

R. Grimm, Chr. Tamm* and J. Mlynek
Institute of Quantum Electronics - Laboratory of Quantum Optics
Swiss Federal Institute of Technology (ETH) Zürich
CH-8093 Zürich, Switzerland

In recent years resonant light pressure effects have attracted increasing attention /1/. While laser cooling of atoms in a beam is an example of special practical interest, resonant light pressure also affects the atomic velocity distribution in a *Doppler-broadened gaseous sample*. Only recently, however, this simple but basic situation has been analyzed in some more detail: it has been found that the effect of radiation pressure on the atomic velocity distribution can strongly modify the *nonlinear susceptibility* of a gas giving rise to new phenomena in its optical response /2,3/; most importantly, a *nonlinear dispersion* has been predicted that displays an *even symmetry* with respect to the optical Doppler detuning /3/. Here, we give a brief theoretical outline of this *light pressure-induced nonlinear dispersion*; in addition, we propose a simple experiment to measure this novel effect.

We calculate the optical response of a Doppler-broadened medium under conditions of narrow bandwidth laser excitation taking into account photon momentum transfer. The atomic medium is modeled as an ensemble of two-level atoms with a Lorentzian velocity distribution. Using an appropriate density matrix formalism, a perturbative treatment yields the following result for the total index of refraction n /4/:

$$n - 1 \sim \frac{-\delta + \varepsilon_r \tau \, r^2 (1+2r^2)^{-3/2}}{\delta^2 + 1}$$

Here $\hbar\varepsilon_r = \hbar^2 k^2/2m$ is the photon recoil energy, r denotes the optical Rabi frequency normalized to the homogeneous optical linewidth Γ, τ is the effective transit time of the atoms through the laser beam and δ describes the detuning of the light field in units of the Doppler width ku. A representative plot of this equation is shown in Fig. 1; quite obviously resonant light pressure can drastically modify the dispersion of an atomic sample. We note that the even symmetry of the

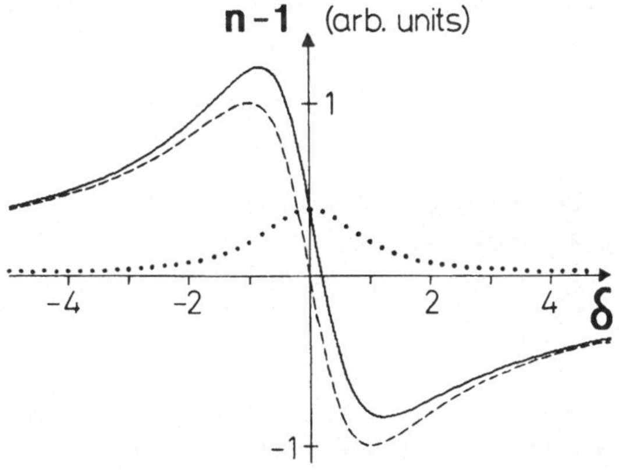

Fig. 1: Dispersion curves showing the total dispersion (——) as the sum of the nonlinear light pressure-induced part (······) and the ordinary linear dispersion (- - -) for $\varepsilon_r \tau = 1$ and $r = 1$.

nonlinear dispersion feature (Fig. 1, ······) results from the fact that the spontaneous scattering force does not depend on the sign of the laser detuning from line center.

For the observation of this new phenomenon, we propose an experiment based on FM spectroscopy /5/. With the use of this simple and sensitive technique (Fig. 2) a phase shift of the strong carrier with respect to the weak sidebands can be measured. Here, for a modulation frequency $\Gamma \ll \omega_m \ll ku$, the weak sidebands can act as a phase reference for the light pressure induced nonlinear dispersion of the carrier (Fig. 2b); this dispersion feature is reflected directly in the in-phase modulation component of the transmitted laser intensity. The corresponding signal shows an even symmetry with respect to the laser detuning δ; moreover, its strength depends on the Rabi frequency and, via the atomic transit time τ, also on the laser beam diameter /4/.

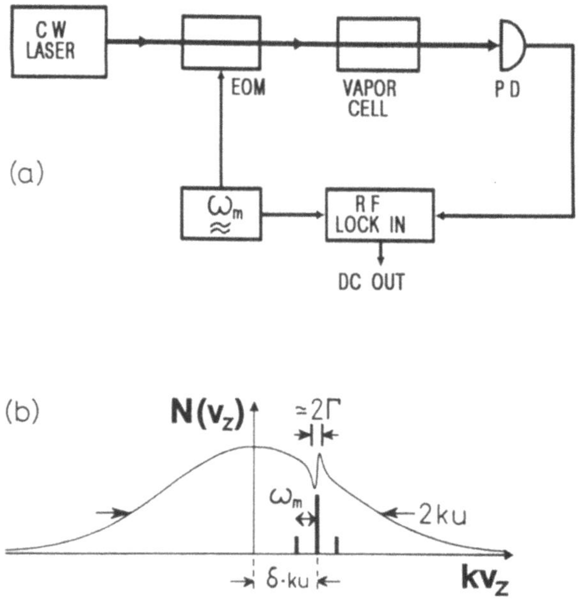

Fig. 2:

(a) Proposed experimental scheme to observe the light pressure-induced nonlinear dispersion: EOM, electro-optic phase modulator; PD, photo detector.

(b) Schematic of the FM laser field (modulation frequency ω_m, propagation in z-direction) interacting with the Doppler-broadened sample. The strong carrier distorts the velocity distribution $N(v_z)$ by its light pressure. This gives rise to a phase shift with respect to the weak sidebands that shows up in the in-phase modulation component of the transmitted light intensity.

Experiments to verify these predictions are currently in preparation. As an atomic sample the Ytterbium λ=555.7nm $^1S_0$-$^3P_1$ line might be an interesting candidate: Here $\varepsilon_r \tau = 1$ can be fulfilled using a beam diameter of about 1cm and r=1 can be obtained easily with laser intensities of less than 1 mW/cm^2.

References

/1/ See, e.g., Jl. Opt. Soc. Am. B2, 1705-1860 (1985) and references therein.

/2/ R. Grimm, Diploma thesis (Hannover University, 1986), unpublished.

/3/ A.P. Kazantsev, G.I. Surdutovich, V.P. Yakovlev, JETP Lett. 43, 281 (1986).

/4/ R. Grimm and J. Mlynek, to be published.

/5/ G.C. Bjorklund, Opt. Lett. 5, 15 (1980).

* Present address: Beijing Institute of Opto-Electronic Technology, Beijing, China.

UNSTABLE PERIODIC ATOMIC ORBITALS

Hubert Klar
Fakultät für Physik, Universität Freiburg
Hermann Herder Str. 3, D-7800 Freiburg

Simple examples of nonintegrable mechanical systems like stadium problems have established the fundamental role of unstable periodic classical trajectories for the structure of the corresponding quantum spectrum (1). These investigations show clearly in a one-to-one correspondence that quantum wave functions are mainly distributed along classical periodic orbits. There is however in general no simple relation between classical solutions and energy positions of individual quantum states except that the statistical distribution of energy levels avoids accumulations because the set of periodic orbits is usually of measure zero with respect to nonperiodic, in general chaotic orbits.

The structure of atomic resonance spectra near thresholds of multiple ionisation is still largely unexplored. The problem is a difficult one, experimentally due to the lack of lasers in the spectral range of interest, and theoretically due to the high degree of non-integrability of the N-body ($N \geq 3$) Coulomb problem. The importance of the problem is augmented by the fact that we expect universal properties for all atoms depending only on the number of highly excited electrons (two in the following for simplicity) because the core is a passive spectator in such a situation. For the reason mentioned above recent theoretical research has focussed to find periodic classical trajectories for two-electron atoms (3-body problem). All these orbits must be unstable because the potential surface has no minimum. Below we report our results.

One class of periodic solutions describes rotating rigid bodies (2). It can be shown that exactly two configurations exist, a linear rotor and a top. In both cases the electrons perform strongly correlated circular orbits with radii equal to each other, see Fig. 1 and 2, such that the centrifugal force cancels all Coulomb forces. The spectrum of local Liapunov exponents identifies these trajectories as hyperbolic fixed points (3).

A second class of periodic solutions describes a "breathing atom". Here each electron performs an ellipse, both ellipses are equal in size and opposite to each other, see Fig. 3. The non-precessing ellipses may be regarded as Kepler ellipses with nuclear charge Z-.25. The straight line limit of the ellipses is known as Wannier solution (4), the obvious generalisation to ellipses seems never have been published (5). Also these solutions are unstable because the motion of the electron pair proceeds along a potential ridge.

A quantisation for unstable periodic orbits is unknown. The identification of such orbits in quantum spectra is simplified in two limiting cases. In the case of a weak instability (small positiv Liapunov exponent) one may regard the orbit as approximately stable, and quantise the degree of freedom according to Bohr-Sommerfeld. This would lead to atomic rotation spectra for the rotor and the top,

$$E_L = I - b/L(L+1)$$

and to a series of ridge riding resonances

$$E_N = I - .5(Z-0.25)^2/(N+2.5)^2$$

Here I is the threshold for double ionisation, L and N are integers >>1 and b is a rotation constant known for rotor and top. In the opposite limit of a large Liapunov exponent the interaction of the periodic orbit with a chaotic background leads to interference pattern. The Fourier transform of excitation cross sections should then spike at the classical recurrence times.

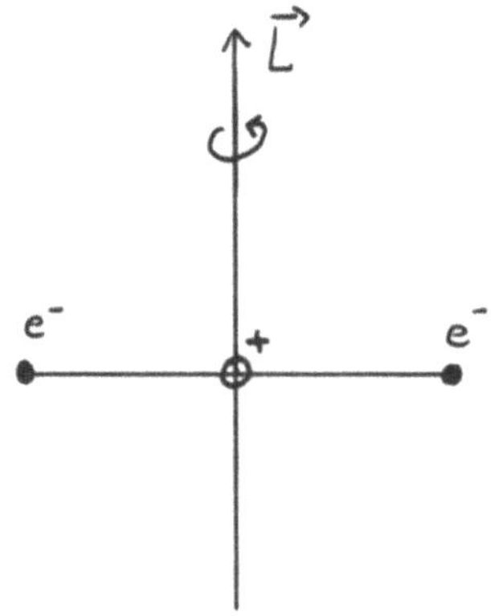

Fig. 1

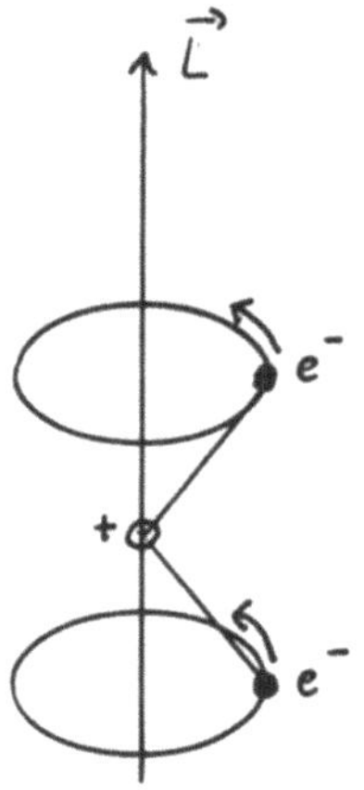

Fig. 2

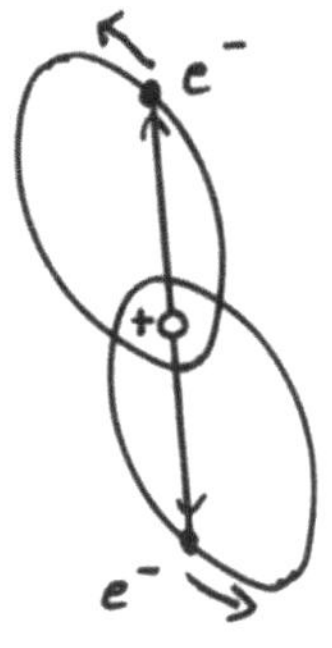

Fig. 3

References

1. E.J. Heller, Lect. Not. Phys. 263, 162 (1986)
2. H. Klar, Phys. Rev. Lett. 57, 66 (1986); Zeit. Phys. D3, 353 (1986)
3. H. Klar, Jour. Opt. Soc. Am. B, in press (1987)
4. G. Wannier, Phys. Rev. 90, 817 (1953)
5. J.H. Macek, priv. comm.

Virtual cloud effects in spontaneous decay

G. Compagno, A. Santangelo
Istituto di Fisica dell' Università, via Archirafi 36
90123 Palermo, Italy

An atom interacting with the e.m. field when is in its ground state is dressed by a cloud of virtual photons. This cloud is connected to the level's radiative shift.[1] Moreover the atom's bare ground state is not asymptotically stationary being its time evolution influenced by the interaction. It was long ago suggested that some processes, like radiative decays, should be considered to occour between dressed states instead than between bare ones. The one-photon decay and resonance scattering from one atom have been studied taking into account the atom dressed ground state.[2] Only the excited levels' decay time has been calculated. Moreover in the one-photon atom's decay the transition amplitudes between dressed states differ from the corresponding ones between bare states only for high order corrections.

It can be shown that starting from a bare state, the time it becomes effectively dressed by a virtual cloud is usually short and at most of the order of the inverse of the transition frequency between the given state an the immediate higher one. This leads one to expect that also for excited states, whenever the dressing time is short compared to the decay time, the dressing should be taken into account. In the two-photon decay from a metastable atomic state the Feynman diagrams describing, at lowest order, dressing are of the same order of the lowest order diagrams for two-photon decay. This implies that, for this process, the differences arising by considering dressed instead than bare states, should be more significant than the corresponding ones in the one-photon decay.

Here we consider a model three-level atom, with bare eigenstates $|i\rangle$,(i=1,2,3) and interacting with the e. m. field, described by the Hamiltonian $H=H_\circ+V$, with

$$H_\circ = \sum_i \omega_i |i\rangle\langle i| + \sum_j \sum_{\underline{k}_j} a^\dagger_{\underline{k}_j} a_{\underline{k}_j} \qquad (j=1,2)$$

and

$$V = \sum_j V_{j3} = \sum_j \sum_{\underline{k}_j} [\epsilon_{\underline{k}_j} (a^\dagger_{\underline{k}_j} + a_{\underline{k}_j}) |3\rangle\langle j| + \text{H.c.}] \qquad (1)$$

With this Hamiltonian the intermediate level 2 executes a two-photon decay to level 1 via level 3. It is dressed at the same order by virtual photons emitted in the virtual transition 2↔3. At the same order level 1 is dressed by virtual photons due to transition 1↔3. To take into account the dressing of both initial and final state,we subject the Hamiltonian (1) to a unitary transformation $T=\exp(iS)$ with $S^{+}=S$. So it is obtained the transformed Hamiltonian $\tilde{H}=THT^{-1}$. By developing S in powers of the coupling constant ϵ as $S=\sum S_n$, where S_n O(n) we obtain

$$\tilde{H} = T H T^{-1} = \tilde{H}_o + \sum \tilde{V}_n \quad , \tag{2}$$

where $\tilde{V}_n$ is the O(n) effective interaction. The S_n can be choosen so that in $\tilde{V}_n$ there are no terms that,at order n, give rise to unobservables virtual transitions. $\tilde{H}_o$ contains only diagonal terms at all orders. Up to second order we obtain:

$$\tilde{H}_o = \sum_i \tilde{\omega}_i |i\rangle\langle i| + \sum_j \sum_{\underline{k}_j} \omega_{\underline{k}_j} a^{+}_{\underline{k}_j} a_{\underline{k}_j} + \tilde{H}'_o \tag{3}$$

where $\tilde{\omega}_3 = \omega_3$, and $\tilde{\omega}_j = \omega_j - \sum_{\underline{k}_j} |\epsilon_{\underline{k}_j}|^2/(\omega_{3j} + \omega_{\underline{k}_j})$ are the dressed atomic frequencies while $\omega_{3j} = \omega_3 - \omega_j$ and $\tilde{H}'_o$ contains the effects of dressing on photon states and does not give a contribution in our case. The effective interaction terms are:

$$\tilde{V}_1 = \sum_j \sum_{\underline{k}_j} (\epsilon_{\underline{k}_j} a_{\underline{k}_j} |j\rangle\langle 3| + \text{H.c.})$$

$$\tilde{V}_2 = -\sum_{\underline{k}_1} \sum_{\underline{k}_2} (\epsilon_{k_1} \epsilon^{*}_{k_2} + \epsilon^{*}_{k_1} \epsilon_{k_2})(a_{\underline{k}_1} a_{\underline{k}_2} |1\rangle\langle 2| + \text{H.c.}) + \tilde{V}'_2 \tag{4}$$

$\tilde{V}'_2$ contains terms which are effective only if photons are present both in the initial and final states. The relevant interaction terms have a r.w. form, also in the second order $\tilde{V}_2$ which now couples directly levels 2 and 1. Now it is possible to use the transformed Hamiltonian in the resolvent $\tilde{G}(z)=1/(z-\tilde{H})$ in order to get the transition amplitudes between dressed states, which are now eigenstates of (3). The persistence amplitude inthe dressed state 2 is $\tilde{I}_{2,2} = \exp(-i\,\tilde{\omega}_2 t - \Gamma_2 t/2)$ with a decay time

$$\tilde{\tau}_2 = 1/\Gamma_2 = 2\pi \sum_{\underline{k}_1} \sum_{\underline{k}_2} |\epsilon_{k_1}|^2 |\epsilon_{k_2}|^2 (\omega_{32} + \omega_{\underline{k}_2})^{-1} \delta(\tilde{\omega}_2 - \tilde{\omega}_1 - \tilde{\omega}_{\underline{k}_1} - \tilde{\omega}_{\underline{k}_2}), \tag{6}$$

It differs from the bare decay time by the replacement of dressed atomic and photon energies in place of the bare ones. The dressed lineshape is given by the square of amplitude:

$$\tilde{I}_{\underline{k}_1 \underline{k}_2} = \left[|\epsilon_{\underline{k}_2}|^2 (\omega_{32} + \omega_{\underline{k}_2})^{-2}\right] |\epsilon_{\underline{k}_1}|^2 \left[(\tilde{\omega}_{21} - \tilde{\omega}_{\underline{k}_1} - \tilde{\omega}_{k_2})^2 + \Gamma_2^2/4\right]^{-1} \quad . \tag{7}$$

In (7) the first factor in the R.H.S. can be associated to the virtual photon distribution in the virtual cloud arond the excited level 2. Eq. (7) differs from the bare amplitude for the change of the denominator of the first R.H.S. factor by $\left[(\omega_{31} - \omega_{\underline{k}_1})^2 + \Gamma_3^2/4\right]^{-1}$.So there is a relevant difference between dressed and bare lineshapes.

References.

1) F. Persico, G. Compagno, R. Passante; Quantum Optics IV, J. D. Harvey and D. F. Walls eds. , Springer-Verlag 1986, and references therein.

2) H. M. Nussenzweig, Quantum Electrodynamics and Quantum Optics, A. O. Barut ed., Plenum Press 1984, and references therein.

LIST OF PARTICIPANTS

N.B. Abraham	Bryn Mawr College, Bryn Mawr, USA
G. Alber	Univ. Innsbruck, Innsbruck, Austria
S.M. Barnett	Imperial College, London, GB
J. Bergou	Univ. of New Mexico, Albuquerque, USA
Z. Białynicka-Birula	Polish Academy of Sciences, Warsaw, Poland
R. Blatt	Univ. Hamburg, Hamburg, FRG
G. Compagno	Univ. Palermo, Palermo, Italy
J. Dalibard	Ecole Normale Superieure, Paris, France
F. Ehlotzky	Univ. Innsbruck, Innsbruck, Austria
W. Ertmer	Univ. Bonn, Bonn, FRG
E. Giacobino	Ecole Normale Superieure, Paris, France
R. Graham	Univ.-Gesamthochschule, Essen, FRG
R. Grimm	ETH, Zürich, Switzerland
R. Grobe	Univ.-Gesamthochschule, Essen, FRG
F. Haake	Univ.-Gesamthochschule, Essen, FRG
S. Haroche	Ecole Normale Superieure, Paris, France Yale Univ.,New Haven, USA
T. Haslwanter	Univ. Innsbruck, Innsbruck, Austria
H.-P. Helm	SRI International, Menlo Park, USA
W. Henle	Univ. Innsbruck, Innsbruck, Austria
M. Höhnerbach	Univ.-Gesamthochschule, Essen, FRG
S. Isermann	Univ.-Gesamthochschule, Essen, FRG
J. Javanainen	Univ. of Rochester, Rochester, USA
H.J. Kimble	Univ. of Texas at Austin, Austin, USA
H. Klar	Univ. Freiburg, Freiburg, FRG
P.L. Knight	Imperial College, London, GB
C. Leubner	Univ. Innsbruck, Innsbruck, Austria
M. Merz	Univ.-Gesamthochschule, Essen, FRG
P. Meystre	Univ. of Arizona, Tucson, USA
J. Mlynek	ETH, Zürich, Switzerland
J.M. Raimond	Ecole Normale Superieure, Paris, France
H. Risken	Univ. Ulm, Ulm, FRG
H. Ritsch	Univ. Innsbruck, Innsbruck, Austria
T. Sauter	Univ. Hamburg, Hamburg, FRG
R. Scharf	Univ.-Gesamthochschule, Essen, FRG
A. Schenzle	Univ.-Gesamthochschule, Essen, FRG
P.E. Toschek	Univ. Hamburg, Hamburg, FRG
J.R. Tredicce	Drexel Univ., Philadelphia, USA
K. Vogel	Univ. Ulm, Ulm, FRG
G. Wagner	MPQ, Garching, FRG
W. Winkler	MPQ, Garching, FRG
P. Zoller	Univ. Innsbruck, Innsbruck, Austria

Lecture Notes in Physics

Vol. 257: Statistical Mechanics and Field Theory: Mathematical Aspects. Proceedings, 1985. Edited by T.C. Dorlas, N.M. Hugenholtz and M. Winnink. VII, 328 pages. 1986.

Vol. 258: Wm. G. Hoover, Molecular Dynamics. VI, 138 pages. 1986.

Vol. 259: R.F. Alvarez-Estrada, F. Fernández, J.L. Sánchez-Gómez, V. Vento, Models of Hadron Structure Based on Quantum Chromodynamics. VI, 294 pages. 1986.

Vol. 260: The Three-Body Force in the Three-Nucleon System. Proceedings, 1986. Edited by B.L. Berman and B.F. Gibson. XI, 530 pages. 1986.

Vol. 261: Conformal Groups and Related Symmetries – Physical Results and Mathematical Background. Proceedings, 1985. Edited by A.O. Barut and H.-D. Doebner. VI, 443 pages. 1986.

Vol. 262: Stochastic Processes in Classical and Quantum Systems. Proceedings, 1985. Edited by S. Albeverio, G. Casati and D. Merlini. XI, 551 pages. 1986.

Vol. 263: Quantum Chaos and Statistical Nuclear Physics. Proceedings, 1986. Edited by T.H. Seligman and H. Nishioka. IX, 382 pages. 1986.

Vol. 264: Tenth International Conference on Numerical Methods in Fluid Dynamics. Proceedings, 1986. Edited by F.G. Zhuang and Y.L. Zhu. XII, 724 pages. 1986.

Vol. 265: N. Straumann, Thermodymamik. VI, 140 Seiten. 1986.

Vol. 266: The Physics of Accretion onto Compact Objects. Proceedings, 1986. Edited by K.O. Mason, M.G. Watson and N.E. White. VIII, 421 pages. 1986.

Vol. 267: The Use of Supercomputers in Stellar Dynamics. Proceedings, 1986. Edited by P. Hut and S. McMillan. VI, 240 pages. 1986.

Vol. 268: Fluctuations and Stochastic Phenomena in Condensed Matter. Proceedings, 1986. Edited by L. Garrido. VIII, 413 pages. 1987.

Vol. 269: PDMS and Clusters. Proceedings, 1986. Edited by E.R. Hilf, F. Kammer and K. Wien. VIII, 261 pages. 1987.

Vol. 270: B.G. Konopelchenko, Nonlinear Integrable Equations. VIII, 361 pages. 1987.

Vol. 271: Nonlinear Hydrodynamic Modeling: A Mathematical Introduction. Edited by Hampton N. Shirer. XVI, 546 pages. 1987.

Vol. 272: Homogenization Techniques for Composite Media. Proceedings, 1985. Edited by E. Sanchez-Palencia and A. Zaoui. IX, 397 pages. 1987.

Vol. 273: Models and Methods in Few-Body Physics. Proceedings, 1986. Edited by L.S. Ferreira, A.C. Fonseca and L. Streit. XIX, 674 pages. 1987.

Vol. 274: Stellar Pulsation. Proceedings, 1986. Edited by A.N. Cox, W.M. Sparks and S.G. Starrfield. XIV, 422 pages. 1987.

Vol. 275: Heidelberg Colloquium on Glassy Dynamics. Proceedings, 1986. Edited by J.L. van Hemmen and I. Morgenstern. VIII, 577 pages. 1987.

Vol. 276: R.Kh. Zeytounian, Les Modèles Asymptotiques de la Mécanique des Fluides II. XII, 315 pages. 1987

Vol. 277: Molecular Dynamics and Relaxation Phenomena in Glasses. Proceedings, 1985. Edited by Th. Dorfmüller and G. Williams. VII, 218 pages. 1987.

Vol. 278: The Physics of Phase Space. Proceedings, 1986. Edited by Y.S. Kim and W.W. Zachary. IX, 449 pages. 1987.

Vol. 279: Symmetries and Semiclassical Features of Nuclear Dynamics. Proceedings, 1986. Edited by A.A. Raduta. VI, 465 pages. 1987.

Vol. 280: Field Theory, Quantum Gravity and Strings II. Proceedings, 1985/86. Edited by H.J. de Vega and N. Sánchez. V, 245 pages. 1987.

Vol. 281: Ph. Blanchard, Ph. Combe, W. Zheng, Mathematical and Physical Aspects of Stochastic Mechanics. VIII, 171 pages. 1987.

Vol. 282: F. Ehlotzky (Ed.), Fundamentals of Quantum Optics II. Proceedings, 1987. X, 289 pages. 1987.